Biostimulants for sustainable crop production

It is widely recognised that agriculture is a significant contributor to global warming and climate change. Agriculture needs to reduce its environmental impact and adapt to current climate change whilst still feeding a growing population, i.e. become more 'climate-smart'. Burleigh Dodds Science Publishing is playing its part in achieving this by bringing together key research on making the production of the world's most important crops and livestock products more sustainable.

Based on extensive research, our publications specifically target the challenge of climate-smart agriculture. In this way we are using 'smart publishing' to help achieve climate-smart agriculture.

Burleigh Dodds Science Publishing is an independent and innovative publisher delivering high quality customer-focused agricultural science content in both print and online formats for the academic and research communities. Our aim is to build a foundation of knowledge on which researchers can build to meet the challenge of climate-smart agriculture.

For more information about Burleigh Dodds Science Publishing simply call us on +44 (0) 1223 839365, email info@bdspublishing.com or alternatively please visit our website at www.bdspublishing.com.

Related titles:

Understanding and improving crop root function
Print (ISBN 978-1-78676-360-0); Online (ISBN 978-1-78676-362-4, 978-1-78676-363-1)

Achieving sustainable crop nutrition
Print (ISBN 978-1-78676-312-9); Online (ISBN 978-1-78676-314-3, 978-1-78676-315-0)

Improving organic crop cultivation
Print (ISBN 978-1-78676-184-2); Online (ISBN 978-1-78676-186-6, 978-1-78676-187-3)

Chapters are available individually from our online bookshop:
https://shop.bdspublishing.com

climate•SMART•publishing
IN AGRICULTURAL SCIENCE

BURLEIGH DODDS SERIES IN AGRICULTURAL SCIENCE
NUMBER 84

Biostimulants for sustainable crop production

Edited by Youssef Rouphael, Patrick du Jardin, Patrick Brown, Stefania De Pascale and Giuseppe Colla

burleigh dodds
SCIENCE PUBLISHING

Published by Burleigh Dodds Science Publishing Limited
82 High Street, Sawston, Cambridge CB22 3HJ, UK
www.bdspublishing.com

Burleigh Dodds Science Publishing, 1518 Walnut Street, Suite 900, Philadelphia, PA 19102-3406, USA

First published 2020 by Burleigh Dodds Science Publishing Limited

Library of Congress Control Number: 2020936883

British Library Cataloguing in Publication Data
A catalogue record for this book is available from the British Library

ISBN 978-1-78676-336-5 (Print)
ISBN 978-1-78676-339-6 (PDF)
ISBN 978-1-78676-338-9 (ePub)
ISSN 2059-6936 (print)
ISSN 2059-6944 (online)

DOI 10.19103/AS.2020.0068

Typeset by Deanta Global Publishing Services, Dublin, Ireland

Contents

Part 2 Non-microbial and microbial categories of biostimulants

Series list

Title	Series number
Achieving sustainable cultivation of maize - Vol 1 From improved varieties to local applications Edited by: Dr Dave Watson, CGIAR Maize Research Program Manager, CIMMYT, Mexico	001
Achieving sustainable cultivation of maize - Vol 2 Cultivation techniques, pest and disease control Edited by: Dr Dave Watson, CGIAR Maize Research Program Manager, CIMMYT, Mexico	002
Achieving sustainable cultivation of rice - Vol 1 Breeding for higher yield and quality Edited by: Prof. Takuji Sasaki, Tokyo University of Agriculture, Japan	003
Achieving sustainable cultivation of rice - Vol 2 Cultivation, pest and disease management Edited by: Prof. Takuji Sasaki, Tokyo University of Agriculture, Japan	004
Achieving sustainable cultivation of wheat - Vol 1 Breeding, quality traits, pests and diseases Edited by: Prof. Peter Langridge, The University of Adelaide, Australia	005
Achieving sustainable cultivation of wheat - Vol 2 Cultivation techniques Edited by: Prof. Peter Langridge, The University of Adelaide, Australia	006
Achieving sustainable cultivation of tomatoes Edited by: Dr Autar Mattoo, USDA-ARS, USA & Prof. Avtar Handa, Purdue University, USA	007
Achieving sustainable production of milk - Vol 1 Milk composition, genetics and breeding Edited by: Dr Nico van Belzen, International Dairy Federation (IDF), Belgium	008
Achieving sustainable production of milk - Vol 2 Safety, quality and sustainability Edited by: Dr Nico van Belzen, International Dairy Federation (IDF), Belgium	009
Achieving sustainable production of milk - Vol 3 Dairy herd management and welfare Edited by: Prof. John Webster, University of Bristol, UK	010
Ensuring safety and quality in the production of beef - Vol 1 Safety Edited by: Prof. Gary Acuff, Texas A&M University, USA & Prof. James Dickson, Iowa State University, USA	011
Ensuring safety and quality in the production of beef - Vol 2 Quality Edited by: Prof. Michael Dikeman, Kansas State University, USA	012
Achieving sustainable production of poultry meat - Vol 1 Safety, quality and sustainability Edited by: Prof. Steven C. Ricke, University of Arkansas, USA	013
Achieving sustainable production of poultry meat - Vol 2 Breeding and nutrition Edited by: Prof. Todd Applegate, University of Georgia, USA	014

Acknowledgements

We wish to acknowledge the following for their help in reviewing particular chapters:

- Chapter 5: Professor Michela Schiavon, University of Padua, Italy
- Chapter 7: Professor Maurizio Ruzzi, Tuscia University, Italy
- Chapter 8: Professor Manuela Giovannetti, University of Pisa, Italy
- Chapter 11: Professor Yang Yang, Director of Purdue Phenotyping Center–Purdue University, USA

Introduction

The EU Fertilizer Regulation 2019/1009 defines a plant biostimulant as a *'fertilizing product the function of which is to simulate plant nutrition process independently of the product's nutrient content'*. Plant biostimulants stimulate natural processes in crops to enhance nutrient uptake, nutrient use efficiency (NUE), resistance to abiotic stress and quality traits, as well as increasing the availability of nutrients in the soil or rhizosphere. They offer the opportunity to enhance fertilizer use and thus contribute to more sustainable crop production. This collection reviews key advances in understanding and using biostimulants. Part 1 reviews ways of classifying microbial and non-microbial biostimulants, types of bioactive compound and ways of evaluating biostimulants. Part 2 surveys the various types of biostimulant, from humic substances and seaweed extracts to protein hydrolysates, silicon, plant growth-promoting rhizobacteria (PGPR) and arbuscular mycorrhizal fungi (AMF). Part 3 discusses advances in designing second-generation biostimulants and their practical application.

Part 1 Introduction and biostimulant characterization

Chapter 1 sets the scene by discussing plant biostimulants as a new paradigm for the sustainable intensification of crop production. Biostimulants are applied to crop plants as a way of modifying plant physiological functions and of increasing crop productivity or quality. They may be regarded as 'functional ingredients' in plant nutrition, distinct from fertilizers and plant protectants (such as insecticides or fungicides). Although biostimulants such as seaweed extracts and humic acids have been used in agriculture for decades, they have only recently been recognized by regulations governing fertilizing products. Biostimulant products placed on the market are identified by claims describing their intended effects on crops such as improved nutrient use efficiency and tolerance to abiotic stress. These effects contribute to the sustainable intensification of crop production. However, their further development requires an improved knowledge of their bioactive effects on plants and associated microorganisms, the responsiveness of recipient plants and environments to biostimulant activity, and their interactions with fertilizers and other agricultural inputs.

As Chapter 2 points out, regulations governing the placement of biostimulants on the market still vary widely across countries and regions. A key factor in all pre-market approval systems governing biostimulants are specifications for the data required for authorization. Most regulations share a focus on composition declarations and, in particular, efficacy claims. A combination of laboratory trials and dedicated ad hoc field studies are recommended to address the definition of both mode(s) of action and effect(s) of plant biostimulants. Traditionally, in vitro assays (e.g. hormone-like activity tests) have been used to support the investigation of potential biostimulant activity. Recent advances in analytics, technology and big data management have raised the potential of -omic sciences in understanding, screening and evaluation of the mode of action for biostimulants. In particular metabolomics and phenotyping are attracting growing interest.

Part 2 Non-microbial and microbial categories of biostimulants

Humic substances (HS) are among the most established biostimulants used in agriculture because they have been shown to significantly improve plant growth, directly or indirectly, as well as improve soil properties and fertility. As Chapter 3 shows, HS affect many agronomic, environmental and geochemical processes that interact with plant growth such as soil structure and porosity, water infiltration rate and moisture-holding capacity of soils as well as affecting the diversity and activity of soil micro-organisms. In addition, HS influence plant physiology by interacting with plant biochemical and physiological processes, stimulating growth and increasing the uptake of nutrients by roots. There is now an extensive body of research that has shown, both under laboratory and field conditions, that HS can have a positive effect on plant growth in terms of increases in biomass of shoots and roots, chlorophyll concentration, and number of lateral roots. Chapter 3 reviews the range of research on key aspects of HS: production and characteristics, biological activities, effects on soil and plant nutrition, and the use of commercial humates in agriculture.

Seaweed extracts are a widely used class of biostimulant. Chapter 4 reviews research on their mechanism of action with a particular focus on primary and secondary metabolites which act as growth stimulating and protecting factors as well as antibacterial agents. Seaweed extracts also modulate the growth of rhizosphere microbial populations. They directly affect plant physiology, in particular the germination of seeds, growth of shoots and roots, improvement of fruit set, as well as improving the quality of food crops. Seaweed extracts can also improve crop abiotic stress tolerance. Although much has been achieved, further research is needed to more fully understand their mechanisms of action.

Protein hydrolysates (PH) are a category of plant biostimulants containing a mixture of polypeptides, oligopeptides and amino acids that are manufactured from animal or plant derived-protein sources using partial hydrolysis. Protein hydrolysates are used for foliar applications and, to a lesser extent, as soil and seed coating applications to promote crop performance in both open field and greenhouse conditions. Chapter 5 provides an overview of the characteristics and production of vegetal and animal-based protein hydrolysates and their beneficial effects on nutrient use efficiency, crop tolerance to abiotic stress and production traits. The chapter reviews recent research on understanding the mode of action and physiological and molecular mechanisms of protein hydrolysates and ways of optimizing the timing and rate of application.

Silicon (Si) is a mineral element that is well known to protect many crops against a range of abiotic stresses, including osmotic and nutritional imbalances. Whilst its exact mechanism of action is still to be fully understood, research highlights the role of silicon in maintaining internal homeostasis in plants. Chapter 6 begins by assessing silicon availability in fertilizers and growing media and then summarises what we know about plant accumulation, transport and deposition of silicon. It then reviews research on ways silicon can enhance plant abiotic defences against drought and salinity, tolerance to heavy metals and other environmental stresses. It also discusses the role of silicon in enhanced crop growth.

Chapter 7 reviews what we know about plant growth-promoting rhizobacteria (PGPR) as plant biostimulants in agriculture. Plant growth-promoting rhizobacteria can improve growth under stressful growing conditions by inducing abiotic stress tolerance via production of antioxidant enzymes, altering plant metabolism, affecting the rate of photosynthesis and shifting osmolyte concentration in plant tissues. These bacteria also help plants resist biotic stress by competing against other microbes for niche space and nutrients, producing antibiotic compounds, and inducing systemic resistance by producing microbe-to-plant signal molecules. However, our understanding of the mechanisms of action of plant growth-promoting rhizobacteria is still relatively new. The importance of factors such as root exudates and intermicrobial signaling needs to be the focus of future research.

Chapter 8 focuses on arbuscular mycorrhizal fungi (AMF) as biostimulants for sustainable crop production. The chapter reviews the functions and benefits of AMF. As research shows, the basis of AMF symbiosis is a bidirectional exchange of nutrients between the plant and AMFs. Host plants provide a physical support and a favourable metabolic framework for the obligate biotrophic lifestyle of AMF. AMF receive carbon fixed by the host plant's photosynthesis in exchange for mineral nutrients that they provide to the host plant via the fungal mycelial network. The chapter also assesses what we know about the requirements for successful implementation of AMF in sustainable plant production. Research suggests that several aspects help determine successful application of AMF. Firstly, the sites of plant production and their conditions must be advantageous for mycorrhizal functioning. Secondly, the genotype of the plant must support the positive functions of the symbiosis. Inocula should also be targeted to particular conditions by 'training' or by combining them with other beneficial microorganisms. The chapter also assesses the current market for mycorrhizal products.

Part 3 Innovation and practical applications

Chapter 9 provides the first overview of the optimal design and formulation of microbial and non-microbial biostimulants. The chapter provides an innovative discussion of the circular production process for the development of plant biostimulants, including i) process development, ii) elucidation of the mode of action (by combining plant phenotyping and omics science), iii) quality control, iv) field trial validation, v) regulation and vi) industrialization/commercialization. The chapter includes two successful industrial case studies of microbial (mycorrhizal inoculants) and non-microbial (vegetal-derived protein hydrolysates) biostimulant products that have been successfully developed and commercialised.

Chapter 10 reviews the effects of humic and fulvic substances, microbial biostimulants, seaweeds and algae as well as protein hydrolysates (PH). It assesses the evidence of for the effects of biostimulants on both agronomic and internal nutrient use efficiency. Improving NUE is of great practical value as it allows for the greater exploitation of added fertilizers and improved recovery of residual nutrients. Regulatory guidelines in the EU and other jurisdictions emphasize that biostimulants can be identified by claims including improved nutrient use efficiency with the goal of enhancing cropping system efficiency. A considerable body of research demonstrates that many biostimulants improve 'agronomic' nutrient use efficiency by enhancing root growth and soil exploration, increasing solubilization of soil nutrients or upregulating nutrient uptake processes, thereby enabling a greater amount of the total soil nutrient reserve to be acquired by the plant. There is, however, much less evidence to demonstrate that biostimulants alter the internal nutrient use efficiency of plants by increasing the productivity of a crop for a given quantity of acquired nutrient. This is a key area for future research.

Precision site-specific application of biostimulants is a great opportunity for optimizing biostimulant efficacy and returns. Chapter 11 looks at the available tools and emerging technologies for the monitoring and management of soil and crops in order to address spatial and temporal variability and inform site-specific management strategies. The chapter assesses methods for site-specific management based on identifying management zones for targeted treatment. The potential of adopting precision agricultural techniques for the use of biostimulants is discussed, focusing on the targeted application of biostimulants in viticulture for mitigation of abiotic stresses such as water, nitrogen and phosphorus deficiency.

Youssef Rouphael, University of Naples, Italy
Patrick du Jardin, University of Liège, Belgium
Patrick Brown, University of California-Davis, USA
Stefania de Pascale, University of Naples, Italy
Giuseppe Colla, University of Tuscia, Italy

Part 1

Introduction and biostimulant characterization

Chapter 1

Plant biostimulants: a new paradigm for the sustainable intensification of crops

Patrick du Jardin, Gembloux Agro-Bio Tech – University of Liège, Belgium

1 Introduction

The definition of plant biostimulants (PBs) has been intensively discussed over the last years, mainly for regulatory purposes (Yakhin et al., 2016; du Jardin, 2015; Caradonia et al., 2019). In Europe, a consensus was reached by a recent regulation on fertilizing products (FPs), a milestone in recognition of the concept and the future harmonization of marketed products. In its Regulation (EU) 2019/1009 of the European Parliament and of the Council of 5 June 2019, laying down rules on the making of European Union (EU) FP available on the market, PBs are defined as follows (EU, 2019):

> A plant biostimulant shall be an EU fertilising product the function of which is to stimulate plant nutrition processes independently of the product's nutrient content with the sole aim of improving one or more of the following characteristics of the plant or the plant rhizosphere:
>
> (a) nutrient use efficiency,
> (b) tolerance to abiotic stress,
> (c) quality traits, or
> (d) availability of confined nutrients in the soil or rhizosphere.

http://dx.doi.org/10.19103/AS.2020.0068.01

The main elements regarding the nature and action of biostimulants contained by this definition were initially proposed by both the industry (EBIC, at http://www.biostimulants.eu) and academic experts (du Jardin, 2012, 2015; Traon et al., 2014) in the preparation of the regulation.

First, PBs are anchored in plant nutrition, and it is acknowledged by the regulation that FPs not only cover nutrient-supplying fertilizers but also products which help the plant make better use of those fertilizers. In improving plant nutrition, the FP may act on the plant, on biotic components of the plant environment (e.g. soil microorganisms) and possibly on abiotic components (like soil physico-chemical properties, possibly covered by the last part of the definition). In the European regulation, PBs are regarded as one of the functional categories of FPs, primarily defined by their intended effects on cultivated plants. Liming materials, inorganic and organic fertilizers and soil improvers are other examples of 'Product Function Categories' listed by the regulation.

Second, the intended functions of PBs are defined as a limited number of claims, which are (a) improvement of nutrient use efficiency (NUE), (b) tolerance to abiotic stress, (c) quality traits and (d) availability of confined nutrients in the soil or rhizosphere. The important factor here is that the placing of biostimulant products on the EU market will depend on the capacity of the company to substantiate the claimed effect of its product. For doing so, the approach of the European legislation is to define EU-harmonized standards, bearing on principles, methods and protocols to which companies will refer to when developing the arguments validating the claims. Compliance to the EU standards when generating data on the products will be the best way to support the claims and access the European market, yet it will not be mandatory as alternatives might be proposed by the companies. Today, the implementation of the adopted EU regulation awaits the setting of standards, a process driven by a dedicated agency (CEN, European Committee for Standardization) and fueled by technical expertise from both the private and public sectors. For such a novel category of FPs as PBs, the way is expected to be long and difficult. The designation and role of 'notified bodies' for claim validation is another aspect to be considered in the near future.

Third, the definition says very little about the composition of PBs. In the regulation, the composition of FPs is described by a separate typology, defining 'Component Material Categories', parallel to the 'Product Function Categories' discussed so far. Accordingly, biostimulant products can be composed of substances or microorganisms in the limitations set by the regulation and framed by implementing the standards that are yet to be adopted. Biostimulant products marketed in the short term under the new European regulation are expected to include substances like seaweed extracts and humic acids, as well as microorganisms with a history of safe use like rhizobia and mycorrhiza. In the current status of the regulation, substances which undergo chemical or enzymatic modification will need registration under the Registration, Evaluation,

Authorisation and Restriction of Chemicals (REACH) regulation (Regulation (EC) No 1907/2006), and this will delay and might even hamper the CE marking of biostimulants by dissuading companies to follow the European track for the placing of their products on the market. Regarding the microbial biostimulants, a limited list of eligible taxa is currently laid down by the regulation, as it will be discussed later in Section 5. In order to anticipate the main biostimulant products marketed under the new Regulation (EU) 2019/1009, the limitations and opportunities set by the future conformity assessment procedure, which will use yet-to-define EU-harmonized standards, need to be clarified. Furthermore, how the European Commission will exercise its power to adopt delegated acts (set by Article 42 of the regulation) to move forward on issues like the limited positive list of microbial biostimulants, or the status of biological polymers including protein hydrolysates, an important category of biostimulants, is uncertain.

One point related to the composition, and that is mentioned in the EU definition, is that a biostimulant is a fertilizing 'product', that is, as supplied to the grower; it is not an ingredient, that is, an isolated compound or microorganism used to develop FPs. The consequence of this is important. On the one hand, the system aims at validating the claimed effects of the product as supplied to the user, and this can be translated into adequate labeling provisions, which seems to be the best way for grower protection. On the other hand, many biostimulants will be composed of mixtures of substances and/or microorganisms, while the scientific research on the mechanisms of biostimulation tends to use single substances (which can be composed of many constituents but are defined by their single origins, for example, an extract of the brown alga *Ascophyllum nodosum*) or single microorganisms. This creates a gap between the practice and research, which challenges efforts to better understand how biostimulant products actually work.

A fourth point to be considered is that, in practice, PBs are often added to macro- and/or micronutrients, or with other ingredients, to give a blended FP combining different materials and different effects on plants, converging to result in improved plant nutrition and higher crop yield and quality. Validation of the claimed agricultural effects is increasingly difficult when relying on interactions between multiple components.

So far, we have discussed the main characteristics of biostimulants laid down by the European regulation, but what about other regions of the world and the United States in particular? The status of biostimulants has made significant progress in the United States over the recent times as well (see http://www.biostimulantcoalition.org/ for updates), under two processes, the 2018 Farm Bill and the 2019 EPA's Guidance for Plant Regulator Label Claims, Including Biostimulants, in an effort to clarify the applicability of the Federal Insecticide, Fungicide and Rodenticide Act (FIFRA) rules on plant regulators (Neuschafer and Paisner, 2019; draft guidance available at https://www.regulations.gov/documen

t?D=EPA-HQ-OPP-2018-0258-0002). Indeed, whether biostimulant products are subject to FIFRA and are regulated as 'plant regulators' by the federal agency EPA, or considered as fertilizers, soil amendments and other products that are not captured by FIFRA and regulated by the state departments of agriculture are important for the pre-market assessment and marketing of the products. Although at the time of writing this chapter, public consultation of the EPA draft guidance text is still ongoing, a definition of biostimulants is proposed:

> a naturally-occurring substance or microbe that is used either by itself or in combination with other naturally-occurring substances or microbes for the purpose of stimulating natural processes in plants or in the soil in order to, among other things, improve nutrient and/or water use efficiency by plants, help plants tolerate abiotic stress, or improve the physical, chemical, and/or biological characteristics of the soil as a medium for plant growth.

This definition may be compared to that adopted by the Congress in the 2018 Farm Bill:

> a substance or micro-organism that, when applied to seeds, plants or the rhizosphere, stimulates natural processes to enhance or benefit nutrient uptake, nutrient efficiency, tolerance to abiotic stress, or crop quality and yield.

These definitions can be compared not only with each other but also with the European definition commented before. Most important is the fact that the three definitions are based on claims. Claims common to all definitions are improvements in plant nutrition (nutrient uptake and use) and tolerance to abiotic stress. Hence, they should be regarded as the cornerstones of the concept and regulation of biostimulants.

But differences between definitions also point to gray areas. The Farm Bill's definition indicates higher yield among the claims, which is not found in other definitions but seems implicit as increased agronomic efficiency of fertilizers and enhanced tolerance to abiotic stress are expected to translate into higher yields. Another point is that the EPA's guidance definition extends the action of biostimulants to soil characteristics, including physical properties, which may be covered by the European definition. Indeed, although a late amendment brought to the definition and adopted by the final regulatory text talks about improved 'availability of confined nutrients in the soil or the rhizosphere', which seems to expand the perimeter of biostimulants to products that would influence some soil properties, clarification is needed and awaits the EU-harmonized standards mentioned before. Furthermore, the EPA's definition talks about 'naturally occurring substance or microbe', but there is no such restriction about the natural origin of biostimulant products in the other definitions. Typically, most biostimulants are of natural origins, like seaweed extracts, protein hydrolysates or humic acids, but another thing is to impose that they are of natural origin, and the EU regulation does not make this step, leaving the possibility of chemical derivatives of natural compounds or synthetic compounds to be used as biostimulants. A requisite will be their registration under the European REACH regulation (Regulation (EC) No 1907/2006 of the European Parliament and the Council concerning the Registration, Evaluation, Authorisation and Restriction of Chemicals).

Finally, although the placing of biostimulants on the market will be based on marketing claims and make use of corresponding labeling provisions, the EPA's guidance pays attention to the active ingredients as well and lists those identifying the products as plant regulators that are captured by FIFRA and regulated as pesticides. The bioactive ingredients listed as examples by this text include plant hormones (e.g. cytokinins, jasmonates) and also substances which are important sources of biostimulants today, like seaweed extracts and humic/fulvic acids.

The reader may refer to a recent review by Caradonia et al. (2019) for further information on the regulation of biostimulants in other countries.

2 The establishment of the term 'biostimulant'

The word 'biostimulant' appeared when it became evident that some products applied to plants were able to stimulate growth at low doses, which could neither be explained by the supply of nutrients nor by some plant protection against pests and pathogens. The pioneering work of two research teams should be mentioned here.

In the 1980s to early 1990s, at the School of Forestry and Environmental Studies of the Yale University, Professor G. P. Berlyn and his team studied the response of woody and grass species to bioactive substances - seaweed extracts, humic acids and vitamins - combined in a proprietary mixture named Roots™. Improvements in root and shoot growth, drought resistance and nitrogen use efficiency were reported (Russo and Berlyn, 1991). There are two remarkable things to be pointed out in this paper. First its title, 'The Use of Organic Biostimulants to Help Low Input Sustainable Agriculture', which, to the best of our knowledge, is the first to use the word 'biostimulant' in a peer-reviewed article. The scope of using biostimulants in agriculture is also far-reaching: low-input agriculture. The second thing is how the authors describe the action of their biostimulant product. After listing the bioactive ingredients of Roots, they propose that 'the innovation of mixing them and capitalizing on their synergistic effects is a real contribution in terms of agricultural production'. Whether the unique properties of biostimulant products rely on synergistic and/or emerging properties of blended bioactive compounds is an issue which we will cover later in this chapter. In a later article describing the effects of Roots on beans (Russo and Berlyn, 1992), the authors define biostimulants as 'nonnutritional products that may reduce fertilizer use and increase yield and resistance to water and temperature stresses', also saying that they 'stimulate plant growth in relatively small amounts'.

In the same period, another team led by Professor R. E. Schmidt - at Department of Crop and Soil Environmental Sciences, Virginia Polytechnic Institute and State University - performed similar studies on growth and stress responses of turfgrass treated by seaweed- and humic acids-based biostimulants. Although the word 'biostimulant' was used and defined by these authors in a popular web magazine in 1997 (du Jardin, 2015), they preferred using the wording 'metabolic enhancers' and 'hormone-containing products' in their peer-reviewed articles (Zhang and Schmidt, 2000). Regarding the action mechanisms of biostimulants, these authors concluded that PBs modify hormone homeostasis and stimulate antioxidant activities which play a major role in environmental stress responses (Zhang and Schmidt, 2000).

In a review paper dated 1993, Kinnersley introduces biostimulants as a category of phytochelates, which are organic compounds and substances promoting plant growth by chelating micro- and macronutrients (Kinnersley, 1993). Humic acids, components of many biostimulant products since that time and still today, enhance metal uptake - for example, magnesium as a constituent of chlorophyll, increasing photosynthetic capacity - and also protect the plant against metal toxicity, which are typical behaviors of phytochelates. The carboxyl groups of humic acids are proposed to be responsible for their chelating activity by Kinnersley (Kinnersley, 1993). The hypothesis of an iron-chelating activity of seaweed extracts was also one the reasons for the interest of Professor Schmidt and colleagues toward biostimulants, in the context of their work on iron nutrition in creeping bentgrass (R. Schmidt, personal communication).

In conclusion, this early literature on biostimulants paved the way for the later research and development, by highlighting some of their main characteristics: their action on plant homeostasis at low doses, their modulation of the plant stress response, their role in plant nutrition and growth promotion by acting on nutrients availability and uptake and the synergistic effects resulting from the combination of bioactive compounds from the same or different substances. These topics are still the agenda of many research teams studying and developing biostimulants (Calvo et al., 2014; du Jardin, 2015; Brown and Saa, 2015; Yakhin et al., 2016).

3 Plant biostimulants as functional ingredients of fertilizing products

Today, PBs are applied to a wide range of crops, and in different ways, for the purpose of improving yield and quality, saving nutrients and water and making the crop more resilient to environmental stress (Fig. 1).

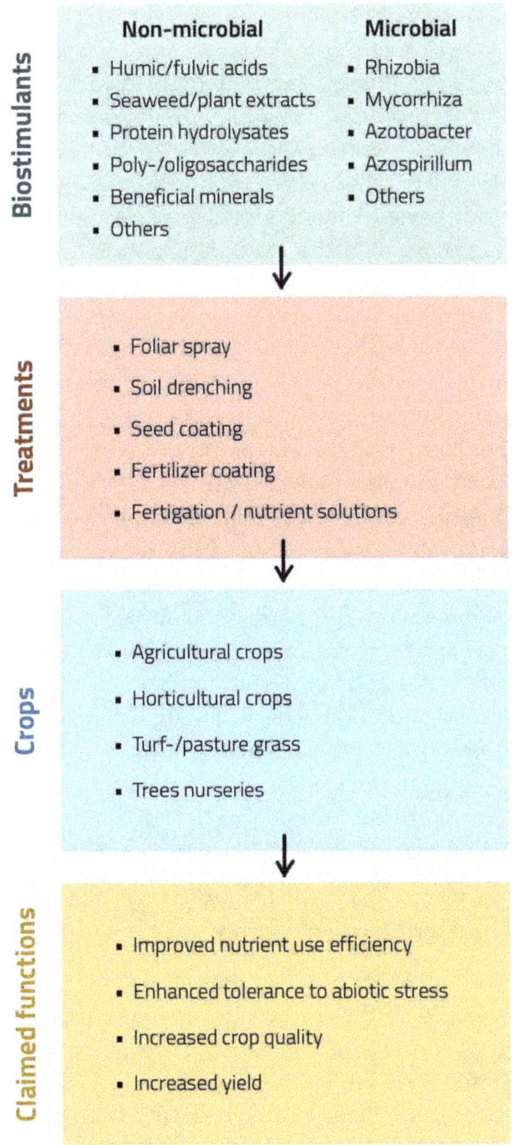

Figure 1 Application of biostimulant products to crop plants and claimed agricultural functions.

Biostimulants introduce a new paradigm in plant production. They appear as a relatively new category of agricultural inputs, which aims at acting on the plant's physiology in ways which are distinct from supplying nutrients or protecting the plants against potentially harmful organisms. Changes in plant physiology and development are targeted by breeding techniques and farming practices, and it is now realized that many plant traits are also amenable to modification through the external application of bioactive substances. In this respect, biostimulant products resemble existing technologies based on plant growth regulators, marketed as pesticides, but most biostimulants are biosourced materials with complex compositions, and their activity is deemed to be supported by many interacting bioactive constituents.

There is some analogy in the way plant nutrition and human nutrition are considered today. Human diets not only provide the body with the nutrients needed for growth and maintenance, but they also supply bioactive constituents which target specific physiological mechanisms, enhancing cell and organ homeostasis. This is achieved via protection against environmental damage and sources of, for example, antioxidants (including vitamins C and E, but also many phenolics), which are examples of such 'functional ingredients' of foods ameliorating human health and welfare. Once the activities of specific food constituent are identified, new products can be developed and marketed which combine relatively high amounts of the bioactive substances and are taken by human consumers as food supplements.

In a way, biostimulant products are food supplements of cultivated plants. But as with food supplements consumed by humans, the 'fertilizer supplements', or biostimulants, will be used by growers if the expected benefits are clear and supported by fair marketing practices, including adequate labeling, and by a sound regulatory framework. This emphasizes the importance of the claims as defined by the regulation in order to inform growers about the intended uses of biostimulant products. Let us now describe in more detail the claims included in the regulatory definition of biostimulants in Europe.

3.1 Improving nutrient use efficiency

Nutrient use efficiency (NUE) has been defined in many ways, depending on the crop and the scope of the study, but the word 'efficiency' always refers to a ratio between outputs and inputs. Both the outputs and inputs can be expressed using different variables: outputs can be grain yield or the amount of exported nitrogen, while the inputs can be the total nutrient content of the soil, its bioavailable fraction or the applied fertilizer. A review of available definitions and calculation methods of NUE can be found in Ussiri and Lal (2013).

In order to analyze the effectiveness of biostimulants in improving NUE, the following function is useful, as it dissects NUE into two components which are controlled by different plant traits and can be influenced by biostimulants according to distinct mechanisms:

$$\text{Nutrient Use Efficiency} = \text{Uptake Efficiency} * \text{Utilization Efficiency}$$

where:

- NUE is the dry mass (or yield) produced per nutrient input (gram total plant biomass or harvested biomass per gram nutrient input).
- Uptake efficiency is the amount of nutrient taken up by the plant per nutrient input (gram nutrient uptake per gram nutrient input).
- Utilization efficiency is the dry mass produced per nutrient taken up by the plant (gram total plant biomass or harvested biomass per gram nutrient uptake).

Taking phosphorus (P) as an example, three strategies for high P use efficiency have been distinguished (Richardson et al., 2011): 'root foraging strategy', that is, efficient capture of P by developing roots even in low-P soils; 'root mining strategy', that is, enhancing the desorption, solubilization or mineralization of P from sparingly available pools; and improved 'internal P-utilization efficiency' (higher yield per unit of P uptake). Each of the described strategies is related to specific plant traits, amenable to modification using biostimulants: upregulation of genes encoding membrane transporters for the uptake of nutrients (Ertani et al., 2017), promotion of root growth (Barone et al., 2018) and of root hairs development (Canellas et al., 2010) and control of the partitioning of nutrients between plant parts via cell membrane transporters (Billard et al., 2014).

Despite the evidence that biostimulants can modify several plant traits involved in NUE, difficulties persist. First, we still largely ignore which phenotypes best serve NUE, at both the developmental and biochemical levels, not to forget the progress made in some crop/nutrient models (Richardson et al., 2011; Lynch, 2013). In this research field, 'phenes' are defined as discrete units of the phenome, making up the phenotype, like genes are the discrete units of the genome making up the genotype. Advances in plant phenome research will benefit the development of biostimulant products, as they will improve the way in which biostimulants can be screened in the laboratory and moved to practical applications on the basis of known mechanisms of action. Second, it is likely more important from an NUE perspective to promote the responsiveness of plants and their roots to changing concentrations of nutrients in their environments, in time and space, than to promote any 'ideal' root architecture. This is more difficult to address in the laboratory, but the interplay of biostimulants with the regulatory gene networks well characterized in model plants is worth studying.

The third and last point is that few field experiments on biostimulants measuring NUE are described in the public literature (Storer et al., 2016). Indeed, most field studies only report on yield and other characteristics directly valued by the market (like grain protein contents). Yet the positive effect of some biostimulants on plant growth and tissue composition when applying a reduced amount of fertilizers (Carillo et al., 2019; Mola et al., 2019; Nguyen et al. 2019a,b) encourages their consideration for a 'high-output-low-input' agriculture.

3.2 Improving tolerance to abiotic stress

Tolerance to abiotic stress is a major concern when aiming at high crop productivity on a warming planet. In all cropping systems of the world, reducing the yield gap between the achievable and actual yields will depend on how farming practices and the genetic properties of plants can be improved to cope with the environmental stresses, like drought, salinity and heat. The magnitude of this yield gap is much different from one farmland to another across the world (Van Ittersum et al., 2013), but, whatever it is, improving tolerance to abiotic stress will be key to increasing crop productivity.

When the yield is threatened by pathogens and pests, farmers know how to deal with them by applying plant protection products, whether they are synthetic chemicals, natural substances or living organisms. The action of these products can be by direct toxicity toward the pathogen or its biological vectors, or by triggering the plant's immunity. Before the advent of PBs, there were few, if at all, solutions to reduce the abiotic stress suffered by the plant, at least in non-controlled growing conditions. Drought mitigation by irrigation is an important exception, but it cannot always be put in practice at affordable costs. Biostimulants appear as a breakthrough in crop protection against abiotic stress, and analyzing their potential in this respect seems to be of paramount importance.

Increased tolerance to abiotic stress is a claim of biostimulant products in all regulatory norms so far. However, quantifying their actual effects on abiotic stress is challenging. When dealing with stress response of an organism, three phases need to be considered: before, during and after the stress episode. Of special interest is the capacity of the organism to be 'primed' against abiotic stressors, which means that some physical, chemical or biological treatment 'before' stress improves the responsiveness of the organism toward subsequent stress application. Although such priming effects have been described since long in the context of both environmental and pathogen stresses, with concepts like hardening and immunity (Hilker et al., 2016), it is only recently that it was realized that the 'priming stimulus' and the 'trigger stimulus' (i.e. the stressor) could be the same or different (Jakab et al., 2005). Interestingly, some chemical treatments are able to prime the organism against a pathogen or abiotic stress, despite any resemblance of the priming molecules to known physiological effectors of the plant (Savvides et al., 2016). When such compounds, substances or microorganisms are developed into commercial products ameliorating plant tolerance to abiotic stress, they may be referred to as PBs. In fact, this appears as a most promising marketing opportunity for PBs, as it complements all other technologies aiming at increasing crop productivity under stress and provides growers with a tool to reduce yield penalties whenever stressful events can be anticipated, like heat and drought episodes.

As an interesting case study, the priming activity of three types of biostimulants in the protection of *Arabidopsis thaliana* against drought stress was studied by Fleming et al. (2019), using phenotypic, biochemical and gene expression analyses. This study emphasizes the importance of applying the biostimulant at the right time with respect to the stress episode and at right doses, showing enhancement of drought tolerance in specific conditions only. Most notably, seaweed extracts proved to be effective as priming stimulus, when applied one week before stress, and ineffective when the stressor and the biostimulant were applied together or when the biostimulant was applied after the onset of stress. In contrast, an amino acid-based biostimulant tested by the same study showed a less stringent timing requirement for efficacy and did help the plants even when applied together with or during the stress (Fleming et al., 2019).

The timing of application of the biostimulants is critical, and their roles in stress tolerance might be described as 'defense primers' (acting before stress), 'rescuers' (acting during stress) or 'restorers' (acting after stress). As an example of enhanced restoring activity, Dalal et al. (2019) have shown the contrasting responses of two biostimulants derived from seaweed extracts and amino acids in pepper and have shown that the latter did improve the resilience of the plants subjected to drought, indicated by the faster water reabsorption activity of roots after stress withdrawal.

Although such studies are scarce, they can explain the 'inconsistent' efficacy of biostimulants observed in the field by some growers. 'Inconsistency' actually reflects insufficient knowledge of the biostimulation mechanisms as well as of the stress experienced by the recipient plants. This shortcoming prevents from delivering sound advice to growers on the application conditions of biostimulants, raising skepticism among them.

Furthermore, molecular studies have shown that plants respond to the combination of stress factors, like drought and heat, in a unique way, different from the addition of the responses to each individual stress (Mittler and Blumwald, 2010). Since plants experience multiple stressors simultaneously in the field, the extrapolation of laboratory experiments applying single stress to the performance of crops in their actual growing conditions can be problematic.

In conclusion, biostimulation is a promising technology to increase yield in adverse environments, but, as with gene technologies, our limited understanding of plant stress physiology, due to the complexity of the plant's response and the resulting methodological challenges, seems to be a major bottleneck to implement efficient strategies in crops. Our capacity to monitor abiotic stress in the field needs to be improved as well in order to determine the suitable timings of application and to quantify the benefits of biostimulant products.

3.3 Improving crop quality

Crop quality refers to specific characteristics of agricultural and horticultural products, which impact their market value and are targeted by breeding and growing practices. Some of the quality traits are closely related to the nutritional characteristics of the plant, such as the protein content in grains under the control of N nutrition, or to abiotic stress tolerance, such as post-harvest chilling injury in fruits controlled by antioxidant metabolism. Hence, there is some overlap between the first three claims introduced by the European regulation on FPs: improvements in NUE, tolerance to abiotic stress and crop quality.

Considering the many cellular targets of PBs and the cross talk between biostimulants, phytohormones and other plant gene regulators, a wide range of effects in a wide range of crops can be anticipated. For example, the effects of the foliar application of biostimulants on fruit tree physiology have been reviewed by Tanou et al. (2017). In vegetable and fruit crops, contents in secondary metabolites are important quality traits. Applications of *A. nodosum* seaweed extracts have been shown to increase the tissue contents of phenolics and flavonoids in woody and non-woody crops (Frioni et al., 2018; Lola-Luz et al., 2013, 2014a,b).

Although post-harvest biology is an important aspect of the market value of vegetable and fruit crops, little attention has been paid so far to the capacity of biostimulants to improve the shelf-life of crop products. An exception to this is chitosan, a deacetylated derivative of chitin, which has been shown to protect fruits from post-harvest damage, resulting from both biotic and abiotic stress, from ageing caused by the oxidative attack and from deterioration due to water loss (Romanazzi et al., 2017; Sharma et al., 2016). However, for chitosan, like for other biopolymers used as biostimulants, the molecular mass profile of the breakdown products of the polymer has a strong influence on their bioactivity, and film-forming properties of high molecular mass chitosan need to be distinguished from the cellular effects of oligochitosans (Malerba and Cerana, 2016; Kerch, 2015; Dzung et al., 2017). Processing of the biopolymer needs to be tailored to its marketing claim.

3.4 Improving the availability of confined nutrients in the soil or rhizosphere

A fourth claim was added to the definition of biostimulants by the European regulation on FPs during the parliamentary process, which is an improvement in the 'availability of confined nutrients in the soil or rhizosphere'. It is so far unclear which mechanisms will be referred to by this claim, and clarification is expected from the future European standards supporting the implementation of the regulation. Anyway, there seems to be a clear overlap with the first claim of the definition bearing on NUE, as increased bioavailability of nutrients in the soil and rhizosphere can be regarded as one of the components of the uptake efficiency of nutrients, which is itself a component of NUE, as explained before.

As a matter of fact, both microbial and non-microbial biostimulants are known to influence the bioavailability of nutrients, via their chelating/complexing activities and via the biochemical activities of microbes releasing sequestered soil nutrients. Phosphate-solubilizing bacteria and fungi are important examples of such microorganisms, which liberate soluble phosphates from insoluble organic and inorganic forms of phosphate, accounting for most of the total P content of the soil (Richardson et al., 2011; Owen et al., 2015).

4 Identifying the bioactive constituents of plant biostimulants

Considering the complex nature of most biostimulants, the identification of their bioactive constituents is both important and challenging. It is important for the development and marketing of the product, allowing the industry to control and monitor the production of effective biostimulants. Pointing the important compositional characteristics of the biostimulant drives product optimization and helps secure the intellectual property on both the production process and the biostimulant product. However, it is challenging to resolve biostimulants into discrete bioactive components, and it might be impossible in case biostimulation is an 'emerging' property of the molecular assembly, as it has been suggested by Yakhin et al. (2016). As defined by Lüttge (2012), 'emergence is the inevitable unfolding of new functions and structures of a system on a higher scalar integrative level'. In other words, it is hard or impossible - both interpretations can be found in the context of biology - to predict the properties of the system from those of its modular components. Although there is little doubt that the properties of complex biostimulants rest on additions of and interactions between their constituents, whether biostimulation is an emerging property of the molecular assembly, that is, a unique and holistic characteristic of the system as a whole, remains elusive.

No generalization is permitted a priori, and a case-by-case study of the different biostimulants is necessary in order to understand their active constituents. A brief journey through the bioactivity of three main biostimulants will illustrate this point.

4.1 Humic and fulvic acids

Humic and fulvic acids, and humic-like substances derived from composted plant biomass, a most popular and diversified category of biostimulants, offer an interesting example of the uniqueness of the action mechanisms of some biostimulants, which can hardly be extrapolated to other categories (Canellas et al., 2015; Savy et al., 2017). Humic acids are dynamic supramolecular structures containing unbound and weakly (non-covalent) and tightly (ester and ether) bound molecules of plant and bacterial origins. Bioactivity was shown to rest on the unbound and weakly bound components, which are mainly hydrophobic and of bacterial origin. Furthermore, there is evidence that organic acids exuded by plant roots contribute to the release of bioactive constituents by destabilizing the weak chemical bonds of the humic suprastructure. In their turn, the released humic constituents influence root biochemical activity by stimulating their proton and calcium ion membrane transporters. Indeed, only low-molecular-mass molecules released from humic materials are expected to penetrate the root apoplasm and interact with membrane proteins. The word 'humeomics' has been coined to describe the sequential fractionation of humic and humic-like materials and their chemical and functional characterization (Nebbioso and Piccolo, 2011).

4.2 Seaweed extracts

The many actions of seaweed extracts on cell and plant physiology can be explained by their complex composition (Khan et al., 2009; Stirk et al., 2020). Categories of compounds common to seaweeds and higher plants, like phytohormones, and molecules specific to algae, like alginates, are both involved in bioactivity. Regarding plant hormones, the main classes have been found in seaweed extracts - cytokinins, auxins, gibberellins, abscisic acid, brassinosteroids and the ethylene precursor ACC - but the current view is that the action of biostimulants is due to both the regulation of the endogenous production of hormones and to a direct supply of plant growth regulators. Other stress-related compounds like betaines, polyamines and phlorotannins play protective roles in the algae and might do the same when applied to crop plants. Finally, polysaccharides from the seaweed cell walls, for example, alginates, or from their carbon reserves, for example, laminarin, show bioactivities which depend on their chemical forms and hence from the liquefaction process used to produce the seaweed extracts. In the end, multiple sources of variation in the composition of the extracts should be considered: the species, the geographical provenance and the ecology of the populations, as the seaweeds developing on rocky coasts accumulate anti-stress compounds in a way that fluctuates with the daily changes in tide height and depends on their location within the intertidal zone (Connan et al., 2007; Flores-molina et al., 2014). These natural and unavoidable sources of variation are superimposed over the varying industrial processes used to make to biostimulant products. As a consequence, it is both challenging and important for the companies producing biostimulants to monitor and control the content of bioactive constituents in their products. On the other hand, this demand opens opportunities to differentiate the products placed on the market and to build intellectual property on the uniqueness of processes and products.

4.3 Protein hydrolysates

Protein hydrolysates contain amino acids and peptides, but also potentially, and actually, many other substances depending on the sources of materials and their processing. Protein hydrolysates point to a paradox of biostimulant within a circular economy. Indeed, turning industrial wastes into value FPs is challenged by two contrasting demands: the cost-effectiveness of waste processing on the one hand and the efficacy and safety of the derived products, on the other hand, which depend on the chemical fractionation and characterization of the processed materials.

Amino acids and peptides applied to plants modify their nitrogen metabolism and, consequently, the assimilation of carbon and plant growth (Colla et al., 2017). These broad effects are shown by the many genes regulated by protein hydrolysates when applied as biostimulants (Ertani et al., 2017). However, the role of specific peptides acting on specific plant receptors should be considered as well. Endogenous plant peptides with hormone-like activities have been described, and genome-wide analysis in the model plant *Arabidopsis thaliana* has identified about 7000 small coding gene candidates (Takahashi et al., 2019; Hirakawa and Sawa, 2019). All aspects of plant development and metabolism are potentially involved, and the signaling role of peptides circulating in the xylem sap for nutrient homeostasis has been documented (Okamoto et al., 2016). In using protein hydrolysates as biostimulants, a root hair-promoting peptide has been obtained from different leguminous species (Colla et al., 2014, 2017; Matsumiya and Kubo, 2011). Matsumiya and Kubo (2011) isolated and sequenced a peptide from soybean meal degraded by a bacterial protease, which stimulates both root hairs and adventitious roots. Comparing this morphogenetic peptide with other classes of phytohormones, the authors concluded that root hair promotion by the peptide resembled the action of ethylene on the same plant materials, but one of the typical ethylene responses of flowering plants (epinasty) was not provoked by the peptide, suggesting different action mechanisms of the peptide on plant development.

From a methodological perspective, the bioactivity of biostimulants can be characterized by 'omics' technologies - transcriptomics, proteomics and metabolomics (Povero et al., 2016; Jannin et al., 2012, 2013; Ertani et al., 2017) - but in order to give sense to this information, a thorough study of plant phenotypes is essential. High-throughput plant phenotyping platforms are being developed with this aim (Rouphael et al., 2018). Once data sets on both the molecular and phenotypic responses can be crossed, hypotheses on the mechanisms of action can be put forward and further tested in field situations. Performance in the field helps refine hypotheses about the metabolic pathways and cellular processes involved, which can be, in turn, evaluated in controlled conditions. Such two-way journeys between the field and the laboratory are likely to be important for the development of efficient biostimulants in the future. Still, our ambitions need to be modest, bearing in mind that even for the most popular phytohormone auxin, the best and the brightest plant biological research over the past decades has failed to deliver a complete picture of its action in the plant (du Jardin et al., 2020).

5 Microbial biostimulants

Biostimulant substances have shown their capacity to positively influence the plant's physiology, but microorganisms living close to or inside the plant body deliver similar effects on crops. Well before the word 'biostimulant' was used in microbiology, beneficial bacteria and fungi were known, and some of them were used as inoculants to enhance crop productivity. Plant growth-promoting rhizobacteria (PGPRs) has become a popular concept and, remarkably, just like biostimulant substances are defined by their claimed effects and not by their chemical nature, PGPRs are defined by their benefits to plants and not by their taxonomic status. Plant growth-promoting microbes (PGPMs) is a useful wording which extends the concept to non-bacterial and to non-rhizospheric microorganisms (Schlaeppi and Bulgarelli, 2015).

The demonstration of the role of the plant microbiota in plant development, nutrition and immunity can be regarded as a groundbreaking event in plant biology over the last decade (Preininger et al., 2018; Lugtenberg, 2015; Berg et al., 2014; Berendsen et al., 2012; Schlaeppi and Bulgarelli, 2015; Kumar et al., 2017). Although the association of microorganisms with plants has been known since quite some time, the metagenomic analyses of the plant microbiome, the advances in the isolation, amplification and formulation of beneficial microbes and the raising efforts of the industry to develop inoculants as commercial products have opened up new horizons in agriculture. The wording 'plant probiotics' is used, paralleling the probiotics in medicine (Kumar et al., 2017), and the analogy is supported by the current knowledge of both plant rhizospheric and human gut microbiota (Berendsen et al., 2012).

However today, full exploitation of beneficial microorganisms in agriculture awaits answers to major questions. The most difficult ones are related to microbial ecology, how bacteria can establish and maintain in microbial communities and how they interplay with the native soil functions and with the plant's immunity. There is a whole continuum in the modes of co-existence between bacteria and plants, ranging from parasitism (adverse effects) to commensalism (neutral) and mutualism (beneficial). Bacteria living within the plant's influence zone shape their ecological niche, by releasing plant growth hormones, for example, auxins which control root development, and by regulating root exudation. Root exudates supply C to the rhizobacteria, which is much higher in the rhizosphere than in the bulk soil. Bacteria are obviously not altruistic organisms, and even popular commercial PGPR strains have been shown to negatively impact plant activities in specific conditions. An example is the downregulation by the *Bacillus velezensis* (formerly *B. amyloliquefaciens*) FZB42 strain of phosphate transporters of the root membranes upon P starvation when upregulation would be expected to help the plant (Talboys et al., 2014). In general, there is a strong interaction between PGPMs and macronutrients in the way they benefit plant growth (Bradakova et al., 2019; Mpanga et al., 2019; Nguyen et al., 2019).

PGPMs have been shown to interfere with nutrition and plan defense mechanisms, but they also alter the plant's response to abiotic stress. An intriguing example is the ACC deaminase activity carried out by some PGPRs and endophytic bacteria, which inhibits the stress-induced production of ethylene in the plant, and by this way, sustains growth upon stress conditions, including salt, drought and flooding stresses (Glick, 2014). This observation is in line with a proposal that higher crop productivity in adverse environments could be obtained by putting a damper on the growth-arresting mechanisms induced by stress. Some of these mechanisms directly interfere with cell cycle progression in plant meristems and are regulated by the 'stress hormone' ethylene (Skirycz et al., 2011). In a way, plants seem to over-respond to at least some types of stress, and downregulating this responsiveness might help sustain productivity in adverse environments.

There are technical issues limiting the marketing of PGPM inoculants (Arora et al., 2011; Kumar et al., 2017), but, most of all, regulatory issues, due to the intermingling of biostimulation and plant protection as soon as plant-associated microbes are involved. Any PGPM interacting with plants will modulate their immunity, although the pathways and magnitude of the response will depend on the host-microbe combinations (Pieterse et al., 2014). Hence, we should not expect any PGPM selected for, for example, improving plant nutrition not to impact the plant's immunity. Besides the induction of defense mechanisms in plants, many rhizobacteria also show antagonistic activity against other microbes, mediated by antibiotics and other secreted allelochemicals (Kumar et al., 2017). In real life, there can be no partition between biostimulation and plant protection, as organisms interacting with plants will *de facto* alter multiple life traits involved in growth, plant health and pathogen response. Despite this biological fact, regulators have attempted to separate biostimulation and plant protection in the scopes of the corresponding regulations. This is why improved tolerance or resistance to 'biotic' stress was excluded from the regulatory definitions of PBs, keeping abiotic stress tolerance only. This explains why the positive list of microorganisms introduced in the new EU regulation on FPs is limited to a few microorganisms, defined as '*Azotobacter*, Mycorrhizal fungi, *Rhizobium* spp., *Azospirillum* spp.'. These are indeed microorganisms with well-established effects on plant nutrition and used since long as biofertilizers, but absence of plant protection effect cannot be claimed on this basis and would simply oppose scientific evidence. In Europe and in the current status of the regulation, any beneficial microorganisms belonging to other taxonomic entities will be placed on the EU market as biopesticides (plant protection products), or follow national laws on fertilizers when the latter authorize PGPMs as biofertilizers.

6 Conclusion

The elements summarized in this chapter help answer the following questions: Are biostimulants a new paradigm in agriculture? Do they fit the general objective of the sustainable intensification of crops in a warming world? Are biostimulants tools of agroecology aiming at strengthening the natural processes of agrosystems in order to enhance crop performance?

PBs appear as a novel category of agricultural inputs, complementing fertilizers and plant protection products: they are 'physiological effectors', that is, substances and microorganisms, the purpose of which is to affect plant functions in a way benefiting the grower, the agrosystem and the environment. Although examples of such physiological effectors are pretty old - for example, plant growth regulators and herbicide safeners have been used since decades, under pesticides regulations - the new situation is the wider acceptance of the concept and the recognition that many natural substances and microorganisms do interplay with the plant's physiology and make it possible to change plant characteristics by pursuing unprecedented technical pathways. Biostimulants do not challenge plant breeding and cropping techniques, but they complete the picture by providing growers with new means of intervention and new commercial products.

PBs have the potential to act at multiple levels of crop productions, from the shaping of the plant microbiome to the reduction of post-harvest wastes, through the mitigation of plant stress and more efficient use of resources (Fig. 2). Despite this encouraging outlook, biostimulant products should never be considered as a 'one-fits-all' solution, and the challenge is to define conditions of use, making specific biostimulants effective on specific plant characteristics.

By targeting NUE and tolerance to abiotic stress, the claims of biostimulant products fit the objectives of the sustainable intensification of crops. However, as said before, despite the published evidence of the many effects of biostimulants on plants and crops, there is a lack of public data quantifying the actual and global benefits of biostimulants in commercial cropping systems at a large scale.

A further contribution to sustainable development is the source of biostimulants, which includes composts, crop bioproducts and digestates. As such, the new EU regulation on FPs is an ambitious expression of the action plan for the circular economy put forward by the EU in 2015 (Bourguignon, 2016) and was actually driven by the objectives of a circular economy.

Are biostimulants part of agroecology? The answer is no, as agroecology can be adopted by growers without any biostimulant products, and biostimulant products can be used in industrial, high-input agriculture as well as in agroecological agriculture. Despite the difficulty to define agroecology, there is a consensus that agroecology gives priority to ecosystem-based approaches to enhance nutrient cycling, increase organic matter in soils, control pests and pathogens and save water. Enhancing these processes will, in turn, improve plant health, soil fertility, crop productivity and ecosystem resilience (Nicholls et al., 2017). The tools of agroecology are those helping the processes listed earlier, and the current knowledge of biostimulants does indicate opportunities in this regard (Gianinazzi et al., 2010).

Agroecology should not be confused with organic farming. Although both share some inspiring principles and cropping practices, organic farming is defined by regulatory norms, while agroecology is not. Many biostimulant products will comply with the requirements imposed to the inputs of organic agriculture (De Pascale et al., 2017), but it will not be automatically the case, as the European way of defining biostimulants is based on functional claims and not on the source and composition of the products. There is a possibility that synthetic molecules, banned from organic agriculture, are proposed as biostimulants on the European market while satisfying other legal requirements, that is, registration in the REACH system. However, it seems more likely that biostimulants will build future market success on their acceptance as 'natural' products for a 'greener' agriculture.

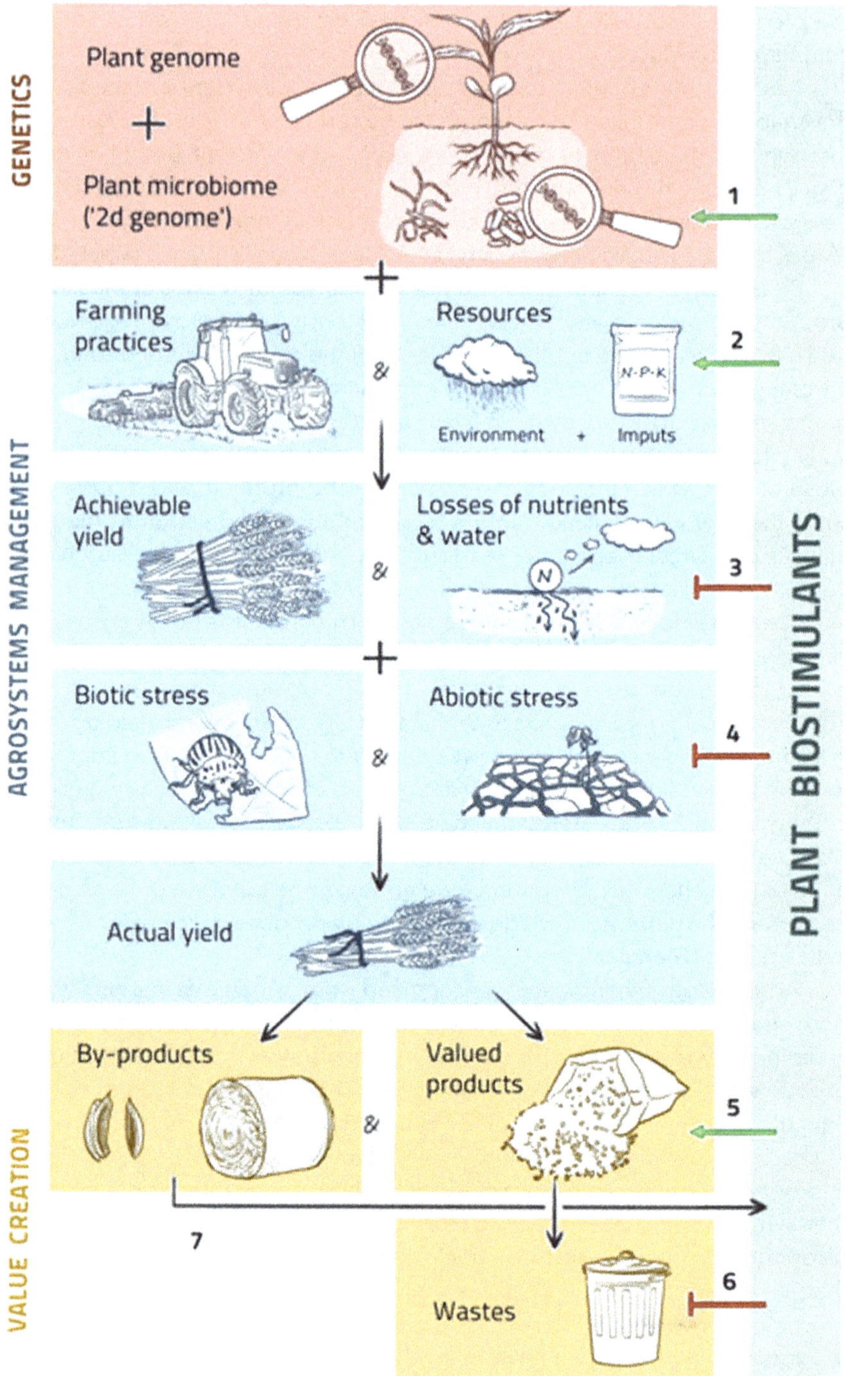
GENETICS
Plant genome
+
Plant microbiome
('2d genome')
+
AGROSYSTEMS MANAGEMENT
Farming practices
&
Resources
Environment + Imputs
N-P-K
Achievable yield
&
Losses of nutrients & water
N
+
Biotic stress
&
Abiotic stress
Actual yield
VALUE CREATION
By-products
&
Valued products
Wastes
1
2
3
4
5
6
7
PLANT BIOSTIMULANTS

◀ **Figure 2** The potential contribution of plant biostimulants to sustainable agriculture and the circular economy. The genetic performance of a plant depends on its genome but also on symbiotic microbiota making the plant microbiome and now considered as its 'second genome'. (1) Biostimulants help shape the microbial communities via inoculants or via substances interacting with the microbiota. (2) Biostimulants increase the bioavailability of nutrients from fertilizers, the soil and the atmosphere and improve their utilization by the plant. (3) Still, not all supplied nutrients are turned into crop products, and a significant part of the surplus is lost to the environment. Biostimulants have the potential to reduce environmental losses by increasing crop nutrients use efficiency. Increased water use efficiency is also expected from the control of root and leaf characteristics of plants by biostimulants. (4) Abiotic stressors, like drought and salinity, compromise yield. Together with pests and pathogens, they turn the achievable yield into the actual yield. Although growers have been using plant protectants for long against biotic stress, few inputs (like plant growth regulators) were available for mitigating 'abiotic' stress before the marketing of biostimulant products. (5) The crop biomass produced can be divided into valued products, by-products and wastes. Biostimulants can be used to improve the quality and economic value of crop products (e.g. tissue contents in nutrients and bioactive molecules), and (6) some of them can reduce wastes caused by the post-harvest damage of crop products. (7) Finally, crop by-products can be used for the production of fertilizing products, including biostimulants, creating value in a circular economy. (Green arrows indicate positive/stimulatory effects, red bars indicate negative/repressive effects; graphic work by Carolina Levicek.)

7 Acknowledgements

The author is grateful to the enlightening discussions about the EU regulation with Dr. Theodora Nikolakopoulou from the European Commission and about the origin of biostimulants research with Dr. Richard Schmidt, Professor Emeritus at Virginia Tech, and with Dr. Acheampong Atta-Boateng, Professor at Oxford University.

8 Where to look for further information

Besides the many review articles mentioned in the text, further information can be found in a special issue of the journal *Scientia Horticulturae* where articles on the different categories of plant biostimulants were published in 2015 (volume 196), and in a book on the Chemical biology of Plant Biostimulants that has been issued recently (Geelen, D., Xu, L. (Eds.), The Chemical Biology of Plant Biostimulants, John Wiley & Sons Ltd). Both the special issue and the book chapters introduce leading research teams in their respective topics, which the reader is invited to follow. Websites of industry coalitions are useful to follow developments in products and regulations (http://www.biostimulants.eu/ and http://www.biostimulantcoalition.org/). A leading company in the development of biostimulants produces educational resources, including podcasts and webinars, to help advance the biostimulant industry (http://www.biostimulant.com/). As research on biostimulants is a meeting point of academia and industry, international congresses specific to the agricultural uses of biostimulants have been sponsored by the industry and organized every second year since 2012 (by New AG International – Informa Connect). More than 1500 participants have contributed to the 2019 edition (see https://informaconnect.com/4th-biostimulants-world-congress/) and both the scientific abstracts and company communications and exhibition materials (see #BIOSTIMULANTSWORLDCONGRESS for examples) give a global picture of biostimulants development.

9 References

Arora, N. K., Khare, E. and Maheshwari, D. K. 2011. Plant growth promoting rhizobacteria: constraints in bioformulation, commercialization, and future strategies. In: Maheshwari, D. K. (Ed.), *Plant Growth and Health Promoting Bacteria*. Microbiology Monographs. Springer, Berlin Heidelberg, Berlin, Heidelberg, pp. 97–116. doi:10.1007/978-3-642-13612-2.

Barone, V., Baglieri, A., Stevanato, P., Broccanello, C., Bertoldo, G., Bertaggia, M., Cagnin, M., Pizzeghello, D., Moliterni, V. M. C., Mandolino, G., Fornasier, F., Squartini, A., Nardi, S. and Concheri, G. 2018. Root morphological and molecular responses induced by microalgae extracts in sugar beet (Beta vulgaris L.). *J. Appl. Phycol.* 30, 1061–71. https://doi.org/10.1007/s10811-017-1283-3.

Berendsen, R. L., Pieterse, C. M. and Bakker, P. A. 2012. The rhizosphere microbiome and plant health. *Trends Plant Sci.* 17(8), 478–86. doi:10.1016/j.tplants.2012.04.001.

Berg, G., Grube, M., Schloter, M. and Smalla, K. 2014. Unraveling the plant microbiome: looking back and future perspectives. *Front. Microbiol.* 5, 148. doi:10.3389/fmicb.2014.00148.

Billard, V., Etienne, P., Jannin, L., Garnica, M., Cruz, F., Garcia-Mina, J.-M., Yvin, J.-C. and Ourry, A. 2014. Two biostimulants derived from algae or humic acid induce similar responses in the mineral content and gene expression of winter oilseed rape (*Brassica napus* L.). *J. Plant Growth Regul.* 33(2), 305–16. doi:10.1007/s00344-013-9372-2.

Bourguignon, D. 2016. Closing the loop - New circular economy package. *Eur. Parliam. Res. Serv, PE* 573, 899.

Bradakova, K., Sittinger, M., Tietz, K., Neuhauser, B., Kandeler, E., Berger, N., Ludewig, U. and Neumann, G. 2019. Maize inoculation with microbial consortia: contrasting effects on rhizosphere activities, nutrient acquisition and early growth in different soils. *Microorganisms* 7(9), 1–16. doi:10.3390/microorganisms7090329.

Brown, P. and Saa, S. 2015. Biostimulants in agriculture. *Front. Plant Sci.* 6, 671. doi:10.3389/fpls.2015.00671.

Calvo, P., Nelson, L. and Kloepper, J. W. 2014. Agricultural uses of plant biostimulants. *Plant Soil* 383(1–2), 3–41. doi:10.1007/s11104-014-2131-8.

Canellas, L. P., Piccolo, A., Dobbss, L. B., Spaccini, R., Olivares, F. L., Zandonadi, D. B. and Façanha, A. R. 2010. Chemical composition and bioactivity properties of size-fractions separated from a vermicompost humic acid. *Chemosphere* 78(4), 457–66. doi:10.1016/j.chemosphere.2009.10.018.

Canellas, L. P., Olivares, F. L., Aguiar, N. O., Jones, D. L., Nebbioso, A., Mazzei, P. and Piccolo, A. 2015. Humic and fulvic acids as biostimulants in horticulture. *Sci. Hortic. (Amsterdam)* 196, 15–27. doi:10.1016/j.scienta.2015.09.013.

Caradonia, F., Battaglia, V., Righi, L., Pascali, G. and La Torre, A. 2019. Plant biostimulant regulatory framework: prospects in Europe and current situation at international level. *J. Plant Growth Regul.* 38(2), 438–48. doi:10.1007/s00344-018-9853-4.

Carillo, P., Colla, G., Fusco, G. M., Dell'Aversana, E., El-Nakhel, C., Giordano, M., Pannico, A., Cozzolino, E., Mori, M., Reynaud, H., Kyriacou, M. C., Cardarelli, M. and Rouphael, Y. 2019. Morphological and physiological responses induced by protein hydrolysate-based biostimulant and nitrogen rates in greenhouse spinach. *Agronomy* 9(8), 1–22. doi:10.3390/agronomy9080450.

Colla, G., Rouphael, Y., Canaguier, R., Svecova, E. and Cardarelli, M. 2014. Biostimulant action of a plant-derived protein hydrolysate produced through enzymatic hydrolysis. *Front. Plant Sci.* 5, 448. doi:10.3389/fpls.2014.00448.

Colla, G., Hoagland, L., Ruzzi, M., Cardarelli, M., Bonini, P., Canaguier, R. and Rouphael, Y. 2017. Biostimulant action of protein hydrolysates: unraveling their effects on plant physiology and microbiome. *Front. Plant Sci.* 8, 2202. doi:10.3389/fpls.2017.02202.

Connan, S., Deslandes, E. and Gall, E. A. 2007. Influence of day-night and tidal cycles on phenol content and antioxidant capacity in three temperate intertidal brown seaweeds. *J. Exp. Mar. Bio. Ecol.* 349(2), 359–69. doi:10.1016/j.jembe.2007.05.028.

Dalal, A., Bourstein, R., Haish, N., Shenhar, I., Wallach, R. and Moshelion, M. 2019. Dynamic physiological phenotyping of drought-stressed pepper plants treated with 'productivity-enhancing' and 'survivability-enhancing' biostimulants. *Front. Plant Sci.* 10, 905. doi:10.3389/fpls.2019.00905.

De Pascale, S., Rouphael, Y. and Colla, G. 2017. Plant biostimulants: innovative tool for enhancing plant nutrition in organic farming. *Eur. J. Hortic. Sci.* 82, 277–85. doi:10.17660/eJHS.2017/82.6.2.

du Jardin, P. 2012. *The Science of Plant Biostimulants – a Bibliographic Analysis*. EU Publication, Brussels, Belgium. Available at https://op.europa.eu/en/publication-detail/-/publication/5c1f9a38-57f4-4f5a-b021-cad867c1ef3c

du Jardin, P. 2015. Plant biostimulants: definition, concept , main categories and regulation. *Sci. Hortic.* 196, 3–14. doi:10.1016/j.scienta.2015.09.021.

du Jardin, P., Xu, L. and Geelen, D. 2020. Agricultural functions and action mechanisms of plant biostimulants (PBS): an introduction. In: Geelen, D. and Xu, L. (Eds), *The Chemical Biology of Plant Biostimulants*. John Wiley & Sons Ltd, Chichester, UK, pp. 1–30.

Dzung, P. D., Phu, D. V., Du, B. D., Ngoc, L. S., Duy, N. N., Hiet, H. D., Nghia, D. H., Thang, N. T., Le, B. V. and Hien, N. Q. 2017. Effect of foliar application of oligochitosan with different molecular weight on growth promotion and fruit yield enhancement of chili plant. *Plant Prod. Sci.* 20(4), 389–95. doi:10.1080/1343943X.2017.1399803.

Ertani, A., Schiavon, M. and Nardi, S. 2017. Transcriptome-wide identification of differentially expressed genes in *Solanum lycopersicon* L. in response to an *alfalfa*-protein hydrolysate using microarrays. *Front. Plant Sci.* 8, 1159. doi:10.3389/fpls.2017.01159.

EU. 2019. Regulation of the European Parliament and of the Council laying down rules on the making available on the market of EU fertilising products and amending regulations (EC) No 1069/2009) and (EC) No 1107/2009 and repealing Regulation (EC) No 2003/2003.

Fleming, T. R., Fleming, C. C., Levy, C. C. B., Repiso, C., Hennequart, F., Nolasco, J. B. and Liu, F. 2019. Biostimulants enhance growth and drought tolerance in *Arabidopsis thaliana* and exhibit chemical priming action. *Ann. Appl. Biol.* 174(2), 153–65. doi:10.1111/aab.12482.

Flores-Molina, M. R., Thomas, D., Lovazzano, C., Nú, A., Zapata, J., Kumar, M., Correa, J. A. and Contreras-Porcia, L., 2014. Desiccation stress in intertidal seaweeds: effects on morphology , antioxidant responses and photosynthetic performance 113, 90-9. doi:10.1016/j.aquabot.2013.11.004.

Frioni, T., Sabbatini, P., Tombesi, S., Norrie, J., Poni, S., Gatti, M. and Palliotti, A. 2018. Effects of a biostimulant derived from the brown seaweed *Ascophyllum nodosum* on ripening dynamics and fruit quality of grapevines. *Sci. Hortic. (Amsterdam)* 232, 97–106. doi:10.1016/j.scienta.2017.12.054.

Gianinazzi, S., Gollotte, A., Binet, M. N., van Tuinen, D., Redecker, D. and Wipf, D. 2010. Agroecology: the key role of arbuscular mycorrhizas in ecosystem services. *Mycorrhiza* 20(8), 519–30. doi:10.1007/s00572-010-0333-3.

Glick, B. R. 2014. Bacteria with ACC deaminase can promote plant growth and help to feed the world. *Microbiol. Res.* 169(1), 30–9. doi:10.1016/j.micres.2013.09.009.

Hilker, M., Schwachtje, J., Baier, M., Balazadeh, S., Bäurle, I., Geiselhardt, S., Hincha, D. K., Kunze, R., Mueller-Roeber, B., Rillig, M. C., Rolff, J., Romeis, T., Schmülling, T., Steppuhn, A., van Dongen, J., Whitcomb, S. J., Wurst, S., Zuther, E. and Kopka, J. 2016. Priming and memory of stress responses in organisms lacking a nervous system. *Biol. Rev.* 91(4), 1118–33. doi:10.1111/brv.12215.

Hirakawa, Y. and Sawa, S. 2019. Diverse function of plant peptide hormones in local signaling and development. *Curr. Opin. Plant Biol.* 51, 81–7. doi:10.1016/j.pbi.2019.04.005.

Jakab, G., Ton, J., Flors, V., Zimmerli, L., Métraux, J. P. and Mauch-Mani, B. 2005 Enhancing Arabidopsis salt and drought stress tolerance by chemical priming for its abscisic acid responses. *Plant Physiol.* 139(1), 267–74. doi:10.1104/pp.105.065698.

Jannin, L., Arkoun, M., Ourry, A., Laîné, P., Goux, D., Garnica, M., Fuentes, M., Francisco, S. S., Baigorri, R., Cruz, F., Houdusse, F., Garcia-Mina, J., Yvin, J.-C. and Etienne, P. 2012. Microarray analysis of humic acid effects on *Brassica napus* growth: involvement of N, C and S metabolisms. *Plant Soil* 359(1–2), 297–319. doi:10.1007/s11104-012-1191-x.

Jannin, L., Arkoun, M., Etienne, P., Laîné, P., Goux, D., Garnica, M., Fuentes, M., Francisco, S. S., Baigorri, R., Cruz, F., Houdusse, F., Garcia-Mina, J., Yvin, J.-C. and Ourry, A. 2013. Brassica napus growth is promoted by *Ascophyllum nodosum* (L.) Le Jol. seaweed extract: microarray analysis and physiological characterization of N, C, and S metabolisms. *J. Plant Growth Regul.* 32(1), 31–52. doi:10.1007/s00344-012-9273-9.

Kerch, G. 2015. Chitosan films and coatings prevent losses of fresh fruit nutritional quality: a review. *Trends Food Sci. Technol.* 46(2), 159–66. doi:10.1016/j.tifs.2015.10.010.

Khan, W., Rayirath, U. P., Subramanian, S., Jithesh, M. N., Rayorath, P., Hodges, D. M., Critchley, A. T., Craigie, J. S., Norrie, J. and Prithiviraj, B. 2009. Seaweed extracts as biostimulants of plant growth and development. *J. Plant Growth Regul.* 28(4), 386–99. doi:10.1007/s00344-009-9103-x.

Kinnersley, A. M. 1993. The role of phytochelates in plant growth and productivity. *Plant Growth Regul.* 12(3), 207–18. doi:10.1007/BF00027200.

Kumar, V., Kumar, M., Sharma, S. and Prasad, R. 2017. *Probiotics and Plant Health.* doi:10.1007/978-981-10-3473-2.

Lola-Luz, T., Hennequart, F. and Gaffney, M. 2013. Enhancement of phenolic and flavonoid compounds in cabbage (*Brassica oleraceae*) following application of commercial seaweed extracts of the brown seaweed (*Ascophyllum nodosum*). *AFSci* 22(2), 288-95. doi:10.23986/afsci.7676.

Lola-Luz, T., Hennequart, F. and Gaffney, M. 2014a. Effect on health promoting phytochemicals following seaweed application, in potato and onion crops grown under a low input agricultural system. *Sci. Hortic. (Amsterdam)* 170, 224–7. doi:10.1016/j.scienta.2014.03.022.

Lola-Luz, T., Hennequart, F. and Gaffney, M. 2014b. Effect on yield, total phenolic, total flavonoid and total isothiocyanate content of two broccoli cultivars (*Brassica oleraceae* var. *italica*) following the application of a commercial brown seaweed extract (*Ascophyllum nodosum*). *AFSci* 23(1), 28–37. doi:10.23986/afsci.8832.

Lugtenberg, B. 2015. *Principles of Plant-Microbe Interactions - Microbes for Sustainable Agriculture*. Springer.

Lüttge, U. 2012. Modularity and emergence: biology's challenge in understanding life. *Plant Biol.* 14(6), 865–71. doi:10.1111/j.1438-8677.2012.00659.x.

Lynch, J. P. 2013. Steep, cheap and deep: an ideotype to optimize water and N acquisition by maize root systems. *Ann. Bot.* 112(2), 347–57. doi:10.1093/aob/mcs293.

Malerba, M. and Cerana, R. 2016. Chitosan effects on plant systems. *Int. J. Mol. Sci.* 17(7), 1–15. doi:10.3390/ijms17070996.

Matsumiya, Y. and Kubo, M. 2011. Soybean peptide: novel plant growth promoting peptide from soybean. In:. El-Shemy, H. (Ed.), *Soybean and Nutrition*. Tech Europe Publisher, Rijeka, Croatia, pp. 215–30.

Mittler, R. and Blumwald, E., 2010. Genetic engineering for modern agriculture: challenges and perspectives. *Annu. Rev. Plant Biol.*, 61(1), 443–462. https://doi.org/10.1146/annurev-arplant-042809-112116

Mola, I. Di, Cozzolino, E., Ottaiano, L., Giordano, M., Rouphael, Y., Colla, G. and Mori, M. 2019. Effect of vegetal- and seaweed extract-based biostimulants on agronomical and leaf quality traits of plastic tunnel-grown baby lettuce under four regimes of nitrogen fertilization. *Agronomy* 9(10). doi:10.3390/agronomy9100571.

Mpanga, I. K., Nkebiwe, P. M., Kuhlmann, M., Cozzolino, V., Piccolo, A., Geistlinger, J., Berger, N., Ludewig, U. and Neumann, G. 2019. The form of N supply determines plant growth promotion by P-solubilizing microorganisms in maize. *Microorganisms* 7(2), 38. doi:10.3390/microorganisms7020038.

Nebbioso, A. and Piccolo, A. 2011. Basis of a humeomics science: chemical fractionation and molecular characterization of humic biosuprastructures. *Biomacromolecules* 12(4), 1187–99. doi:10.1021/bm101488e.

Neuschafer, B. and Paisner, B. 2019. Plant biostimulants and plant regulators: EPA publishes draft guidance on the applicability of FIFRA. *Plant Coalit*. Web Site. WWW Document.

Nguyen, M. L., Glaes, J., Spaepen, S., Bodson, B., du Jardin, P. and Delaplace, P. 2019a. Biostimulant effects of *Bacillus* strains on wheat from in vitro towards field conditions are modulated by nitrogen supply. *J. Plant Nutr. Soil Sci.* 182(3), 325–34. doi:10.1002/jpln.201700610.

Nguyen, M. L., Spaepen, S., du Jardin, P. and Delaplace, P. 2019b. Biostimulant effects of rhizobacteria on wheat growth and nutrient uptake depend on nitrogen application and plant development. *Arch. Agron. Soil Sci.* 65(1), 58–73. doi:10.1080/03650340.2018.1485074.

Nicholls, C., Altieri, M. and Vazquez, L. 2017. Agroecological principles for the conversion of farming systems. In: Wezel, A. (Ed.), *Agroecological Practices for Sustainable Agriculture*. World Scientific Publishing Europe Ltd., London, UK, pp. 1–18.

Okamoto, S., Tabata, R. and Matsubayashi, Y. 2016. Long-distance peptide signaling essential for nutrient homeostasis in plants. *Curr. Opin. Plant Biol.* 34, 35–40. doi:10.1016/j.pbi.2016.07.009.

Owen, D., Williams, A. P., Griffith, G. W. and Withers, P. J. A. 2015. Use of commercial bio-inoculants to increase agricultural production through improved phosphrous acquisition. *Appl. Soil Ecol.* 86, 41–54. doi:10.1016/j.apsoil.2014.09.012.

Pieterse, C. M. J., Zamioudis, C., Berendsen, R. L., Weller, D. M., Van Wees, S. C. M. and Bakker, P. A. H. M. 2014. Induced systemic resistance by beneficial microbes. *Annu. Rev. Phytopathol.* 52, 347–75. doi:10.1146/annurev-phyto-082712-102340.

Povero, G., Mejia, J. F., Tommaso, D. and Piaggesi, A. 2016. A systematic approach to discover and characterize natural plant Biostimulants. *Front. Plant Sci.* 7, 1–9. https://doi.org/10.3389/fpls.2016.00435.

Preininger, C., Sauer, U., Bejarano, A. and Berninger, T. 2018. Concepts and applications of foliar spray for microbial inoculants. *Appl. Microbiol. Biotechnol.* 102(17), 7265–82. doi:10.1007/s00253-018-9173-4.

Richardson, A. E., Lynch, J. P., Ryan, P. R., Delhaize, E., Smith, F. A., Smith, S. E., Harvey, P. R., Ryan, M. H., Veneklaas, E. J., Lambers, H., Oberson, A., Culvenor, R. A. and Simpson, R. J. 2011. Plant and microbial strategies to improve the phosphorus efficiency of agriculture. *Plant Soil* 349(1–2), 121–56. doi:10.1007/s11104-011-0950-4.

Romanazzi, G., Feliziani, E., Baños, S. B. and Sivakumar, D. 2017. Shelf life extension of fresh fruit and vegetables by chitosan treatment. *Crit. Rev. Food Sci. Nutr.* 57(3), 579–601. doi:10.1080/10408398.2014.900474.

Rouphael, Y., Spíchal, L., Panzarová, K., Casa, R. and Colla, G. 2018. High-throughput plant phenotyping for developing novel biostimulants: from lab to field or from field to lab? *Front. Plant Sci.* 9, 1–6. https://doi.org/10.3389/fpls.2018.01197

Russo, R. O. and Berlyn, G. P. 1991. The use of organic biostimulants to help low input sustainable agriculture. *J. Sustain. Agric.* 1(2), 19–42. doi:10.1300/J064v01n02_04.

Russo, R. O. and Berlyn, G. P. 1992. Vitamin-humic-algal root biostimulant increases yield of Green bean. *HortScience* 27(7), 847. doi:10.21273/HORTSCI.27.7.847.

Savvides, A., Ali, S., Tester, M. and Fotopoulos, V. 2016. Chemical priming of plants against multiple abiotic stresses: mission possible? *Trends Plant Sci.* 21(4), 329–40. doi:10.1016/j.tplants.2015.11.003.

Savy, D., Canellas, L., Vinci, G., Cozzolino, V. and Piccolo, A. 2017. Humic-like water-soluble lignins from giant reed (*Arundo donax* L.) display hormone-like activity on plant growth. *J. Plant Growth Regul.* 36(4), 995–1001. doi:10.1007/s00344-017-9696-4.

Schlaeppi, K. and Bulgarelli, D. 2015. The plant microbiome at work. *Mol. Plant Microbe Interact.* 28(3), 212–7. doi:10.1094/MPMI-10-14-0334-FI.

Sharma, S., Barman, K. and Siddiqui, M. W. 2016. Chitosan: properties and roles in postharvest quality preservation of horticultural crops. In: Siddiqui, M. W. (Ed.) *Eco-Friendly Technology for Postharvest Produce Quality*. Elsevier Inc. doi:10.1016/B978-0-12-804313-4.00009-8.

Skirycz, A., Claeys, H., de Bodt, S., Oikawa, A., Shinoda, S., Andriankaja, M., Maleux, K., Eloy, N. B., Coppens, F., Yoo, S. D., Saito, K. and Inzé, D. 2011. Pause-and-stop: the effects of osmotic stress on cell proliferation during early leaf development in

Arabidopsis and a role for ethylene signaling in cell cycle arrest. *Plant Cell* 23(5), 1876–88. doi:10.1105/tpc.111.084160.

Stirk, W. A., Rengasamy, K. R. R., Kulkarni, M. G. and van Staden, J. 2020. Plant biostimulants from seaweed: an overview. In: Geelen, D. and Xu, L. (Eds), *The Chemical Biology of Plant Biostimulants*. John Wiley & Sons Ltd, Chichester, UK, pp. 33–56.

Storer, K., Kendall, S., White, C., Roques, S. and Berry, P. 2016. *A Review of the Function, Efficacy and Value of Biostimulant Products Available for UK Cereals and Oilseeds*. Agriculture and Horticulture Development Board (ADHB). Kenilworth, UK.

Takahashi, F., Hanada, K., Kondo, T. and Shinozaki, K. 2019. Hormone-like peptides and small coding genes in plant stress signaling and development. *Curr. Opin. Plant Biol.* 51, 88–95. doi:10.1016/j.pbi.2019.05.011.

Talboys, P. J., Owen, D. W., Healey, J. R., Withers, P. J. A. and Jones, D. L. 2014. Auxin secretion by *Bacillus amyloliquefaciens* FZB42 both stimulates root exudation and limits phosphorus uptake in *Triticum aestivium*. *BMC Plant Biol.* 14, 51. doi:10.1186/1471-2229-14-51.

Tanou, G., Ziogas, V. and Molassiotis, A. 2017. Foliar nutrition, biostimulants and prime-like dynamics in fruit tree physiology: new insights on an old topic. *Front. Plant Sci.* 8, 75. doi:10.3389/fpls.2017.00075.

Traon, D., Amat, L., Zotz, F. and du Jardin, P. 2014. *A Legal Framework for Plant Biostimulants and Agronomic Fertiliser Additives in the EU*. EU Publication, Brussels, Belgium. Available at https://op.europa.eu/en/publication-detail/-/publication/5c1f9a38-57f4-4f5a-b021-cad867c1ef3c.

Ussiri, D. and Lal, R. 2013. *Soil Emission of Nitrous Oxide and Its Mitigation*. Springer, Dordrecht, Heidelberg, New York, London. doi:10.1007/978-94-007-5364-8.

Van Ittersum, M. K., Cassman, K. G., Grassini, P., Wolf, J., Tittonell, P. and Hochman, Z. 2013. Yield gap analysis with local to global relevance: a review. *Crop. Res.* 143, 4–17. doi:10.1016/j.fcr.2012.09.009.

Yakhin, O. I., Lubyanov, A. A., Yakhin, I. A. and Brown, P. H. 2016. Biostimulants in plant science: a global perspective. *Front. Plant Sci.* 7, Article 2049. doi:10.3389/fpls.2016.02049.

Zhang, X. and Schmidt, R. E. 2000. Hormone-containing products' impact on antioxidant status of tall fescue and creeping bentgrass subjected to drought. *Crop Sci.* 40(5), 1344–9. doi:10.2135/cropsci2000.4051344x.

Chapter 2

Bioactive compounds and evaluation of biostimulant activity

Luigi Lucini and Begoña Miras-Moreno, Università Cattolica del Sacro Cuore, Italy; and Andrea Ertani, Università degli Studi di Padova, Italy

1 Introduction

The concept of biostimulant activity is inevitably connected with current and future regulations and regulatory prescriptions regarding the placement on the market of biostimulants. Currently, regulatory schemes are very variable and range from free access to the market to pre-market approval processes that include data requirements. However, in most cases, the identification/definition/characterization requirements of such regulatory schemes do not make it possible to fully characterize substances. In the United States, biostimulants are regulated in fertilizer legislation (www.aapfco.org) at state agricultural department level, since there is not a federal law. Of course, the complexity of the regulatory dossier and the inherent costs are dramatically affected by the regulation scheme to be considered. A general scheme of the different authorization processes is summarized in Fig. 1.

Regardless of the type of regulation (which may be harmonized in the near future), key considerations to address are efficacy claims and content declarations.

According to the existing regulations, there is still no agreement on the terms, since the use of terms like 'material' and 'substance' can create possible

http://dx.doi.org/10.19103/AS.2020.0068.02

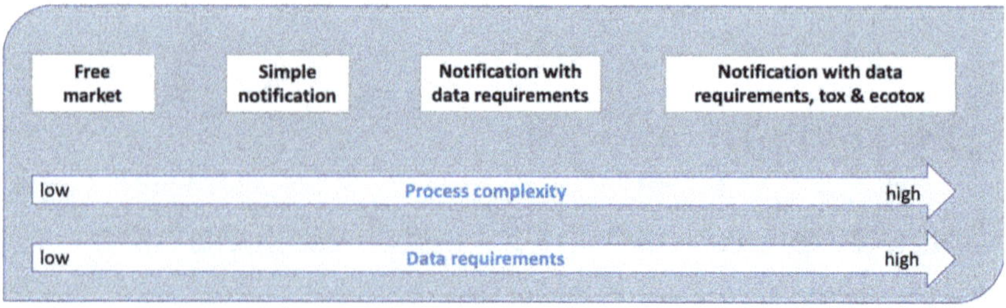

Figure 1 Different authorization processes.

confusion. Starting from that, it is important to distinguish between regulatory schemes that cover the products as supplied to the users (formulations) rather than the components of the product.

Speaking about biostimulants, also the terms 'plant' and 'rhizosphere' can create some confusion, as they appear too restrictive (the former not explicitly including seeds, the second often being too limited in scope).

Another definition, for which there is no agreement (despite being one of the claims related to biostimulants), is the concept of crop/fruit quality. Potentially, this is a very broad concept overlapping with other claims related to plant protection products and growth regulators. Therefore, the definition of fruit quality is a major challenge in the future regulation(s) of plant biostimulants.

Nonetheless, it is also important to keep in mind that the actual composition of a biostimulant product is very complex (since they arise from natural products), that they trigger a wide variety of physiological triggers with many cellular targets. This makes the molecular/physiological characterization of biostimulant action very challenging but still a very exciting goal with potentially large benefits. Furthermore, since plant biostimulants include a wide diversity of products ranging from inorganic and organic components, up to microorganisms, the data required to support authorization processes should be probably focused and declined according to the nature of the biostimulant.

Within this framework, and given the relatively new introduction of biostimulants, there are no clear definitions about the active components (biostimulants are currently defined based on their activity) and even less about regulatory needs and experimental trials that might be necessary for market regulation purposes. Therefore, the current chapter aims to address this gap by providing a framework regarding the definition of active components, the mode(s) of action and the current regulation, as well as by highlighting the most recent approaches being proposed by the scientific community to investigate such complex mode(s) of action.

2 Active components

Several of the existing regulations refer to a 'material', speaking about biostimulant products. This term is rather vague, and we should better refer to substance(s), microorganism and their combinations. This implies that we should intend biostimulants in the form as they are supplied to the final users.

Within this concept, we consider substance(s) in the same way as fertilizers or plant protection products, that is, meaning chemical element(s) or compound(s) as they are present naturally or by manufacture (thus including any side-component or impurity inevitably resulting from the manufacturing process).

The following categories of substances can be distinguished:

- **Substances from synthesis**: Substances which are produced by chemical reactions under defined processes, hence resulting in repeatable profiles and purity ranges.
- **Natural or botanical substances:** Substances spontaneously occurring in nature either unprocessed or processed. Typically, gravitational or mechanical processes, water dissolution or extraction (eventually under heating), distillation and/or concentration, can occur.
- **Microorganisms:** Any microbiological entity qualitatively identified by the species and the strain, and quantitatively defined in terms of viable cells or spores per mass. With this regard, the definition of strain can be taken from the OECD document on microbials for pest control (OECD No. 43, ENV/JM/MONO(2008)36): 'a strain is a population of organisms that descends from a single cell or a pure culture isolate. Typically, it is the result of a succession of cultures ultimately deriving from an initial single colony.'

The complete definition of a given substance implies that we have to include not only its identification and characterization, but also relevant and physicochemical properties and the eventual presence of contaminants, the definition of manufacturing process and quality control, as well as the identification of the related analytical method(s).

This means that the constituent(s) contributing to the claimed biostimulant effects should be identified whenever possible. However, especially in the case of natural substances, this wide definition of substance can raise up the demand for a challenging analytical characterization. Therefore, despite leaving the space to more subjective definitions, we must take into account that such complex definition is not possible in all the cases. Furthermore, in several cases there might be several components synergically exerting the biostimulant effect to the plant. These multiple challenges related to the definition and identification of the components being responsible of the biological activity has led to a limited agreement about the actual bioactive compound(s). Roughly, we can identify either chemical substances or microorganisms. Among the former, few classes are widely recognized to play a pivotal role in determining the actual biostimulant effect, namely humic substances (HS), protein hydrolysates (PH), seaweed extracts (SE) and chitosan (CH). Several other chemicals have been ascribed among biostimulant compounds, although in some cases there is a lower agreement among scientists rather than preliminary evidences.

Humic substances (HS) are a component of the soil organic matter, resulting from the biochemical and microbial decomposition of plant, animal and microbial residues. HS structure is complex and has been proposed to be a supramolecular association of heterogeneous compounds. HS chemistry has been proposed as a self-assembled supramolecular association of small heterogeneous molecules whose structure is assembled mainly because of weak hydrophobic interactions. Their composition has been traditionally proposed according to their solubility at different pH values.

Among HS, fulvic acids are associations of small hydrophilic molecules rich in acid groups, thus remaining dispersed in solution at any pH, while humic acids are made of hydrophobic compounds such as polymethylenic chains, fatty acids and steroids compounds. Unlike fulvic acids, humic acids are stable only at neutral pH, since they are held by hydrophobic dispersive forces. Indeed, humic acids aggregate at lower pH because of intermolecular hydrogen bonds being progressively formed until humic matter flocculates.

The colloidal nature of such supramolecular compounds shows complex association/dissociation dynamics as a function of soil characteristics and root exudates in the rhizosphere. Notwithstanding, HS show a wide chemical diversity as a function of the source considered, the pedo-climatic conditions, the interaction with plant(s) and the time of residence into soil.

Another important class of biostimulants is represented by **protein hydrolysates** (PHs), mixtures of free amino acids and peptides that are obtained by chemical or enzymatic hydrolysis of proteins from either plant or animal origin. Chemical hydrolysis of proteins is done under strong acid or alkaline conditions, that is, a very aggressive process carried out at high temperature and pressure. Such strong conditions lead to a high degree of hydrolysis and the corresponding products show a high content of free amino acids rather than peptides. Moreover, chemical hydrolysis leads to the destruction of several amino acids (total degradation of tryptophan under acid hydrolysis; partial degradation of cysteine, serine and threonine; conversion of asparagine and glutamine into their acidic forms) as well as to racemization. However, hydrolysis using proteolytic enzymes of animal rather than plant origin, is milder than chemical hydrolysis, usually targets specific peptide bonds and does not require high temperatures. Therefore, PH from enzymatic hydrolysis are typically a mixture of amino acids and peptides of different lengths, as a function to the conditions used.

Besides PH, other nitrogen-containing molecules include **betaines and polyamines**, whose degree of biological activity is somehow under debate. Among betaines, glycine betaine (trimethyl glycine) is an amino acid derivative being widely recognized as a stress-modulating compound. Interestingly, glycine betaine has been identified in PH where it might play a synergistic activity, together with amino acids and peptides.

As far as concerned, **seaweed extracts** (SE), polysaccharides, are major components and have been proposed among bioactive compounds. Indeed, polysaccharides may account for up to 30–40% of the extract on a dry weight basis. Brown algae are the most commonly used seaweeds, with *Ascophyllum*, *Fucus* and *Laminaria* as main genera. The most common polysaccharides found in brown seaweed extracts include fucoidans, alginates, laminarins and fucose containing glucans. However, red seaweeds (which correspond to a distinct phylogenetic line) have been reported to be rich in carrageenans.

The season of harvest, the extraction conditions from the raw material, the actual algal species determine the actual composition of polysaccharides. Fucoidans are fucose polymers differing in structures for their degree of branching, degree of methylation or sulfation. Alginates are polymers of guluronic and mannuronic acids, whereas laminarins are beta-(1,3)-linked glucans with 6-*O* branching and beta-(1,6) intrachain links. Carrageenans are high-molecular-weight linear polysaccharides made of repeating galactose and 3,6-anhydrogalactose, both sulfated and non-sulfated, linked by alpha-(1,3)- and beta-(1,4) linkages.

Seaweeds are also rich in phenolic compounds including phlorotannins (complex polymers of phloroglucinol), compounds presenting a higher number of phenolic rings in their structure, thus showing radical scavenging activities. Phytohormones including auxins, cytokinins, gibberellins, abscisic acid and brassinosteroids have been also reported in SE, as well as the polyol mannitol and several betaines and betaine analogues. The dominant betaines in *A. nodosum* are γ-aminobutyric acid betaine (ABAB), glycine betaine, δ-aminovaleric acid betaine, γ-aminobutyric acid betaine and laminine.

The last class of widely used biostimulants is represented by CH. Chitin and **chitosan** are both copolymers of glucosamine and *N*-acetyl-glucosamine, differing in the ratio of each monomer of the polymer chain. The monosaccharide residues in both chitin and chitosan are linked together via beta-(1,4) linkages. However, chitin is typically made by more than 95% of *N*-acetyl-glucosamine. Chitosan is derived from chitin and can be quite different as a function of the degree of polymerization and the degree of deacetylation. Such parameters strongly affect the physical-chemical properties and the biological activity of chitosan. Treatments with NaOH are usually adopted to modulate the degree of deacetylation, whereas chemical and/or enzymatic treatments are used to modulate the degree of polymerization.

Finally, **inorganic compounds** represent another class of biostimulants. They include silicates and carbonates playing distinct roles from their function as nutrients. Silicon is also gaining popularity among biostimulants; this element is mainly present in soil solution as monosilicic acid (H_4SiO_4). When silicic acid solubility limit is exceeded, for example, during evapotranspiration, precipitation to colloidal silicic acid and conversion to amorphous silica ($SiO_2 \cdot nH_2O$) occur.

3 Data requirements

The definitions used in pre-market approval systems that specify the data required for authorization are a key factor affectingthe market for biostimulants. The schemes designed for synthetic chemicals and plant protection products in general include risk assessment procedures that lead to very high costs for applicants and which are difficult to apply to natural products, as most biostimulants are. Other existing schemes, such as the REACH regulation at the EU level (for chemicals manufactured or supplied at or above 1 ton/year), do not consider microorganisms.

Therefore, it is evident that the market of biostimulants requires a dedicated regulation focused on the inherent nature of biostimulants. To date, there is no general agreement about the data required to support the commercialization of biostimulants. Although efficacy is a prerequisite and must be demonstrated in all countries, few authorization schemes require e-complex dossier that includes also toxicological and ecotoxicological data.

The abovementioned wide diversity of products, ranging from inorganic and organic components up to microorganisms, should also be considered when the data required for supporting authorization processes are to be produced.

As a reference framework here, the principal authorization schemes implemented at the EU level are summarized below.

Chemicals (REACH): Manufacturers or importers are required to provide a registration for all chemical compounds commercialized at or above 1 ton per year. A specified set of properties of these substances (proportionally to the amounts manufactured) must be made available to assess the hazards and risks that a substance may pose and the consequent measures of control.

Plant protection products: The regulation (EC) No 1107/2009 defines a set of information that must be made available for both the active substance(s) and all formulated products derived from it. These information include tox and ecotox data, physical-chemical information, efficacy and residues in the targeted commodities (including the methods for post-authorization control). As a general concept underlying the authorization, health and environment protection predominate over agricultural production.

Biocides: Biocides require a pre-market authorization based on the regulation (EC) No 528/2012. The regulation introduces formal exclusion and substitution criteria in case of carcinogenicity, reprotoxicity, endocrine disruptors or persistent, bioaccumulative and toxic substances.

It becomes clear that none of the existing schemes can adequately support the use of biostimulants. Ideally, a 'dossier' on biostimulant products should cover at least:

- The identification and characterization of the material, including the absence of contaminants or hazardous components/contaminants (e.g. heavy metals, polycyclic aromatic hydrocarbons, polychlorinated biphenyls, polychlorinated dibenzodioxins, potential to produce toxins and pathogenic potential for microorganisms).
- The biological and physicochemical properties of the product.
- Information on the manufacturing process and the related quality control.
- The demonstration of agronomic efficiency and the related claim(s).

Some other information, such as the mode of action and eventually also toxicology and environmental impact of the product, are advisable as well. In this case, the concepts of technical equivalence, data sharing and data bridging (taken from the regulations for plant protection products) could be used in order to avoid repeating studies, especially when animal testing is involved. However, it is questionable whether data justification based on history of safe use (rather than qualified presumptions of safety for microorganisms) shall be fully considered in the authorization process. In all the cases, the authority in charge of conducting risk assessment should be the sole party providing a scientific opinion, in order to keep separate the responsibilities for risk assessment and risk management.

Concerning the **agronomic efficiency**, it is essential to specify the dosages, crop growth stage(s) and/or environmental conditions that maximize the chance of having a biostimulant effect. In the absence of common agreement, the conduct of 3–6 field trials should be considered as minimal investigation to support biological efficacy in agreement with the European and Mediterranean Plant Protection Organization on good agricultural practice. Such trials must consider the specific environmental, pedological and climatic conditions, together with plant health conditions.

A reliable definition of agronomic efficiency is defined by the balance between positive aspects (e.g. improved yield, lower chemical inputs) and eventual negative aspects (e.g. phytotoxicity, detrimental effects on the microbial population). In other words, efficacy should provide an overall improvement in crop production that justifies the use of the biostimulant product.

4 Mode of action

The definition of a mode of action is of enormous utility to support agronomic efficiency rather than a given claim for biostimulants. With this regard, it becomes important to clearly distinguish between 'mode of action' and 'effect'. The former refers to a biological activity related to biochemical (chemical or enzymatic), physiological or cellular phenomena. However, the effect is defined as the proven result of a given mode of action and allows the definition of claims.

Generally, we can assume that the level of evidence for a specific effect allows the users to verify it. Overall, effects typically related to plant biostimulants are tightly connected to the definition of biostimulants, and include:

- increased yield (biomass, fruits) or quality of the products,
- increased nutrient-use efficiency,
- increased plant functions (hormone-like effects, improved photosynthetic efficiency), and
- improved tolerance to abiotic stress (salinity, drought, temperature).

Some of these effects are well-established concepts in agronomy and plant science, this is the case of yield and nutrient-use efficiency. The latter refers to the availability and uptake, transport and/or assimilation of the nutrients.

Some other concepts are vague and still need a precise definition. This is the case of 'crop and fruit quality', generally referring to desired traits or attributes. This concept is broad and includes functional quality (e.g. increased benefits to human health), sensorial quality (organoleptic properties), technological quality (e.g. fruit size, postharvest performance) and environmental quality (e.g. increased sustainability). Remarkably, the evaluation of fruit quality improvements is also very complicated and multifactorial. As an example, it can be noted that amino acids from protein hydrolysates can be biosynthetic precursors contributing to aroma (e.g. alanine, leucine, isoleucine, valine), color (phenylalanine for anthocyanins) and taste (arginine, alanine, glycine and proline).

Biostimulants are often a mixture of different bioactive substances, likely acting in a synergic way. This means that multiple modes of action can be identified, thus complicating the elucidation of the processes involved in plant response to biostimulants. Preliminary evidences about the effects of biostimulants in plants can be gained from the analysis of scientific literature, eventually making reference to 'gray' literature and/or history of use. However, care must be taken in this case, because sometimes information regarding the exact definition/characterization of the material tested is missing, and because very few works deepened their investigation until the identification of the mode(s) of action of the active substance(s). Furthermore, controlled conditions (often not representative of open-field real conditions) or model plants (in place of crops) are used.

Therefore, dedicated ad hoc field studies are strongly recommended to better address the definition of both mode(s) of action and effect(s) for plant biostimulants. These trials should be conducted according to the guidelines provided by manufacturers and include:

- multi locations and multi-year experimental designs (e.g. at least 3–4 locations over a minimum period of 2 crop cycles),
- a preliminary definition of the most appropriate rates and timings of application,
- the evaluation of a dose-response relationship,
- tests either under optimal plant conditions or under stress conditions (as a function of the claims to be supported), and
- the concurrent investigation of any unintended effects (e.g. phytotoxicity).

Although the adoption of proper statistic treatments is a common prerequisite in all cases, a reliable assessment of yield improvements is not an easy task. Some authors proposed that small plot experiments are not rugged enough in detecting significant yield differences of less than 0.5 t/ha, even though smaller improvements are likely to be involved (at least under optimal plant growth conditions). On the other hand, it is also important to note that plant biostimulants are often designed to prevent yield losses under limiting conditions, rather than improve yields under optimal conditions. All these considerations should be carefully taken into account when designing the experimental work. An alternative to classic approaches is to use plant phenotyping approaches. Plant phenotyping refers to a quantitative description of the plant's anatomical, morphological and biochemical properties by using optical sensors. This high-throughput non-destructive approach has been made possible, thanks to the advance in image-analysis-based characterization of plant traits. Plant phenotyping makes it possible to monitor the actual growth performance, morphological and architectural parameters using multispectral image analysis, thermographical analysis, reflectance and fluorescence analysis. The high throughput of the approach is ensured by the possibility to use automated facilities in greenhouses and growth chambers, thus allowing for the comparison of several hundred plants per day.

Noteworthy, the evaluation of mode(s) of action, following the conduct of tailored trials, is also very complicated. Indeed, plant metabolism is very complicated, with more than 200 000 different metabolites reported in a single cell, and pathways are interconnected one to each other. Furthermore, several compounds or protein functions still need to be elucidated, and no reference materials are available for most of them. The comprehension of such complex

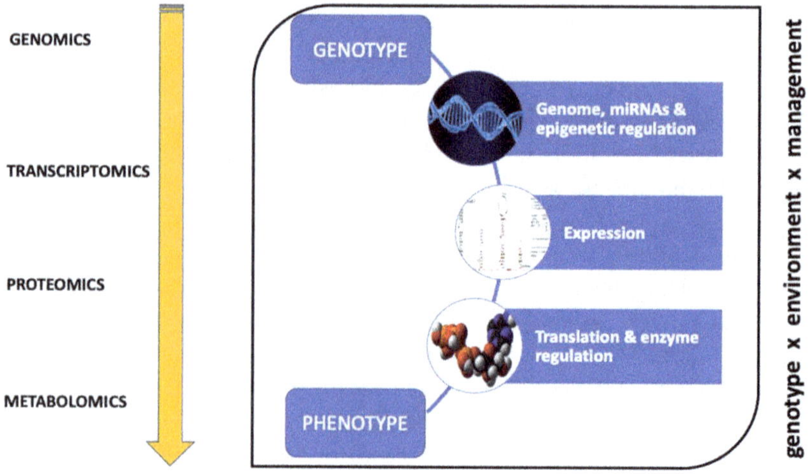

Figure 2 Molecular processes in plant responses to biostimulants.

molecular interplay and regulation of cellular functions at different levels can greatly benefit from the so-called omics sciences including genomics, transcriptomics, proteomics and metabolomics. The untargeted, holistic, no hypothesis-driven nature of omics sciences allows providing a huge amount of information (the so-called big data) that, opportunely managed through multivariate statistics, can allow to shed light onto the molecular processes involved in plant response to biostimulants (Fig. 2).

The most powerful approach, still complicated and not routinely available, is the systems biology approach, according to which a computational analysis and modeling of molecular processes is provided by the integration between different omics outputs within biological systems.

5 The omics approach

Although phenotypical approaches are required for assessing the successful effect of a biostimulant on plant crops, other holistic strategies should be taken into consideration to elucidate their mode of action. The omics approach is a global assessment of a set of genes, proteins, metabolites or phenotype parameters in an organism. Plant omics approach implies a holistic understanding of data collections to comprehend the complex plant system (Gemperline et al., 2016). In recent years, the development of new technologies such as second- and third-generation nucleic acid sequencing, second-generation peptide-sequencing platforms together with bioinformatics and statistical methodologies have allowed the emergence of -omics sciences. In a broad perspective, the principal -omics are genomics, transcriptomics, proteomics and metabolomics although more specific subcategories have been arising that include transposomics, glycomics and lipidomics (Hakeem et al., 2016). Annotation data drives to highlight the evolutionary relationships as well as to identify genes and elucidate the molecular basis of agronomically important traits, such as yield and tolerance to abiotic and biotic stresses (Hakeem et al., 2016). For this purpose, several studies revealed that -omics strategies imply potent tools to explain the plant response to biostimulant application.

Particularly, the study of genome and transcriptome, the total DNA and RNA, respectively, of a cell or organism, have been applied to this field as an upstream approach. For instance, microarrays analysis supported by physiological evidence pointed out that a plastid division regulator was involved in the plant responses activated in *Brassica napus* after seaweed extract (Jannin et al., 2013). Some differences in gene expression involved in plant stress response and senescence were found after humic acid application (Jannin et al., 2012). In addition, Ertani et al. (2017) used cDNA microarrays to better understand the responses of tomato plants stimulated by alfalfa-derived PH. They showed a connection between PH and upregulation of genes involved in essential processes such as carbon and nitrogen metabolism or photosynthesis, as well as genes involved in secondary pathways such as phenol and terpene-related gene expression. Moreover, mRNA sequencing allowed Wilson et al. (2015) to insight the mode of action of gelatin seed treatment, which increases plant growth in cucumber probably by enhancing the expression levels of related genes.

As well as the study of plant genome and transcriptome are crucial to understand the plant behavior after stress or biostimulant application, the comprehension of protein involvement in plant response contributes to broaden their knowledge on biological processes at molecular level. Proteomics is the study of the protein set included in a tissue, cell or subcellular compartment, which plays an essential role in plant survival and adaptation to external stresses. Although more studies are necessary to enhance existing knowledge about protein response to biostimulants, the application of seaweed extract, fungal extract or amino acid-based product to wheat crop has been studied at the proteome level by several authors (Pichereaux et al., 2019; Martínez-Esteso et al., 2016). These studies highlighted the implication of proteins related to storage, regulation and defense in the mechanism of reaction to fungal and marine biostimulants (Pichereaux et al., 2019). The results obtained by Martínez-Esteso et al. (2016) suggest an increase of wheat productivity supported by changes in CO_2 fixation protein synthesis and oxidative stress proteins.

Although genomics and proteomics are essential tools for understanding plant physiology and biological activities, plant metabolomics emerges as an approach much closer to phenotypes and therefore, suitable for the study of environmental and ecological interactions. Metabolomics provides a metabolite profiling and allows comparing the plant behavior under both normal and stressed conditions (Hong et al., 2016; Hakeem et al., 2016). In this sense, PH have been studied by using metabolomics-untargeted approach to highlight their mode of action (Lucini et al., 2015; Paul et al., 2019a,b). Specifically, plant PH applied to lettuce under stress condition provoked changes in the oxidative stress and osmotic adjustment, hormone profiling and secondary metabolites (Lucini et al., 2015). Similarly, a metabolic/proteomic reprograming was found in tomato in the presence or absence of abiotic stress (Paul et al., 2019a,b) and following mycorrhiza colonization under drought (Bernardo et al., 2017, 2019). Different product-based biostimulants were tested to highlight their mode of action by studying the metabolic profiling changes as in the case of biopolymer in melon (Lucini et al., 2018) and a lipo-complex formulation in pepper plants formulation (Barrajón-Catalán et al., 2020). In addition, metabolomics has been used for plant functional and nutraceutical properties, food quality or traceability, among others (Lucini et al., 2016; Rocchetti et al., 2018a,b; Hakeem et al., 2016).

Besides laboratory analyses, plant phenotyping has also emerged in the last years. Plant phenotyping platforms provide automated measure of several traits related to plant phenotype over a time-course, using sensors. This technique is non-invasive and can be used indoor (where the response of individual plants can be restricted by pot growth) or outdoor (in this case we measure canopy, and the restriction is that experiments are season dependent). The use of growth-chambers allows experiments to be conducted under controlled conditions of temperature, light/dark cycle and photosynthetically active radiation (PAR) light intensity. When other factors are included in plant experiments (e.g., time/dosages of application, presence/absence of stresses, etc.), then the real throughput of phenomics depends on the level of automation and platform capacity (Rouphael et al., 2018). Phenotyping platforms larger than growth chambers allow integrating watering and weighing and even controlled nutrient delivery to plant. Field facilities for phenomics have been also proposed, with the aim of achieving very realistic plant growth conditions. With this regard, fixed platforms (in which the sensors do not move, such as for towers, tripods and wireless sensor networks) and mobile platforms (ranging from fixed rails to tractor-based systems, buggies, rovers) have been developed.

Considering that the plant phenotype is the result of a complex genome-environment-management interaction, it is evident that the approach can be successfully used to investigate or to validate biostimulant activity. The image analysis of plant development rather than the whole canopy is typically done using a red, green and blue (RGB) camera, whereas a dedicated camera can investigate chlorophyll fluorescence. Morphometric parameters can be collected continuously and then used to calculate germination rates and to infer projected canopy height, a parameter to follow plant growth kinetics.

Among others, phenotyping has been successfully proposed to investigate the different tolerance to drought (Marchetti et al., 2019) or other abiotic stress conditions (De Diego et al., 2017; Danzi et al., 2019). Recently, phenomics has been successfully proposed for screening potential biostimulants to study the biostimulant effects (Povero et al., 2016, Ugena et al., 2018; Paul et al., 2019a,b; Briglia et al., 2019).

Starting from the most recent works, a combined approach that integrates different omics (the so-called system biology perspective) is being proposed also in plant science, including the study of biostimulants. Indeed, the integration of data from phenomics with other omics may be very useful to provide complementary information thus gaining insight on key processes related to molecular, biochemical and/or physiological mechanisms of plant response to biostimulant application (Danzi et al., 2019; Paul et al., 2019b).

6 Hormone-like activity and in vitro assays

The extreme heterogeneity of biostimulant products and their several influences on plants have attracted the attention of researchers during the years (Thalheimer and Paoli, 2002; Schiavon et al., 2008; Ertani et al., 2009). Biostimulants can influence plant metabolism and morphology by interacting with a variety of biochemical mechanisms and physiological processes, thereby stimulating growth and increasing the amount of nutrients taken up by the plants (Colla and Rouphael, 2015; Caruso et al., 2019; Carillo et al., 2019, Ertani et al., 2019a). A considerable amount of literature shows that under laboratory and field conditions, biostimulants can have a beneficial effect on plants in terms of increases in lengths, fresh and dry weights of shoots and roots, leaf chlorophyll concentration, number of lateral root initials and many other biological parameters (Paul et al., 2019a,b; Lucini et al., 2018; Ertani et al., 2019b). The ecological roles of biostimulants are important because they are able to interact with plants, thanks to their hormone-like activity, as demonstrated by several studies (Nardi et al., 2018; Palumbo et al., 2018; Ertani et al., 2018a.b; Scaglia et al., 2015). This property involves all the activities of biostimulants in regulating plant biochemical and physiological processes (Adams-Phillips et al., 2004; Khan et al., 2009; Nardi et al., 1988, 1991, 2000, 2016). However, the reported hormonal effects did not solely correlate to the amount of IAA detected in the biostimulants (Ertani et al., 2013), but also due to the presence of different compounds such as phenols or amino acids.

The researches on seaweed-based biostimulants have demonstrated a wide range of positive effects on plants, such as enhanced crop performance and yield (Ertani et al., 2018a) and increased resistance to stress (Hankins and Hockey, 1990; Blunden, 1991; Norrie and Hiltz, 1999). These characteristics were imputable to plant growth regulators (Crouch and van Staden, 1993), osmolites (e.g. betaines), amino acids and other compounds. The biological action of these compounds was evaluated using bioassays and their presence was confirmed by gas chromatography/mass spectrometry and nuclear magnetic resonance (Crouch and van Staden, 1993). The determination of the hormone-like activity of biostimulants is a key to understand the mechanisms of action and the physiological responses potentially induced in plants. A bioassay is a laboratory method which, by using portions of plants or whole plants, allows the evaluation of quality and intensity of the effects induced by a substance, in order to allow its inclusion in a class of compounds capable of eliciting similar responses.

A compound possesses hormone-like activity when it induces effects comparable to those induced by the hormones. These tests are based on the physiological responses and allow to compare the effects induced by substances of different origin and chemical characteristics, with hormones such as auxins or gibberellins. In the case of auxins, although the path of transduction of the auxinic signal still needs to be completely clarified, the cytological response to the hormone is cellular distension, which occurs thanks to the acidification of the apoplast and consequent reorganization of the cell wall (Teale et al., 2006).

Gibberellins are involved in the regulation of numerous physiological processes that span all stages of plant development, from promoting seed germination to fruiting (Gupta and Chakrabarty, 2013). The processes of response to gibberellins better understood, however, are those related to the growth of the stem. Over the years, various products have been subjected to these bioassays: humic substances (Muscolo et al., 1993; Giannattasio et al., 2014; Palumbo et al., 2018), algae extracts (Ertani et al., 2018b) and protein hydrolysates (Schiavon et al., 2008; Ertani et al., 2009).

Rayorath et al. (2008) proposed the used bioassays to test the bioactivity of seaweed extract and its different fractions. *Arabidopsis thaliana* was selected as a plant for these bioassays as it provides several advantages, such as small size of plant, easy growth, fast life cycle, small genome size and accessibility of large amounts of genetic evidences, making it a perfect laboratory model (Meinke et al., 1998). The bioassays include Arabidopsis root-tip elongation bioassay, Arabidopsis liquid growth bioassay and greenhouse growth bioassay. The root growth elongation assay could be used as an indicator for salt stress conditions (Halfter et al. 2000), but it is not an ideal system for studies under water-deficit and drought conditions. Greenhouse assays could be used to test several parameters of plant shoot growth such as fresh and dry weight.

Alternative bioassay was proposed by Zandonadi et al. (2016) based on the proton pump activity as marker of biostimulant activity. Proton pumps present in the vacuolar and plasma membranes are the main agents responsible for the regulation of the electrochemical gradient that energizes the nutrient uptake system and acid growth mechanism of plant cells. The proposed protocol, based on the root acidification using a pH-sensitive dye and the Micro-Tom tomato as a model plant, is a procedure to evaluate the efficacy of humic substances that are claimed to improve nutrient acquisition and root growth. The authors found a direct correlation between plasmalemma proton-adenosine triphosphatase (H+-ATPase) activation and the amount of rhizosphere acidification observed in the bromocresol gel.

Biological activity of humic substances (HS) are the topic of research in several studies, because these compounds can greatly benefit plant physiology (Nardi et al., 2006, 2017). In these studies, the authors proved the hormone-like activity using bioassays (Audus, 1972) that compare the physiological effects induced by HS with the responses elicited by hormones (indoleacetic acid (IAA) and gibberellic acid (GA)). The IAA-like and GA-like activities were assessed by checking the reduction in the growth of watercress (*Lepidium sativum* L.) roots and the increase in the length of lettuce (*Lactuca sativa* L.) epicotyls, respectively (Audus, 1972). These two bioassays were applied to several biostimulants derived from different matrices, such as protein hydrolysate derived from alfalfa plants (EM) (Ertani et al., 2012), leather of animal origin (MH) (Ertani et al., 2009, 2013), agro-industrial residues (Ertani et al., 2012) or cow manure preparation (Giannattasio et al., 2014; Morau et al., 2020). Also, in paper of Schiavon et al. (2008), the authors studied the biological activity of EM using bioassays. The results have evidenced that EM possess gibberellin- and auxin-like activity. Again, Ertani et al. (2009) demonstrated that a product obtained by fully controlled enzymatic hydrolysis of leather of animal origin elicited both a high gibberellin-like activity and a weak auxin-like one.

Cytokinins (CK) hormones serve many important functions in plant development and morphogenesis because they are involved in the regulation of cell division, in the control of apical dominance and lateral branching and the root-shoot ratio in intact plants and in tissue culture. The cytokinin-like activity in biostimulants was evaluated using the methods proposed by Letham (1971). In specific, the cytokinin-like activity was evaluated by weighing the cotyledon leaves of treated radish and comparing them with radish treated with IPA. Pizzeghello et al. (2013) used this bioassay to check the cytokinin-like activity of humic substances extracted from lignohumates, leonardite and earthworm faeces. The results have shown that IPA as well as HS induced significant dose-dependent responses on the weight of cotyledons.

Given the existence of different techniques and bioassays for the characterization of biostimulants, it must however be emphasized that the extreme heterogeneity and complexity of the matrices and formulations of biostimulants makes it difficult to quantify.

The different bioassays can be used as an effective and sensitive method to evaluate the biostimulant activity of different compounds. The results obtained from the application of these protocols must however be combined with tests in controlled conditions and in the open field.

7 Conclusion and future trends

Despite the fact that plant biostimulants can be considered as the 'newborns' among agricultural inputs, their rapid increase in market penetration has attracted the attention of manufacturers, farmers and plant scientists. For biostimulants, we probably will not be able to define an active substance, as we do for pesticides, but rather active components. This will be the starting point for hypothesis-driven targeted studies, aimed to support one or more of the claims that the regulations have related to biostimulant activity.

The current context in terms of biological activity and comprehension of biostimulants mode of action offers great opportunities but also big challenges.

From one side, it is crucial to identify the active component(s) and the mode(s) of action for a biostimulant product, to support the placement on the market. This is important for all the actors of the chain, starting from manufacturers to regulatory until the end users (i.e. the farmers).

On the other side, we must consider the wide chemical diversity of products on the market, their convoluted composition and the technically complex investigations needed to support the currently defined biostimulant claims.

Within this context, the opportunities relate to the early definition of regulations concerning biostimulant activity. This might retrace what was already done in past decades for fertilizers and pesticides, while taking into account the peculiarities of biostimulants. At the same time, developing and adopting feasible approaches to sustain biostimulant claims will not be an easy task. In fact, the support of recent scientific advances (e.g. -omic sciences) are growing fast, in parallel with plant science related to biostimulants. In the near future, such recent scientific progress can efficiently complement more classical approaches like bioassays and/or field trials.

The combination of technically available and cost-effective scientific approaches, together with an industrial perspective in products development, is likely to guarantee the market of biostimulants in a long-term perspective. In turn, this will guarantee more sustainable agricultural productions, where chemical inputs can be optimized to ensure the expectations of farmers (yield, economic efficiency, quality of the produce) as well as consumers and stakeholders (quality, environmental aspects, safety).

8 References

Adams-Phillips, L., Barry, C. and Giovannoni, J. 2004. Signal transduction systems regulating fruit ripening. *Trends Plant. Sci.* 9(7), 331–8. doi:10.1016/j.tplants.2004.05.004.

Audus, L. J. 1972. Plant growth substances chemistry and physiology. In: Hill, L. (Ed.), *Plant Science Monographs* (vol. 1). University of California, London, UK, pp. 24–65.

Barrajón-Catalán, E., Álvarez-Martínez, F. J., Borrás, F., Pérez, D., Herrero, N., Ruiz, J. J. and Micol, V. 2020. Metabolomic analysis of the effects of a commercial complex biostimulant on pepper crops. *Food Chem.* 310, 125818. doi:10.1016/j.foodchem.2019.125818.

Bernardo, L., Morcia, C., Carletti, P., Ghizzoni, R., Badeck, F. W., Rizza, F., Lucini, L. and Terzi, V. 2017. Proteomic insight into the mitigation of wheat root drought stress by arbuscular mycorrhizae. *J. Prot.* 169, 21–32. doi:10.1016/j.jprot.2017.03.024.

Bernardo, L., Carletti, P., Badeck, F. W., Rizza, F., Morcia, C., Ghizzoni, R., Rouphael, Y., Colla, G., Terzi, V. and Lucini, L. 2019. Metabolomic responses triggered by arbuscular mycorrhiza enhance tolerance to water stress in wheat cultivars. *Plant Physiol. Biochem.* 137, 203–12. doi:10.1016/j.plaphy.2019.02.007.

Blunden, G. 1991. Agricultural uses of seaweeds and seaweed extracts. In: Guiry, M. D. and Blunden, G. (Eds), *Seaweed Resources in Europe: Uses and Potential*. Wiley, Chichester, UK, pp. 65–81.

Briglia, N., Petrozza, A., Hoeberichts, F. A., Verhoef, N. and Povero, G. 2019. Investigating the impact of biostimulants on the row crops corn and soybean using high-efficiency phenotyping and next generation sequencing. *Agronomy* 9(11). doi:10.3390/agronomy9110761.

Carillo, P., Colla, G., Fusco, G. M., Dell'Aversana, E., El-Nakhel, C., Giordano, M., Pannico, A., Cozzolino, E., Mori, M., Reynaud, H., Kyriacou, M. C., Cardarelli, M. and Rouphael, Y. 2019. Morphological and physiological responses induced by protein hydrolysate-based biostimulant and nitrogen rates in greenhouse spinach. *Agronomy* 9(8), 450. doi:10.3390/agronomy9080450.

Caruso, G., Stoleru, V., De Pascale, S., Cozzolino, E., Pannico, A., Giordano, M., Teliban, G., Cuciniello, A. and Rouphael, Y. 2019. Production, leaf quality and antioxidants of perennial wall rocket as affected by crop cycle and mulching type. *Agronomy* 9(4), 194. doi:10.3390/agronomy9040194.

Colla, G. and Rouphael, Y. 2015. Biostimulants in horticulture. *Sci. Hortic.* 196, 1–2. doi:10.1016/j.scienta.2015.10.044.

Crouch, I. J. and van Staden, J. 1993. Evidence for the presence of plant growth regulators in commercial seaweed products. *Plant Growth Regul.* 13(1), 21–9. doi:10.1007/BF00207588.

Danzi, D., Briglia, N., Petrozza, A., Summerer, S., Povero, G., Stivaletta, A., Cellini, F., Pignone, D., De Paola, D. and Janni, M. 2019. Can high throughput phenotyping help food security in the Mediterranean area? *Front. Plant Sci.* 10, 15 doi:10.3389/fpls.2019.00015.

De Diego, N., Fürst, T., Humplík, J. F., Ugena, L., Podlešáková, K. and Spíchal, L. 2017. An automated method for high-throughput screening of *Arabidopsis rosette* growth in multi-well plates and its validation in stress conditions. *Front. Plant Sci.* 8, 1702 doi:10.3389/fpls.2017.01702.

Ertani, A., Cavani, L., Pizzeghello, D., Brandellero, E., Altissimo, A., Ciavatta, C. and Nardi, S. 2009. Biostimulant activity of two protein hydrolyzates in the growth and nitrogen metabolism of maize seedlings. *J. Plant Nutr. Soil Sci.* 172(2), 237–44. doi:10.1002/jpln.200800174.

Ertani, A., Nardi, S. and Altissimo, A. 2012. Review: Long-term research activity on the biostimulant properties of natural origin compounds. *Acta Hortic.* 1009, 181–8.

Ertani, A., Pizzeghello, D., Baglieri, A., Cadili, V., Tambone, F., Gennari, M. and Nardi, S. 2013. Agro-industrial residues and their biological activity on maize (*Zea mays* L.) metabolism. *J. Geochem. Explor.* 129, 103–11.

Ertani, A., Schiavon, M. and Nardi, S. 2017. Transcriptome-wide identification of differentially expressed genes in *Solanum lycopersicon* L. in response to an alfalfa-protein hydrolysate using microarrays. *Front. Plant Sci.* 8, 1159. doi:10.3389/fpls.2017.01159.

Ertani, A., Francioso, O., Ferrari, E., Schiavon, M. and Nardi, S. 2018a. Spectroscopic chemical fingerprint and biostimulant activity of a protein-based product in solid form. *Molecules* 23(5), 1031. doi:10.3390/molecules23051031.

Ertani, A., Francioso, O., Tinti, A., Schiavon, M., Pizzeghello, D. and Nardi, S. 2018b. Evaluation of seaweed extracts from Laminaria and *Ascophyllum nodosum* spp. as biostimulants in *Zea mays* L. using a combination of chemical, biochemical and morphological approaches. *Front. Plant Sci.* 9, 428. doi:10.3389/fpls.2018.00428.

Ertani, A., Nardi, S., Francioso, O., Pizzeghello, D., Tinti, A. and Schiavon, M. 2019a. Metabolite-targeted analysis and physiological traits of *Zea mays* L. in response to application of a leonardite-humate and lignosulfonate-based products for their evaluation as potential biostimulants. *Agronomy* 9(8), 445. doi:10.3390/agronomy9080445.

Ertani, A., Nardi, S., Francioso, O., Sanchez-Cortes, S., Di Foggia, M. D. and Schiavon, M. 2019b. Effects of two protein hydrolysates obtained from chickpea (*Cicer arietinum* L.) and Spirulina platensis on *Zea mays* (L.) plants. *Front. Plant Sci.* 10, 954. doi:10.3389/fpls.2019.00954.

Gemperline, E., Keller, C. and Li, L. 2016. Mass spectrometry in plant-omics. *Anal. Chem.* 88(7), 3422–34.

Giannattasio, M., Vendramin, E., Fornasier, F., Alberghini, S., Zanardo, M., Stellin, F., Concheri, G., Stevanato, P., Ertani, A., Nardi, S., Rizzi, V., Piffanelli, P., Spaccini, R., Mazzei, P., Piccolo, A. and Squartini, A. 2014. Mikrobiologie und Bioaktivität Des biodynamischen Präparats 500. *Lebendige Erde* 3, 42–5.

Gupta, R. and Chakrabarty, S. K. 2013. Gibberellic acid in plant. *Plant Signal. Behav.* 8(9), 9. doi:10.4161/psb.25504.

Hakeem, K. R., Tombuloğlu, H. and Tombuloğlu, G. (Eds). 2016. *Plant Omics: Trends and Applications*. Springer, Basel, Switzerland.

Halfter, U., Ishitani, M. and Zhu, J. K. 2000. The *Arabidopsis* SOS2 protein kinase physically interacts with and is activated by the calciumbinding protein SOS3. *Proc. Natl. Acad. Sci. USA* 97(7), 3735–40. doi:10.1073/pnas.040577697.

Hankins, S. D. and Hockey, H. P. 1990. The effect of a liquid seaweed extract from *Ascophyllum nodosum* (Fucales, Phaeophyta) on the two spotted red spider mite *Tetranychus urticae*. *Hydrobiologia* 204–205(1), 555–9. doi:10.1007/BF00040286.

Hong, J., Yang, L., Zhang, D. and Shi, J. 2016. Plant metabolomics: an indispensable system biology tool for plant science. *Int. J. Mol. Sci.* 17(6), 767. doi:10.3390/ijms17060767.

Jannin, L., Arkoun, M., Ourry, A., Laîné, P., Goux, D., Garnica, M., Fuentes, M., San Francisco, M., Baigorri, R., Cruz, F., Houdusse, F., García-Mina, J. M., Yvin, J. C. and Etienne, P. 2012. Microarray analysis of humic acid effects on Brassica napus growth: involvement of N, C and S metabolisms. *Plant Soil* 359, 297–319.

Jannin, L., Arkoun, M., Etienne, P., Laîné, P., Goux, D., Garnica, M., Fuentes, M., San Francisco, M., Baigorri, R., Cruz, F., Houdusse, F., García-Mina, J. M., Yvin, J. C. and Ourry, A. 2013. *Brassica napus* growth is promoted by *Ascophyllum nodosum* (L.) Le

Jol. seaweed extract: microarray analysis and physiological characterization of N, C, and S metabolisms. *J. Plant Growth Regul.* 32, 31-52.

Khan, W., Rayirath, U. P., Subramanian, S., Jithesh, M. N., Rayorath, P., Hodges, D. M., Critchley, A. T., Craigie, J. S., Norrie, J. and Prithiviraj, B. 2009. Seaweed extracts as biostimulants of plant growth and development. *Plant Growth Regul.* 28(4), 386-99. doi:10.1007/s00344-009-9103-x.

Letham, D. S. 1971. Regulators of cell division in plant tissues. XII. A cytokinin bioassayusing excised radish cotyledons. *Physiol. Plant.* 25(3), 391-6. doi:10.1111/j.1399-3054.1971.tb01462.x.

Lucini, L., Rouphael, Y., Cardarelli, M., Canaguier, R., Kumar, P. and Colla, G. 2015. The effect of a plant-derived biostimulant on metabolic profiling and crop performance of lettuce grown under saline conditions. *Sci. Hortic.* 182, 124-33. doi:10.1016/j.scienta.2014.11.022.

Lucini, L., Kane, D., Pellizzoni, M., Ferrari, A., Trevisi, E., Ruzickova, G. and Arslan, D. 2016. Phenolic profile and in vitro antioxidant power of different milk thistle [*Silybum marianum* (L.) gaertn.] cultivars. *Ind. Crops Prod.* 83, 11-6. doi:10.1016/j.indcrop.2015.12.023.

Lucini, L., Rouphael, Y., Cardarelli, M., Bonini, P., Baffi, C. and Colla, G. 2018. A vegetal biopolymer-based biostimulant promoted root growth in melon while triggering brassinosteroids and stress-related compounds. *Front. Plant Sci.* 9, 472. doi:10.3389/fpls.2018.00472.

Marchetti, C. F., Ugena, L., Humplík, J. F., Polák, M., Ćavar Zeljković, S., Podlešáková, K., Furst, T., De Diego, N. and Spíchal, L. 2019. A novel image-based screening method to study water-deficit response and recovery of barley populations using canopy dynamics phenotyping and simple metabolite profiling. *Front. Plant Sci.* 10, 1252 doi:10.3389/fpls.2019.01252.

Martínez Esteso, M. J., Vilella-Antón, M., Sellés-Marchart, S., Martínez-Márquez, A., Botta-Català, A., Piñol-Dastis, R. and Bru-Martínez, R. 2016. A DIGE proteomic analysis of wheat flag leaf treated with TERRA-SORB® foliar, a free amino acid high content biostimulant. *JIOMICS* 6(1), 9-17. doi:10.5584/jiomics.v6i1.188.

Meinke, D. W., Cherry, J. M., Dean, C., Roundsley, S. D. and Koorneef, M. 1998. *Arabidopsis thaliana*: a model plant for genome analysis. *Science* 282(5389), 662, 679-82. doi:10.1126/science.282.5389.662.

Morau, A., Piepho, H. and Fritz, J. 2020. Growth responses of garden cress (Lepidium sativum L.) to biodynamic cow manure preparation in a bioassay. *Biol. Agric. Hortic.* 36(1), 16-34. doi:10.1080/01448765.2019.1644668.

Muscolo, A., Felicim, M., Concheri, G. and Nardi, S. 1993. Effect of earthworm humic substances on esterase and peroxidase activity during growth of leaf explants of *Nicotiana plumbaginifiolia*. *Biol. Fertil. Soils* 15(2), 127-31. doi:10.1007/BF00336430.

Nardi, S., Arnoldi, G. and Dell'Agnola, G. 1988. Release of the hormone-like activities from *Allolobophora rosea* (Sav.) and *Allolobophora caliginosa* (Sav.) feces. *Can. J. Soil Sci.* 68(3), 563-7. doi:10.4141/cjss88-054.

Nardi, S., Concheri, G., Dell'Agnola, G. and Scrimin, P. 1991. Nitrate uptake and ATPase activity in oat seedlings in the presence of two humic fractions. *Soil Biol. Biochem.* 23(9), 833-6. doi:10.1016/0038-0717(91)90094-Z.

Nardi, S., Pizzeghello, D., Gessa, C., Ferrarese, L., Trainotti, L. and Casadoro, G. 2000. A low molecular weight humic fraction on nitrate uptake and protein synthesis in maize seedlings. *Soil Biol. Biochem.* 32(3), 415-9. doi:10.1016/S0038-0717(99)00168-6.

Nardi, S., Carletti, P., Ertani, A. and Pizzeghello, D. 2006. Conoscere e valutare l'attività biostimolante. *Inf. Agrar.* 45, 2–5.

Nardi, S., Pizzeghello, D., Schiavon, M. and Ertani, A. 2016. Plant biostimulants: physiological responses induced by protein hydrolyzed- based products and humic substances in plant metabolism. *Sci. Agric. (Piracicaba, Braz.)* 73(1), 18–23. doi:10.1590/0103-9016-2015-0006.

Nardi, S., Ertani, A. and Francioso, O. 2017. Soil-root cross-talking: the role of humic substances', *J. Plant Nutr. Soil Sci.* 180(1), 5–13. doi:10.1002/jpln.201600348.

Nardi, S., Pizzeghello, D. and Ertani, A. 2018. Hormone-like activity of the soil organic matter. *Appl. Soil Ecol.* 123, 517–20. doi:10.1016/j.apsoil.2017.04.020.

Norrie, J. and Hiltz, D. A. 1999. Agricultural applications using Ascophyllum seaweed products. *Agro Food Ind. Hi Tec.* 2, 15–8.

Palumbo, G., Schiavon, M., Nardi, S., Ertani, A., Celano, G. and Colombo, C. M. 2018. Biostimulant potential of humic acids extracted from an amendment obtained via combination of olive mill wastewaters (OMW) and a pre-treated organic material derived from municipal solid waste (MSW). *Front. Plant Sci.* 9, 1028. doi:10.3389/fpls.2018.01028.

Paul, K., Sorrentino, M., Lucini, L., Rouphael, Y., Cardarelli, M., Bonini, P., Reynaud, H., Canaguier, R., Trtílek, M., Panzarová, K. and Colla, G. 2019a. Understanding the biostimulant action of vegetal-derived protein hydrolysates by high-throughput plant phenotyping and metabolomics: a case study on tomato. *Front. Plant Sci.* 10, 47. doi:10.3389/fpls.2019.00047.

Paul, K., Sorrentino, M., Lucini, L., Rouphael, Y., Cardarelli, M., Bonini, P., Miras Moreno, M. B., Reynaud, H., Canaguier, R., Trtílek, M., Panzarová, K. and Colla, G. 2019b. A combined phenotypic and metabolomic approach for elucidating the biostimulant action of a plant-derived protein hydrolysate on tomato grown under limited water availability. *Front. Plant Sci.* 10, 493. doi:10.3389/fpls.2019.00493.

Pichereaux, C., Laurent, E. A., Gargaros, A., Viudes, S., Durieu, C., Lamaze, T., Grieu, P. and Burlet-Schiltz, O. 2019. Analysis of durum wheat proteome changes under marine and fungal biostimulant treatments using large-scale quantitative proteomics: a useful dataset of durum wheat proteins. *J. Proteomics* 200, 28–39. doi:10.1016/j.jprot.2019.03.003.

Pizzeghello, D., Francioso, O., Ertani, A., Muscolo, A. and Nardi, S. 2013. Isopentenyladenosine and cytokinin-like activity of four humic substances. *J. Geochem. Expl.* 129, 70–5.

Povero, G., Mejia, J. F., Di Tommaso, D., Piaggesi, A. and Warrior, P. 2016. A systematic approach to discover and characterize natural plant biostimulants. *Front. Plant Sci.* 7, 435, doi:10.3389/fpls.2016.00435.

Rayorath, P., Khan, W., Palanisamy, R., MacKinnon, S. L., Stefanova, R., Hankins, S. D., Critchley, A. T. and Prithiviraj, B. 2008. Extracts of the brown seaweed Ascophyllum nodosum induce gibberellic acid (GA3)-independent amylase activity in barley. *J. Plant Growth Regul.* 27(4), 370–9. doi:10.1007/s00344-008-9063-6.

Rocchetti, G., Chiodelli, G., Giuberti, G. and Lucini, L. 2018a. Bioaccessibility of phenolic compounds following *in vitro* large intestine fermentation of nuts for human consumption. *Food Chem.* 245, 633–40. doi:10.1016/j.foodchem.2017.10.146.

Rocchetti, G., Lucini, L., Gallo, A., Masoero, F., Trevisan, M. and Giuberti, G. 2018b. Untargeted metabolomics reveals differences in chemical fingerprints between

PDO and non-PDO Grana Padano cheeses. *Food Res. Int.* 113, 407–13. doi:10.1016/j.foodres.2018.07.029.

Rouphael, Y., Spíchal, L., Panzarová, K., Casa, R. and Colla, G. 2018. High-throughput plant phenotyping for developing novel biostimulants: from lab to field or from field to lab? *Front.Plant Sci.* 9, 1197 doi:10.3389/fpls.2018.01197.

Scaglia, B., Pognani, M. and Adani, F. 2015. Evaluation of hormonelike activity of the dissolved organic matter fraction (DOM) of compost and digestate. *Sci. Total Environ.* 514, 314–21. doi:10.1016/j.scitotenv.2015.02.009.

Schiavon, M., Ertani, A. and Nardi, S. 2008. Effects of an alfaalfa protein hydrolysate on the gene expression and activity of enzymes of TCA cycle and N metabolism in *Zea mays* L. *J. Agric. Food Chem.* 56(24), 11800–8. doi:10.1021/jf802362g.

Teale, W. D., Paponov, I. A. and Palme, K. 2006. Auxin in action: signaling, transport and the control of plant growth and development. *Nat. Rev. Mol. Cell. Biol.* 7(11), 847–59. doi:10.1038/nrm2020.

Thalheimer, M. and Paoli, N. 2002. Effectiveness of various leaf- applied biostimulants on productivity and fruit quality of apple. *International Symposium on Foliar Nutrition of Perennial Fruit Plants. ISHS Acta Hortic.* 594(594), 335–9. doi:10.17660/ActaHortic.2002.594.41.

Ugena, L., Hýlová, A., Podlešáková, K., Humplík, J. F., Doležal, K., Diego, N. D. and Spíchal, L. 2018. Characterization of biostimulant mode of action using novel multi-trait high-throughput screening of *Arabidopsis* germination and rosette growth. *Front. Plant Sci.* 9, 1327 doi:10.3389/fpls.2018.01327.

Wilson, H. T., Xu, K. and Taylor, A. G. 2015. Transcriptome analysis of gelatin seed treatment as a biostimulant of cucumber plant growth. *Sci. World J.* 2015, 391234. doi:10.1155/2015/391234.

Zandonadi, D. B., Santos, M. P., Caixeta, L. S., Marinho, E. B., Peres, L. E. P. and Façanha, A. R. 2016. Plant proton pumps as markers of biostimulant action. *Sci. Agric. (Piracicaba, Braz.)* 73(1), 24–8. doi:10.1590/0103-9016-2015-0076.

Part 2

Non-microbial and microbial categories of biostimulants

Chapter 3

Humic substances (HS) as plant biostimulants in agriculture

Andrea Ertani, Università degli Studi di Padova and Università degli Studi di Torino, Italy; and Michela Schiavon and Serenella Nardi, Università degli Studi di Padova, Italy

1 Introduction

Ensuring sufficient food production for a developing world population while protecting natural resources represents a great challenge for agricultural science. Intensive cultivation practices have had negative environmental impacts and partially contributed to global climate change. Consequently, new advances in achieve more environmentally friendly agricultural production are needed. One possible approach is the use of biostimulants. In the EU the regulatory definition of biostimulants is: 'CE marked products which stimulate plant physiological processes independently of their nutrient content by improving one or more of the following characteristics of the plant rhizosphere or phyllosphere: nutrient use efficiency, tolerance to abiotic stress, crop quality, availability of confined nutrients in the soil and rhizosphere, humification and degradation of organic compounds in the soil' (European Commission, 2016).

Humic substances (HS) are among the most established biostimulants used in agriculture because they can significantly improve plant growth, directly or indirectly, as well as improving soil properties and fertility (Nardi et al., 2016). HS affect many agronomic, environmental and geochemical processes (Hayes and Swift, 1978; Stevenson, 1982), such as soil structure and porosity, water infiltration rate and moisture-holding capacity of soils as well as the diversity

http://dx.doi.org/10.19103/AS.2020.0068.04

and activity of soil microorganisms. In addition, HS influence plant physiology by interacting with plant biochemical and physiological processes, stimulating growth and increasing the uptake of nutrients by roots (Vaughan and Malcolm, 1985). There is now an extensive body of research that has shown, both under laboratory and field conditions (Table 1), that HS can have a positive effect on plant growth in terms of increases in biomass of shoots and roots, chlorophyll concentration, and number of lateral roots (Nardi et al., 1996).

This chapter will cover different aspects of HS: production and characteristics, biological activities, effects on soil and plant nutrition, and the use of commercial humates in agriculture.

2 Production and characteristics of humic substances

HS are produced by chemical and biological transformations of plant and animal residues through the activity of microorganisms and represent the major pool of organic carbon in soils. Several authors reported the complexity of the structure of HS (Muscolo et al., 2013; Canellas and Olivares, 2014; García et al., 2016). One of the first definitions of HS was proposed by Aiken and Wershaw (1985), who defined humic substances as 'a general category of naturally occurring, biogenic, heterogeneous organic substances that can generally be characterized as being yellow to black in color, of high molecular weight and refractory'. The chemical composition of the vegetable material and the development of humification processes are the major factors determining the properties of HS (Howard et al., 1998; Gonet and Debska, 1999). With this in mind, Piccolo (2001) defined HS as 'heterogeneous supramolecular mixtures of relatively small molecules, held together by no covalent intermolecular interactions (e.g. p-p, CH-p, Vander Waals, charge-transfer, and hydrogen-bonding) between relatively small organic moieties'. The exterior hydrophilic domain comprises polar groups (e.g. carboxylic acids), whereas the interior hydrophobic domain is formed by plant macromolecules (Piccolo and Spiteller, 2003; Smejkalova and Piccolo, 2008).

Humates are metal (mineral) salts of humic acids (HAs) or fulvic acids (FAs). Within any humic substance there are a large number of complex humate molecules. The formation of a humate is based on the ability of the carboxyl (COOH) and hydroxyl (OH) groups (on the outside of the polymers) to dissociate (expel) the hydrogen ion. Once the hydrogen ions are dissociated, a negatively charged anion (COO- or -CO-) results. Two of these anions can bind to positive metal cations, such as iron (Fe++), copper (Cu++), zinc (Zn++), calcium (Ca++). Humate composition is specific for individual HS, resulting to a large variability in molecular composition of different HS.

Among the parameters that change during the humification process are the degree of polydispersity of HS and the amount of hydrophobic and hydrophilic

Table 1 Summary of studies of the effects of HS

Reference	Crop	Effects on plants
Poapst and Schnitzer (1971)	*Phaseolus vulgaris* L.	Increase of root number
Rauthan and Schnitzer (1981)	*Cucurbita pepo* L.	Increase of plant biomass and content of N, P, K, Ca, Mg, Cu, Fe, and Zn
Dell'Agnola et al. (1981)	*Hordeum vulgare* L.	Antidote action of humic substances (HS) on atrazine inhibition of sulphate uptake
Dell'Agnola et al. (1981)	*Avena sativa* L.	Increase of Fe uptake by Ferric- humates
Xudan (1986)	*Triticum durum* Desf.	Increase of roots uptake; partial alleviation of grain yield depression by drought
Maggioni et al. (1987)	*Avena sativa* L.	Increase of ion uptake and (Mg^{2+}, K^{+}) ATPase activity
Albuzio et al. (1986)	*Hordeum vulgare* L.	Increase of nitrate uptake and assimilation
Dell'Agnola and Nardi (1987)	*Avena sativa* L.	Enhancement of nitrate uptake
Nardi et al. (1988)	*Hordeum vulgare* L.	Increased nitrate uptake
Nardi et al. (1991)	*Avena sativa* L.	Stimulation of nitrate uptake and ATPase activity
Ferretti et al. (1991)	*Zea mays* L.	Stimulation of sulphate assimilation
Piccolo et al. (1992)	*Hordeum vulgare* L.	Increase of nitrate uptake
Muscolo et al. (1993)	*Nicotiana plumbaginifolia* L.	Enhancement of peroxidase and esterase activities
Concheri et al. (1994)	*Triticum durum* Desf.	Induction of morphological changes and increase of invertase and peroxidase activities
Nardi et al. (1994)	*Nicotiana plumbaginifolia* L.	Evidence for auxin-like effects of HS in plants
Muscolo et al. (1996)	*Daucus carota* L.	Increase of glutamate dehydrogenase, glutamine synthetase and phosphoenolpyruvate carboxylase activities
Concheri et al. (1996)	*Triticum aestivum* L.	Modification of root morphology and enzymes activity
Nardi et al. (1996)	Review	Biological activity of humus
Concheri et al. (1996)	*Hordeum vulgare* L.	Promotion of nitrate uptake from HS obtained from forest soils
Nardi et al. (1997)	*Zea mays* L.	Influence of maize exudates on HS mobilization in the rhizosphere
Panuccio et al. (1997)	*Pinus laricio*	Stimulation of nitrate uptake
Nardi et al. (1998)	*Pinus Silvestris*	Increase of invertase and peroxidase activities by forest HS

(*Continued*)

Table 1 (Continued)

Reference	Crop	Effects on plants
Muscolo et al. (1998)		IAA detection in HS
Adani et al. (1998)	*Solanum lycopersicon* L.	Increase of root and shoot growth; enhancement of N, P, Fe, and Cu uptake
Pandeya et al. (1998)	*Oryza sativa* L.	Increase of Fe uptake following application of Fe-fulvic acid than $FeCl_3$
Muscolo et al. (1999)	*Daucus carota* L.	Stimulation of cell growth and nitrate metabolism
Nardi et al. (2000a)	*Zea mays* L.	Increase of protein synthesis and enzyme activity
Nardi et al. (2000b)	*Picea abies, Pinus sylvestris*	Mobilization of HS in the rhizosphere by forests root exudates
Panuccio et al. (2000)	*Pinus spp.*	Positive effect of HS on nitrogen uptake and assimilation in two species of Pinus
Sessi et al. (2001)	*Norway Spruce*	HS extracted from pasture and forested soils stimulate RED-OX
Pizzeghello et al. (2001)	*Fagus sylvatica*	HS biological activity was influenced by pedological conditions
Peng et al. (2001)	*Triticum durum* Desf.	Reduced symptoms of Se toxicity with high levels of Se
Nardi et al. (2002)	*Picea abies, Pynus sylvestris, Zea mays* L.	Biological activity of soil organic matter mobilized by root exudates
Canellas et al. (2002)	*Zea mays* L.	Increased secondary roots and root surface area
Sánchez-Sánchez et al. (2002)	*Citrus limon* L.	Increased Fe and Cu in leaves and vitamin C
Pizzeghello et al. (2002)	*Norway Spruce, Picea abies*	Forest ecosystems influences biological activities
Nardi et al. (2002)	Review	Physiological effects of HS on higher plants
Nardi et al. (2003)	Higher plants	Influenced field recruitment
Chen et al. (2004)	*Glycine max* (L.) Merr., *Cucumis melo* L. and ryegrass	Increased SPAD index
Nardi et al. (2005)	*Picea abies, Pynus sylvestris, Zea mays* L.	HS extracted by root exudates influences nitrate metabolisms
Murillo et al. (2005)	Wild olive	Increased nitrogen without increases to phytotoxic levels of metals
Bocanegra et al. (2006)	*Helianthus annuus* L.	Fulvic acid chelated Fe^{3+} for plant uptake
Muscolo et al. (2005)	*Pinus callus*	Influenced sucrose metabolism
Sánchez-Sánchez et al. (2006)	*Vitis vinifera* L.	Increased P and Fe uptake

(Continued)

Table 1 (Continued)

Reference	Crop	Effects on plants
Russell et al. (2006)	*Pisum sativum* L.	Stomatal responses to HS and auxin are sensitive to inhibitors of phospholipase A2
Muscolo et al. (2007)	*Pinus nigra*	HS from forest soils affected sugar content, free amino acids and enzymes in C and N metabolisms
Dobbss et al. (2007)	*Arabidopsis thaliana*, micro-Tom *Solanum lycopersicon*	Increased lateral root growth in Arabidopsis and wild-type micro-Tom tomato
Muscolo et al. (2007)	*Daucus carota* L. cell cultures	The auxin-like activity of HS is related to membrane interactions in carrot cell cultures
Yildrim (2007)	*Solanum lycopersicon* L.	Increased yield, total soluble solids, and ascorbic acid
Nardi et al. (2007)	*Zea mays* L.	Relationship between molecular characteristics of soil humic fractions and glycolytic pathway and Krebs cycle
Costa et al. (2008)	*Lantana camara*	Increase plant biomass, earlier flowering of rooted cuttings
Carletti et al. (2008)	*Zea mays* L.	Protein Expression changes in maize roots
Karakurt et al. (2009)	*Cucurbita pepo* L.	Increased yield, sugars, and chlorophylls
Canellas et al. (2009)	*Zea mays* L.	Stimulated root growth and length
Karakurt et al. (2009)	*Capsicum annuum* L.	Increased yield, early yield, fruit weight, sugars, and chlorophylls
Schiavon et al. (2010)	*Zea mays* L.	Enhancement of secondary metabolism by high molecular size HS
Marino et al. (2010)	*Pyrus communis* L.	Increase of rooting, plant height, chlorophyll content and nutrient uptake
Cimrin et al. (2010)	*Capsicum annuum* L.	Increase of N, P, K, Ca, Mg, S, Mn, and Cu accumulation under salt conditions
Dobbss et al. (2010)	*Arabidopsis thaliana*, *Solanum lycopersicon* L., *Zea mays* L.	Increase of lateral root emergence
Mora et al. (2010)	*Cucurbita pepo* L.	Increase of shoot growth and nitrate assimilation in shoots
Morard et al. (2011)	Several species	Increase of plant biomass, number of flowers, and water-use efficiency
Ertani et al. (2011)	*Zea mays* L.	Positive effect of lignosulfonate-humate on plants' metabolism
Canellas et al. (2011)	*Solanum lycopersicon* L.	Enhancement of lateral roots number
Tahir et al. (2011)	*Triticum durum* Desf.	Increase of plant height, enhancement of N uptake

(Continued)

Table 1 (Continued)

Reference	Crop	Effects on plants
Selim et al. (2012)	*Solanum tuberosum* L.	Increased of yield and ascorbic acid content
Jindo et al. (2012)	*Zea mays* L.	Promotion of root growth; increase of the number of mitotic sites in roots
El-Nemr et al. (2012)	*Cucurbita pepo* L.	Increase of plant growth and uptake of N, P, K, Ca, and Mg
García et al. (2012)	*Oryza sativa* L.	Reduction of oxidative stress of plants grown under water stress
Moghaddam and Soleimani (2012)	*Pistacia vera*	Increase of shoot growth under salinity conditions
Mazhar et al. (2012)	*Chrysanthemum* L.	Increase of plant growth, flowering, and accumulation of total carbohydrates, N, P, and K under salinity stress
Ertani et al. (2013)	*Zea mays* L.	Positive effects on growth and nitrogen assimilation determined by Humic-like substances obtained from agroindustry residues
Pizzeghello et al. (2013)	*Zea mays* L.	Isopentenyladenosine and cytokinin-like activity of different HS
Befrozfar et al. (2013)	*Ocimum basilicum* L.	Increase of oil yield by with humic acid alone and in combination with plant growth promoters
Muscolo et al. (2013)	Review	Deeper information on the relationship between structure and activity of HS
Vaccaro et al. (2015)	*Zea mays* L.	Stimulation of nitrogen assimilation and amino acid metabolism at physiological and molecular level by HS
Nardi et al. (2016)	Review	Overview of: physiological responses induced by HS and protein hydrolysates on plant metabolism
Nardi et al. (2017b)	Review	Overview of the role of HS in the soil-root cross-talking
Palumbo et al. (2018)	*Zea mays* L.	Positive effects of HS extracted from an amendment obtained via combination of olive mill wastewaters (OMW) on plant metabolism

fractions (Aleksandrova, 1980; Debska et al., 2005). Humic substances (HS) possess several functional groups such as carboxyl, phenolic and alcoholic compounds that influence their chemical properties. These groups and their chemical properties account to a large degree for most of the functions ascribed to HS. The relationship between HS structure, functional groups, and activity of the different fractions has been widely studied (Nardi et al., 2002).

The proportions of these fractions influence the solubility and the bioactivity of HS.

HS can be characterized in terms of their solubility in aqueous acids and bases. In particular, humic acids (HA) are soluble in aqueous alkaline solutions but precipitate at pH 1–2, fulvic acids (FA) remain in solution after the aqueous alkaline extracts are acidified, and humin has low solubility in aqueous solution and is not affected by pH conditions (Hayes, 2006). HA, FA, and humin comprise highly complex and heterogeneous mixtures of organic substances. As a result, there is no model compound of identified composition and molecular weight (MW), and significant uncertainty about the sizes and real MW subsists. Studies show that these values may vary from 500 Daltons for some aquatic HS, to more than 106 Daltons for soil HA (Stevenson, 1994).

Organic acids (such as succinic, fumaric, tartaric, malic, and citric acids) exuded by plant roots into the rhizosphere change humic aggregations, releasing smaller size-fractions and single molecules that may bind to cellular receptors on the root surface and promoting biological activity (Nardi et al., 2003; Canellas et al., 2010). In light of these studies, Nardi and co-authors (2016) suggested that HS may act as signaling compounds, perhaps releasing phytohormones at the plant and/or soil biota level. Although transmembrane receptors for HS have yet to be described, plants respond to the presence of HS by exuding several compounds that may interact with HS in the rhizosphere (Nardi et al., 2017a,b).

3 Humic substances and plant nutrition

In order to establish whether a humic substance exerts positive effects on plant growth, it is crucial to define the conditions in which the experiments will take place. In many studies the control growth media consists only of water. In such conditions HS would be expected to enhance plant growth by simply providing nutrients to plants (Vaughan and Malcolm, 1985). The composition of the nutrient solution, as well as its aeration, also influences the effect of HS on plant growth. As an example, the addition of HS to Hoagland No. 2 solution (Hoagland and Arnon, 1950) was found to be more effective in promoting plant growth than when the humic fractions were mixed with quartz (Nardi et al., 2002).

Research has shown a positive effect of HS on plant growth, with a more pronounced effect on root development than on leaf and stem development in plants (Vaughan and Malcolm, 1985). Studies were undertaken with HS labeled with ^{14}C to establish whether the positive effect of these substances on plants was via direct or indirect contact (e.g. superficial physical contact). It was concluded that HS, especially those with low molecular weight (MW), could be absorbed by the plant roots and translocated to the shoot. The positive effects of HS on plant growth relate to their capacity to increase nutrient uptake by plants and depend on their concentration. Variability in effects is likely to be due to differences in HS origin (Lee and Bartlett, 1976) and fractions (Ernst et al., 1987). Different plant species treated with the same humic fraction display different responses (Dormaar, 1975).

Several studies have shown that one of the first effects induced by HS in plants is the modification of root architecture (Fig. 1) and root hair initiation (Nardi et al., 2009; Canellas et al., 2010; Dobbss et al., 2010; Tahir et al., 2011; Canellas and Olivares 2014). The increase in root area results in a better interaction of roots with nutrient elements and enhancement of their uptake (Nardi et al., 2016). The stimulation of root development could be due to hormonal effects triggered by HS (Cacco and Dell'Agnola, 1984; Dell'Agnola and Nardi, 1987; Nardi et al., 1988; Piccolo et al., 1992). HS have been shown to contain auxins and thus exert auxin-like activity. However, it is still not clear if

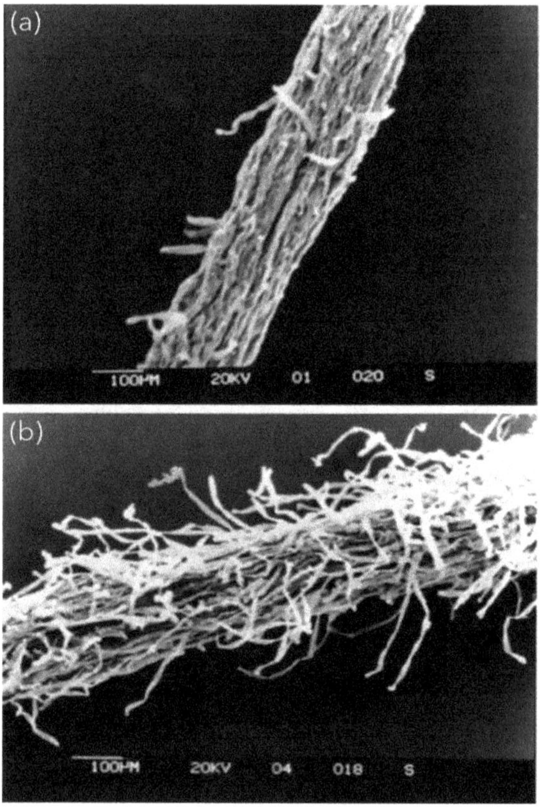

Figure 1 SEM micrographs of the 0–20 mm region behind the root tips of *Z. mays* seedlings surface grown in Hoagland solution and treated for the last 2 days with MH. A = untreated plant; B = plant treated with HS.

this activity is associated with the structure of HS or depends on the presence of hormones of microbial origin enclosed in HS. A number of studies have investigated the hormone-like activity of HS in different plant species and under several environmental conditions (Concheri et al., 1996; Muscolo and Nardi, 1997; Nardi et al., 1999, 2007; Muscolo et al., 1999; Pizzeghello et al., 2001, 2002; Schiavon et al., 2010; Sitzia et al., 2014, 2018).

4 Effects of humic substances on soil

Several studies have shown the positive roles played by HS in improving the chemical and physical properties of soil, including cation exchange capacity (CEC), nitrogen content, and soil microbial activity (Gümüş and Şeker, 2015; Alagöz and Erdem, 2009). Soil structure affects soil water movement and retention, erosion, crusting, nutrient recycling, root penetration, and crop production. Structural and physical soil degradation is frequently related to a reduction in organic matter content. HS have a fundamental role in structural stabilization. Piccolo et al. (1997) reported that HS increase the structural stability of soils via formation of clay-humic complexes through bridging polyvalent cations adsorbed on clays (Piccolo and Mbagwu, 1994).

One of the main properties of HS is the water retention capacity, which is connected with their partial hydrophilicity and porous structure (Jaeger et al., 2010). At low water content, the water molecule bridges (WaMB) connect and stabilize the segments of HS (Aquino et al., 2011; Schaumann and Bertmer, 2008). Increasing water content improves the sorption capacity of HS toward non-polar compounds (Borisover, 2013), highlighting the significance of non-covalent polar links in stabilization of soil organic matter. Soils which contain high amounts of HS hold more water for crop use during periods of drought. This is why growers who apply humate-based fertilizers, and use production practices which preserve HS, can frequently harvest a crop during periods of dry weather.

The effect of HS on soil microbes has been known for a long time (Visser, 1985). It has been shown that the addition of HS to microbial cultures at concentrations of 30 mg L^{-1} resulted in proliferation of soil microbes up to 2000-fold. In the presence of HS, certain microbes also appeared to proliferate on substrates that they could not utilize originally, suggesting that HS can alter microbe metabolism.

5 Biological activities of humic substances

HS include more than 60% of soil organic matter and are major components of organic fertilizers. They contain significant amounts of nutrients (Stevenson, 1994) and biologically active compounds (Nardi et al., 1996). Because of their properties, HS exert both direct and indirect effects on plants (Chen and Aviad, 1990). Indirect effects include improvement of soil properties, such as aggregation, aeration, permeability, water-holding capacity, and nutrient movement (Tan, 2003). Direct effects relate to greater nutrient uptake of HS by plants (Muscolo et al., 2013). However, it is not completely known how HS induce the signaling pathways that activate and/or stimulate different plant metabolic routes in order to produce root modifications (Erturk et al., 2015), build rhizospheres, and microbial interactions (Nardi et al., 1989).

The responses of plants to endogenous and environmental signals produced by HS are associated with the presence of molecules used as chemical messengers (Swarup et al., 2002). Plant hormones such as auxins, gibberellins, cytokinins, and abscissic acid are able to elicit key physiological responses (Marré and Ballarin-Denti, 1985). The auxin-like activity of HS reported by Dell'Agnola and Nardi (1987) was related to the presence of indole acetic acid (IAA) (Muscolo et al., 1998). These authors measured a concentration of IAA in HS equal to 0.5% (w/w) using an enzyme immunoassay, and 2-3.7% using a radioimmunoassay. The presence of IAA in HA has been more recently quantified using chromatography mass spectrometry (Canellas et al., 2002). Schiavon et al. (2010) have detected IAA in HS from earthworm feces.

The hormone-like activity of HS could be related to other compounds, such as polyamines, putrescine, spermidine, and spermine, which have been reported in amounts of 1.54–7.00, 0.39–3.88, 0.48–4.79 nmol g^{-1}, respectively, using high performance liquid chromatography (HPLC) (Young and Chen, 1997). Arancon et al. (2012) measured 185 ng L^{-1} of N-indole-3-acetic acid, cytokinins, and gibberellins in vermicompost tea. Recent studies have shown the presence of cytokinin (Isopentenyladenosine) in physiologically active concentrations in lignosulphonate humates, leonardite humic acid, and HS from earthworm feces, using an ELISA immunoassay (Ertani et al., 2013; Pizzeghello et al., 2013).

There have been a number of studies to evaluate the morphological changes induced by HS using a simplified system such as leaves and cells in culture (Fig. 2) (Ehsanpour and Fatahian, 2003; Vidal et al., 2004). Muscolo and co-authors (1993) used leaf explants of tobacco (*Nicotiana plumbaginifolia*) supplemented with hormones or HS at low and high molecular weight and concentrations (1.0 and 5.0 mg C L^{-1}). The results demonstrated that leaf explants of tobacco culture combined with either indole acetic acid (IAA) and/or HEf (humic matter fraction <3500 Da) displayed a number of morphological changes compared to controls (Nardi et al., 1994).

Schmidt and coauthors (2005) found a substantial proliferation of root hair density in *Arabidopsis thaliana* subjected to HS, suggesting the induction of a 'nutrient acquisition response', which supports the uptake of nutrient elements via an increase in the absorption surface area. After 3 days of HS additin,

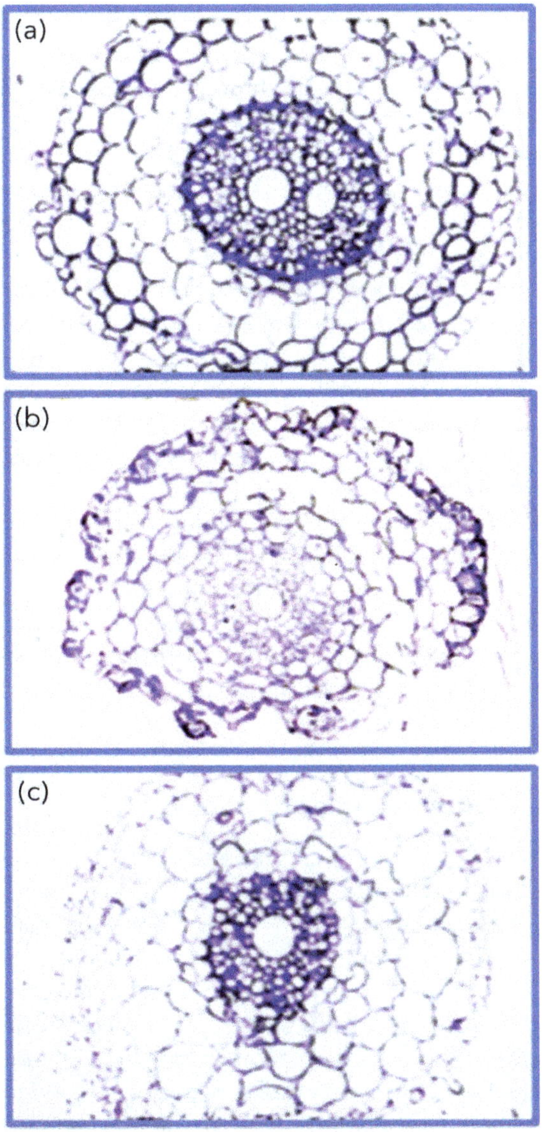

Figure 2 Central cylinder of wheat root of plants treated with HS (A and C) showed a differentiation level higher than the untreated plants (B).

Canellas et al. (2002) confirmed that the treatment stimulated the number of sites of lateral root emergence by 7–12 times compared to control plants. The role of HS in regulation root proliferation is due to signaling compounds such as ABA, nitrate, NO, ROS, and IAA that regulate root physiological modifications through several pathways in response to HS conditions.

The beneficial role of H+-ATPase activity, in plant treated with HS, has been reported in several studies. The function of the plasma membrane H+-ATPase is to promote uptake and efflux of ions through the membrane by producing an electrochemical gradient. Jindo et al. (2012) described how HS stimulated root growth of maize through the release of auxin-like compounds that induce H^+-ATPase activity in the plasma membrane of root cells, causing the acidification of the apoplast and leading to loosening of cell walls and elongation of root cells. Stimulation of H^+-ATPase has been associated with humic acid-induced lateral root development of maize (Nardi et al., 1991; Canellas et al., 2002; Zandonadi et al., 2007; Canellas et al., 2009). Its functionality has been generally associated with nutrient uptake. Pinton et al. (1997) observed an enhanced proton extrusion from roots in *Avena sativa* plants treated with HS compared to control plants. In a study with maize, Pinton et al. (1999) found an increased nitrate uptake prompted by HS, which was related to enhanced plasma membrane H^+-ATPase activity.

In 1997, Macdonald demonstrated the involvement of a phospholipase A2 (PLA2) in auxin signaling leading to activation of H+-ATPase. Russell et al. (2006) studied the effects of two humic fractions of different molecular weight, containing a very low amount of free IAA with auxin-like properties, on stomatal opening of leaves of an Argenteum mutant of pea (*Pisum sativum* L.). Both fractions of IAA caused stomatal opening in the epidermal peels, corroborating the theory that HS possess hormone-like properties.

6 Commercial humates in agriculture

The physicochemical and biological properties of HS, together with changes in the fertilizer market toward greater sustainability, have contributed to the increasing interest in HS. Over recent decades, commercial humic products have found multiple applications in agriculture and environmental technologies. The use of humates derived from different salts of HA, such as ammonium and potassium humates, has been growing (Brownell et al., 1987; Fernandez-Escobar et al., 1996). Research has long highlighted the importance of maintaining the humic acid content of soils to ensure good crop productivity (Sauchelli, 1944). Humates isolated from spent sulphite pulping liquors of lignin (McGovern, 1980; Ertani et al., 2011) are widely used in industrial applications (Gargulak and Lebo, 2000) and as biostimulants in agriculture (Montoneri et al., 2004).

Humates from different mineral deposits have their own unique features. A commercial humate is produced in a controlled process. The process involves a primary raw material (potassium lignosulphonan) homogenized with potassium hydroxide and subjected to hydrolysis with high pressure and specific temperatures. The result is the synthesis of humates from a material that does not primarily contain such substances. In fact, what happens is the initiation and acceleration of a process that would take many years in nature. From an industrial point of view, the efficiency of humate extraction is a key factor affecting the efficiency and economic viability of the process (Ertani et al., 2011). As potential fertilizer components, HA derived from different sources under different conditions need to be well characterized. Although research has elucidated the primary structure of these polymers, few studies have investigated their effects on soil, plant growth, and production. Lignosulphonates seem to show comparable properties to HS in terms of chelation, buffering and cation exchange capacity due to a high amount of carboxylic and phenol groups bonded to the aromatic ring (Lutnaes et al., 2008; Nardi et al., 2009). Commercial humates, especially potassium (K) humates, are extensively used as biostimulants to increase soil organic matter content. They are used to increase crop growth, yield and product quality, increase soil microbial activity, in deactivation of toxic metals, to improve soil structure and water retention capacity, enhance crop nutrient uptake and resistance to stress conditions such as salinity and sodicity, as well as increasing the efficiency of inorganic fertilizers (Kulikova et al., 2005; Billingham, 2012).

Most research on commercial humates report positive effects (Mikkelsen, 2005), which include increased plant growth and yield (Lobartini et al., 1992; Gaur et al., 1971; Mora et al., 2014) and/or helping plants to overcame stress situations, including salinity (Kulikova et al., 2005; Lee and Bartlett, 1976; Sadiq et al., 2014). Imbufe et al. (2005) suggested that potassium humate is potentially effective as a soil conditioner in improving aggregate stability of acidic and sodic soils against the adverse effects of cyclic seasonal wetting and drying. Fe-humate is an effective Fe source for citrus and grape trees grown in alkaline soils, which has been shown to increase plant biomass and fruit production (Alva and Obreza, 1998). HS have been shown to enhance the solubility of soil potassium (P) through complexation/combination of Fe and Al in acid soils and Ca in calcareous soils.

It is important to note that commercial humates derived from the same source and obtained by the same company can differ extensively in composition (Iakimenko, 2005). Evidence for effectiveness in the field (as opposed to controlled conditions) is inconsistent (Billingham, 2012; Lyons and Genc, 2016). Benefits appear to be greater, for example, in hydroponic and sand culture studies under controlled conditions (Ertani et al., 2011). There is currently not enough research to explain possible mechanisms and accurately predict when humic materials might prove beneficial in practice (Mikkelsen, 2005; Billingham, 2012). There needs to be better characterization of the chemistry of particular commercial products and field studies to test their efficacy (Lyons and Genc, 2016). Table 1 summarizes the range of studies on the effects of HS.

7 Conclusions

The main issue encountered while studying HS effects on plants and soils is the fact that they are complex mixtures without a unique formula, which are difficult to chemically characterize in structure. The intricate mechanisms responsible for the effects of HS in plants suggest several interconnected signaling pathways related to both relevant hormonal networks and secondary messengers. Existing evidence suggests that the biostimulant effects of HS are linked to both structural and physiological changes in plants that improve nutrient uptake and assimilation. In addition, HS can induce stimulation of plant primary and secondary metabolism related to abiotic stress-defence, and thus the use of HS in agronomic systems can be a valuable tool for the development of more sustainable agricultural practices. As most HS used in agriculture are currently derived from non-renewable resources like coal and peat, the promotion of new technology to produce HS from other sources, requires further research. The effects of HS in plants could be further investigated by using a combination of new omics techniques, including ionomics, metabolomics, proteomics, and transcriptomics.

8 References

Adani, F., Genevini, P., Zaccheo, P. and Zocchi, G. (1998), 'The effect of commercial humic acid on tomato plant growth and mineral nutrition', *J. Plant Nutr.* 21, 561–75.

Aiken, G. R. and Wershaw, R. L. (1985), 'Molecular size and weight measurements of humic substances', *in* Aiken, G. R., McKnight, D. M., Wershaw, R. L. and MacCarthy, P. (Eds), *Humic Substances in Soil, Sediment, and Water: Geochemistry, Isolation, and Characterization*, John Wiley & Sons, New York, pp. 477–92.

Alagöz, Z. and Erdem, Y. (2009), 'Effects of different sources of organic matter on soil aggregate formation and stability: a laboratory study on a Lithic Rhodoxeralf from Turkey', *Soil Till. Res.* 103, 419–24.

Albuzio, A., Ferrari, G. and Nardi, S. (1986), 'Effects of humic substances on nitrate uptake and assimilation in barley seedlings', *Can. J. Soil Sci.* 66, 731–6.

Aleksandrova, L. N. (1980), *Organic Soil Matter and Processes of Its Transformation* (in Russian), Nauka, Leningrad, p. 288.

Alva, A. K. and Obreza, J. A. (1998), 'By-product iron-humate increases tree growth and fruit production of orange and grapefruit', *Hort. Sci.* 33, 71–4.

Aquino, A., Tunega, D., Pasalic, H., Schaumann, G. E., Haberhauer, G., Gerzabek, M. H. and Lischka, H. (2011), 'Molecular dynamics simulations of water molecule-bridges in polar domains of humic acids', *Environ. Sci. Technol.* 45, 8411–19.

Arancon, N. Q., Pant, A., Radovich, T., Hue, N. V., Potter, J. K. and Converse, C. E. (2012), 'Seed germination and seedling growth of tomato and lettuce as affected by vermicompost water extracts (Teas)', *Hort. Sci.* 47, 1722–8.

Befrozfar, M. R., Habibi, D., Asgharzadeh, A., Sadeghi-Shoae, M. and Tookallo, M. R. (2013), 'Vermicompost, plant growth promoting bacteria and humic acid can affect the growth and essence of basil (*Ocimum basilicum* L.)', *Ann Biol. Res.* 4, 8–12.

Billingham, K. (2012), 'Humic products potential or presumption for agriculture? Can humic products improve my soil?' *in* Harris, C., Lodge, G. and Waters, C. (Eds), *Proceedings of the 27th Annual Conference of the Grassland Society of NSW Inc.*, Orange, NSW, Australia, 24-26 July 2012, Grassland Society of NSW Inc., Wagga Wagga, Australia, pp. 43–50.

Bocanegra, M. P., Lobartini, J. C. and Orioli, G. A. (2006), 'Plant uptake of iron chelated by humic acids of different molecular weights', *Commun. Soil Sci. Plant Anal.* 37, 1-2.

Borisover, M. (2013), 'The effect of organic sorbates on water associated with environmentally important sorbents: estimating and the LFER analysis', *Adsorption* 19, 241-50.

Brownell, J. R., Nordstrom, G., Marihart, J. and Jorgensen, G. (1987), 'Crop responses from two new leonardite extracts', *Sci. Total Environ.* 62, 491-9.

Cacco, G. and Dell'Agnola, G. (1984), 'Plant growth regulator activity of soluble humic complex', *Can. J. Soil Sci.* 62, 306-10.

Canellas, L. P. and Olivares F. L. (2014), 'Physiological responses to humic substances as plant growth promoter', *Chem. Biol. Technol. Agric.* 1, 3.

Canellas, L. P., Olivares, F. L., Okorokova-Façanha, A. L. and Façanha, A. R. (2002), 'Humic acids isolated from earthworm compost enhance root elongation, lateral root emergence, and plasma membrane H^+-ATPase activity in maize roots', *Plant Physiol.* 130, 1951-7.

Canellas, L. P., Spaccini, R., Piccolo, A., Dobbss, L. B., Okorokova-Facanha, A. L Santos, G., Olivares, F. and Façanha, A. (2009), 'Relationships between chemical characteristics and root growth promotion of humic acids isolated from brazilian oxisols', *Soil Sci.* 174, 611-24.

Canellas, L. P., Piccolo, A., Dobbss, L. B., Spaccini, R., Olivares, F. L., Zandonadi, D. B. and Façanha, A. R. (2010), 'Chemical composition and bioactivity properties of size fractions separated from a vermicompost humic acid', *Chemosphere* 78, 457-66.

Canellas, L. P., Dantas, D. J., Aguiar, N. O., Peres, L. E. P., Olivares, F. L., Dobbss, L. B., Façanha, A. R., Nebbioso, A. and Piccolo, A. (2011), 'Probing the hormonal activity of fractionated molecular humic omponents in tomato auxin mutants', *Ann. Appl. Biol.* 159, 202-11.

Carletti, P., Masi, A., Spolaore, B., De Laureto, P. P., De Zorzi, M., Turetta, L., Ferretti, M. and Nardi, S. (2008), 'Protein expression changes in maize roots in response to humic substances', *J. Chem. Ecol.* 34, 804-18.

Chen, Y. and Aviad, T. (1990), 'Effects of humic substances on plant growth', *in* MacCarthy, P., Malcolm, R. L., Clapp, C. E. and Bloom, P. R. (Eds), *Humic Substances in Soil and Crop Science: Selected Readings*, American Society of Agronomy and Soil Science Society of America, Madison, WI, 161-87.

Chen, Y., Clapp, C. E. and Magen, H. (2004), 'Mechanisms of plant growth stimulation by humic substances: the role of organo-iron complexes', *Soil Sci. Plant Nutr.* 50, 1089-95.

Çimrin, K. M., Türkmen, O., Turan, M. and Tuncer B. (2010), 'Phosphorus and humic acid application alleviate salinity stress of pepper seedling', *Afr. J. Biotech.* 9, 5845-51.

Concheri, G., Nardi, S., Piccolo, A., Rascio, N. and Dell'Agnola, G. (1994), 'Effects of humic fractions on morphological changes related to invertase and peroxidase activities in wheat seedlings', *in* Senesi, N. and Miano, T. M. (Eds), *Humic Substances in the Global Environment and Implications on Human Health*, Elsevier Science Publisher B.V., Miano, pp. 257-62.

Concheri, G., Nardi, S. and Dell'Agnola, G. (1996), 'Productivity and biological humus activities in forest soils', *in* Clapp, C. E., Hayes, M. H. B., Senesi, N. and Griffith, S. M. (Eds), *Humic Substances and Organic Matter in Soil and Water Environments: Characterization, Transformations and Interactions*, International Humic Substances Society Inc., St. Paul, MN, pp. 131-8.

Costa, G., Labrousse, P., Bodin, C., Lhernould, S., Carlué, M., Authier, F. and Krausz, P. (2008) 'Effects of humic substances on the rooting and development of woody plant cuttings', *Acta Hortic*. 779, 255-61.

Debska, B., Drag, M. and Andersen D. (2005), 'Properties of organic matter and molecular size distribution of humic acids isolated from forest soil', *Latvijas Uniwersitates Raksti*, 696, 19-28.

Dell'Agnola, G. and Nardi, S. (1987), 'Hormone-like effect of enhanced nitrate uptake induced by depolycondensed humic fractions obtained from *Allolobophora rosea* and *A. caliginosa* faeces', *Biol. Fertil. Soils* 4, 115-18.

Dell'Agnola, G., Ferrari G. and Nardi S. (1981), 'Antidote action of humic substances on atrazine inhibition of sulphate uptake in barley roots', *Pest. Bioch. Physiol*. 15, 101-4.

Dobbss, L. B., Medici, L. O., Peres, L. E. P., Pino-Nunes, L. E., Rumjanek, V. M., Façanha, A. R. and Canellas, L. P. (2007), 'Changes in root development of *Arabidopsis* promoted by organic matter from oxisols', *Ann. Appl. Biol*. 151, 199-211.

Dobbss, L. B., Canellas, L. P., Olivares, F. L., Aguiar, N. O., Pers, L. E. P., Azevedo, M., Spaccini, R., Piccolo, A. and Façanha, A. R. (2010), 'Bioactivity of chemically transformed humic matter from vermicompost on plant root growth', *J. Agric. Food Chem*. 58, 3681-8.

Dormaar, J. F. (1975), 'Effects of humic substances from chenozemic Ah horizons on nutrient uptake by *Phaseolus vulgaris* and *Festuca scabrella*', *Can. J. Soil Sci*. 55, 111-18.

Ehsanpour, A. A. and Fatahian, N. (2003), 'Effects of salt and proline on *Medicago sativa* callus', *Plant Cell. Tissue Organ Cult*. 73, 53-6.

El-Nemr, M. A., El-Desuki, M., El-Bassiony, A. M. and Fawzy, Z. F. (2012), 'Response of growth and yield of cucumber plants (*Cucumis sativus* L.) to different foliar applications of humic acid and bio-stimulators', *Aust. J. Basic Appl. Sci*. 6, 630-7.

Ernst, W. H. O., Kraak, M. H. S. and Stoots, L. (1987), 'Growth and mineral nutrition of *Scrophularia nodos* with various combinations of fulvic and humic acids', *J. Plant Physiol*. 127, 171-5.

Ertani, A., Francioso, O., Tugnoli, V., Righi, V. and Nardi, S. (2011), 'Effect of commercial lignosulfonate-humate on *Zea mays* L. metabolism', *J. Agric. Food Chem*. 59, 11940-8.

Ertani, A., Pizzeghello, D., Baglieri, A., Cadili, V., Tambone, F., Gennari, M. and Nardi, S. (2013), 'Agro-industrial residues and their biological activity on maize (*Zea mays* L.) metabolism', *J. Geochem. Exp*. 129, 103-11.

Erturk, F. A., Agar, G., Arslan, E. and Nardemir, G. (2015), 'Analysis of genetic and epigenetic effects of maize seeds in response to heavy metal (Zn) stress', *Environ. Sci. Pollut. Res*. 22(13), 10291-7.

European Commission (2016), Proposal for a regulation laying down rules on the making available on the market of CE marked fertilizing products and amending Regulations (EC)1069/2009 and (EC)1107/2009, COM(2016) 157.

Fernandez-Escobar, R., Benlloch, M., Barranco, D., Duenas, A. and Guterrez Ganan, J. A. (1996), 'Response of olive trees to foliar application of humic substances extracted from leonardite', *Sci. Hort*. 66, 191-200.

Ferretti, M., Ghisi, R., Nardi, S. and Passera, C. (1991), 'Effect of humic substances on photosynthetic sulphate assimilation in maize seedlings', *Can. J. Soil Sci.* 71, 239-42.
García, A. C., Santos, L. A., Izquierdo, F. G., Sperandio, M. V. L., Castro, R. N. and Berbara, R. L. L. (2012), 'Vermicompost humic acids as an ecological pathway to protect rice plant against oxidative stress', *Ecol. Eng.* 47, 203-8.
García, A. C., de Souza, L. G. A., Pereira, M. G., Castro, R. N., García-Mina, J., Zonta, E., Lisboa, F. J. G. and Berbara, R. L. L. (2016), 'Structure-property-function relationship in humic substances to explain the biological activity in plants', *Sci. Rep.* 6, 20798.
Gargulak, J. D. and Lebo, S. E. (2000), 'Commercial use of lignin-based materials', *in* Glassr W. G., Northey, R. A. and Schultz, T. P. (Eds), *Lignin: Historical, Biological and Haterials Perspectives*, ACS Symposium Series 742, American Chemical Society, Washington DC, pp. 304-20, Chapter 15.
Gaur, A. C., Sadasivam, K. V., Vimal, O. P. and Mathur, R. S. (1971), 'A study on the decomposition of organic matter in an alluvial soil: CO_2 evolution, microbiological and chemical transformations', *Plant Soil.* 35, 17-28.
Gümüş. İ. and Şeker, C. (2015), 'Influence of humic acid applications on soil physicochemical properties', *Solid Earth Discuss.* 7, 2481-500.
Gonet, S. S. and Debska, A. B. (1999), 'Properties of humic acids produced during decomposition of plant residues in soil', *Rostlinná Výroba* 45, 455-60.
Hayes, M. H. B. (2006), 'Solvent systems for the isolation of organic components from soils', *Soil Sci. Soc. Am. J.* 70, 986-94.
Hayes, M. H. B. and Swift, R. S. (1978), 'The chemistry of soil organic colloids', *in* Greenland, D. J. and Hayes, M. H. B. (Eds), *The Chemistry of Soil Constituents*, John Wiley & Sons, Chichester, pp. 179-320.
Hoagland, D. R. and Arnon D. (1950), 'The water culture method for growing plants without soil', Agricultural Experimental Station Circular 347, University of California.
Howard, P. J. A., Howard, D. M. and Lowe, L. E. (1998), 'Effects of tree species and soil physico-chemical conditions on the nature of soil organic matter', *Soil Biol. Biochem.* 30, 285-97.
Iakimenko, O. S. (2005), 'Commercial humates from coal and their influence on soil properties and initial plant development', *in* Perminova, I. V., Hatfield, K. and Hertkorn, N. (Eds), *Use of Humic Substances to Remediate Polluted Environments: From Theory to Practice*, Springer, Dordrecht, The Netherlands, pp. 365-78.
Imbufe, A. U., Patti, A. F., Surapeneni, A., Jakson, R. and Webb, A. J. (2005), 'Effects of brown coal derived materials on pH and electrical conductivity of an acidic vineyard soil', *Super Soil: 3rd Australian New Zealand Soils Conference*, 5-9 December 2004, University of Sydney, Australia.
Jaeger, F., Shchegolikhina, A., van As, H. and Schaumann, G. E. (2010), 'Proton NMR relaxometry as a useful tool to evaluate swelling processes in peat soils', *Open Magn. Reson. J.* 3, 27-45.
Jindo, K., Martim, S. A., Navarro, E. C., Pérez-Alfocea, F., Hernandez, T., Garcia, C., Aguiar, N. O. and Canellas, L. P. (2012), 'Root growth promoting by humic acids from composted and non-composted urban organic wastes', *Plant Soil* 353, 209-20.
Karakurt, Y., Unlu, H. and Padem, H. (2009), 'The influence of foliar and soil fertilization of humic acid on yield and quality of pepper', *Acta Agric. Scand. Sect. B.* 59, 233-7.
Kulikova, N. A., Stepanova, E. V. and Koroleva, O. V. (2005), 'Mitigating activity of humic substances: direct influence on biota', *in* Perminova, I. V., Hatfield, K. and Hertkorn, N. (Eds), *Use of Humic Substances to Remediate Polluted Environments: From Theory to Practice*, Springer, Dordrecht, The Netherlands, pp. 285-309.

Lee, Y. S. and Bartlett, R. J. (1976), 'Stimulation of plant growth by humic substances', *Soil Sci. Soc. Am. J.* 40, 876–9.
Lobartini, J. C., Tan, K. H., Rema, J. A., Gingle, A. R., Pape, C. and Himmelsbach, D. S. (1992), 'The geochemical nature and agricultural importance of commercial humic matter', *Sci. Total Environ.* 113, 1–15.
Lutnaes, B. F., Myrvold, B. O., Lauten, R. A. and Endeshaw, M. M. (2008), '1H and 13C NMR data of benzylsulfonic acids model compounds for lignosulfonate', *Magn. Reson. Chem.* 46, 299–305.
Lyons, G. and Genc, Y. (2016), 'Commercial humates in agriculture: real substance or smoke and mirrors?' *Agronomy* 6, 50–8.
Macdonald, H. (1997), 'Auxin perception and signal transduction', *Physiol. Plant.* 100, 423–30.
Maggioni A., Varanini Z., Nardi S. and Pinton R. (1987), 'Action of soil humic matter on plant roots: stimulation of ion uptake and effects on (Mg^{2+} + K^{+}) ATPase activity', *Sci. Total Environ.* 62, 355–63.
Marino, G., Cellini, A., Masia, A., Francioso, O. and Gessa, C. (2010), '*In vitro* treatment with a low molecular weight humic acid can improve growth and mineral uptake of pear plantlets during acclimatization', *Acta Hortic.* 884, 565–72.
Marré, E. and Ballarin-Denti, A. (1985), 'The proton pumps of the plasmalemma and the tonoplast of higher plants', *J. Bioen. Biomemb.* 17, 1–21.
Mazhar, A. A. M., Shedeed, S. I., Abdel-Aziz, N. G. and Mahgoub, M. H. (2012), 'Growth, flowering and chemical constituents of *Chrysanthemum indicum* L. plant in response to different levels of humic acid and salinity', *J. Appl. Sci. Res.* 8, 3697–706.
McGovern, J. N. (1980), 'Pulping', *in* Casey, J. P. (Ed.), *Pulp and Paper: Chemistry and Chemical Technology* (3rd edn.), Wiley, New York, Chapter 4.
Mikkelsen, R. L. (2005), 'Humic materials for agriculture', *Better Crop* 89, 6–10.
Moghaddam, A. R. L. and Soleimani, A. (2012), 'Compensatory effects of humic acid on physiological characteristics of pistachio seedlings under salinity stress', *Acta Hortic.* 940, 252–55.
Montoneri, E., Adani, F., Genevini, P. L., Ricca, G., Cherubini, S. and Spitaleri, C. (2004), 'Polyalkylphenyl-sulphonic acids with acid groups of variable strength from animal-vegetable wastes', *Waste Manag.* 24, 513–22.
Mora, V., Bacaicoa, E., Zamarrenõ, A. M., Aguirre, E., Garnica, M., Fuentes, M. and Garcıá-Mina, J. M. (2010), 'Action of humic acid on promotion of cucumber shoot growth involves nitrate-related changes associated with the root-to-shoot distribution of cytokinins, polyamines and mineral nutrients', *J. Plant Physiol.* 167, 633–42.
Mora, V., Bacaicoa, E., Baigorri, R., Zamarreno, A. M. and Garcia-Mina, J. M. (2014), 'NO and IAA key regulators in the shoot growth promoting ction of humic acid in *Cucumis sativus* L', *J. Plant Growth Regul.* 33, 430–9.
Morard, P., Eyheraguibel, B., Morard, M. and Silvestre, J. (2011), 'Direct effects of humic-like substance on growth, water, and mineral nutrition of various species', *J. Plant Nutr.* 34, 46–59.
Muscolo, A., Sidari, M., Attinà, E., Francioso, O., Tugnoli, V. and Nardi S. (2007), 'Biological activity of humic substances is related to their chemical structure', *Soil Sci. Soc. Am. J.* 71, 75–85.
Murillo, J. M., Madejón, E., Madejón, P. and Cabrera, F. (2005), 'The response of wild olive to the addition of a fulvic acid-rich amendment to soils polluted by trace elements (SW Spain)', *J. Arid Environ.* 63, 284–303.

Muscolo, A. and Nardi, S. (1997), 'Auxin or auxin-like activity of humic matter', *in* Drozd, J., Gonet, S. S., Senesi, N. and Weber, J. (Eds), *The Role of Humic Substances in the Ecosystems and Environmental Protection*, Polish Society of Humic Substances, Wroclaw, Poland, pp. 987–92.

Muscolo, A., Felici, M., Concheri, G. and Nardi, S. (1993), 'Effect of humic substances on peroxidase and esterase patterns during growth of leaf explants of *Nicotiana plumbaginifolia*', *Biol. Fertil. Soils* 15, 127–31.

Muscolo, A., Panuccio, M. R., Abenavoli, M. R., Concheri, G., and Nardi, S. (1996), 'Effect of molecular complexity and acidity of earthworm faeces humic fractions on glutamate dehydrogenase, glutamine synthetase and phosphoenolpyruvate carboxylase in *Daucus carota* II cells', *Biol. Fertil. Soils* 22, 83–8.

Muscolo, A., Cutrupi, S. and Nardi, S. (1998), 'IAA detection in humic substances', *Soil Biol. Biochem.* 30, 1199–201.

Muscolo, A., Bovalo, F., Gionfriddo, F. and Nardi, S. (1999), 'Earthworm humic matter produces auxin-like effects on *Daucus carota* cell growth and nitrate metabolism', *Soil Biol. Biochem.* 31, 1303–11.

Muscolo, A., Panuccio, M. R., Sidari, M. and Nardi, S. (2005). The effects of humic substances on Pinus callus are reversed by 2,4-dichlorophenoxy acetic acid', *J. Chem. Ecol.* 31, 577–90.

Muscolo, A., Sidari, M. and Nardi S. (2013), 'Humic substance: relationship between structure and activity. Deeper information suggests univocal findings', *J. Geochem. Explor.* 129, 103–11.

Nardi, S., Arnoldi, G. and Dell'Agnola, G. (1988), 'Release of the hormone-like activities from *Allolobophora rosea* and *A. caliginosa* faeces', *Can. J. Soil Sci.* 68, 563–7.

Nardi, S., Arnoldi, G. and Dell'Agnola, G. (1989), 'Effect of actynomicete metabolites on ion absorption by rice seedlings', *Sci. Total Environ.* 81-82, 675–8.

Nardi, S., Concheri, G., Dell'Agnola, G. and Scrimin, P. (1991), 'Nitrate uptake and ATPase activity in oat seedlings in the presence of two humic fractions', *Soil Biol. Biochem.* 23, 833–6.

Nardi S., Panuccio M. R., Abenavoli M. R. and Muscolo A. (1994), 'Auxin-like effect of humic substances extracted from faeces of *Allolobophora caliginosa* and *Allolobophora rosea*', *Soil Biol. Biochem.* 26, 1341–6.

Nardi, S., Concheri, G. and Dell'Agnola, G. (1996), 'Biological activity of humus', *in* Piccolo, A. (Ed.), *Humic Substances in Terrestrial Ecosystems*, Elsevier, The Netherlands, pp. 361–406.

Nardi, S., Reniero, F. and Concheri, G. (1997), 'Soil organic matter mobilization by root exudates of three maize hybrids', *Chemosphere* 35, 2237–44.

Nardi, S., Pizzeghello, D., Muscolo, A., Dalla Vecchia, F. and Concheri, G. (1998), 'Effects of forest humus on biological activity in roots of pinus sylvestris related to chemical humus fraction characteristics', *Fres. Environ. Bull.* 7, 203–8.

Nardi, S., Pizzeghello, D., Reniero, F. and Muscolo, A. (1999), 'Biological activity of humic substances extracted from soils under different vegetation cover', *Commun. Soil Sci. Plant Anal.* 30, 621–34.

Nardi S., Pizzeghello D., Gessa C., Ferrarese L., Trainotti L. and Casadoro G. (2000a), 'A low molecular weight humic fraction on nitrate uptake and protein synthesis in maize seedlings', *Soil Biol. Biochem.* 32, 415–19.

Nardi, S., Concheri, G., Pizzeghello, D., Sturaro, A., Rella, R. and Parvoli, G. (2000b), 'Soil organic matter mobilization by root exudates', *Chemosphere* 41, 653–8.

Nardi, S., Pizzeghello, D., Muscolo, A. and Vianello, A. (2002), 'Physiological effects of humic substances on higher plants', *Soil Biol. Biochem.* 34, 1527–36.
Nardi, S., Pizzeghello, D., Bragazza, L. and Gerdol, R. (2003), 'Low-molecular-weight organic acids and hormone-like activity of dissolved organic matter in two forest soils in N Italy', *J. Chem. Ecol.* 29, 1549–64.
Nardi, S., Tosoni, M., Pizzeghello, D., Provenzano, M. R., Cilenti, A., Sturaro, R. and Vianello, A. (2005), 'A comparison between humic substances extracted by root exudates and alkaline solution', *Soil Sci. Soc. Am. J.* 69, 2012–19.
Nardi, S., Muscolo, A., Vaccaro, S., Baiano, S., Spaccini, R. and Piccolo, A. (2007), 'Relationships between molecular characteristics of soil humic fractions and glycolytic pathway and kerb cycle in maize seedlings', *Soil Biol. Biochem.* 39, 3138–46.
Nardi, S., Carletti, P., Pizzeghello, D. and Muscolo, A. (2009), 'Biological activities of humic substances', *in* Senesi, N., Xing, B. and Huang, P. M. (Eds), *Volume 2-Biophysico-Chemical Processes Involving Natural Nonliving Organic Matter in Environmental Systems. PART I. Fundamentals and Impact of Mineral-organic-biota Interactions on the Formation, Transformation, Turnover, and Storage of Natural Nonliving Organic Matter (NOM)*, John Wiley & Sons, Hoboken, NJ, pp. 301–35.
Nardi, S., Pizzeghello, D., Schiavon, M. and Ertani, A. (2016), 'Plant biostimulants: physiological responses induced by protein hydrolyzed-based products and humic substances in plant metabolism', *Sci. Agr.* 73, 18–23.
Nardi, S., Pizzeghello, D. and Ertani, A. (2017a), 'Hormone-like activity of the soil organic matter', *Appl. Soil Ecol.* 123, 517–20.
Nardi, S., Ertani, A. and Francioso, O. (2017b), 'Soil-root cross-talking: the role of humic substances', *J. Plant Nutr. Soil Sci.* 180, 5–13.
Palumbo, G., Schiavon, M., Nardi, S., Ertani, A., Celano, G. and Colombo, C. M. (2018), 'Biostimulant potential of humic acids extracted from an amendment obtained via combination of olive mill wastewaters (OMW) and a pre-treated organic material derived from municipal solid waste (MSW)', *Front. Plant Sci.* 9, 1028.
Pandeya, S. B., Singh, A. K. and Dhar P. (1998), 'Influence of fulvic acid on transport of iron in soils and uptake by paddy seedlings', *Plant Soil* 198, 117–25.
Panuccio, M. R., Muscolo, A. and Nardi S. (1997), 'Effect of humic substances on the nitrogen uptake in *Pinus laricio* (P., spp. Calabrica) seedlings', *J. Biol. Res.* 73, 9–10.
Panuccio, M. R., Muscolo, A. and Nardi, S. (2000), 'Effect of humic substances on nitrogen uptake and assimilation in two species of pinus', *J. Plant Nutr.* 24, 693–704.
Peng, A., Xu, Y. and Wang, Z. J. (2001), 'The effect of fulvic acid on the dose effect of selenite on the growth of wheat', *Biol. Trace Elem. Res.* 83, 275–9.
Piccolo, A. (2001), 'The supramolecular structure of humic substances', *Soil Sci.* 166, 810–32.
Piccolo, A. and Mbagwu, J. S. C. (1994). 'Humic substances and surfactants effects on the stability of two tropical soils', *Soil Sci. Soc. Am. J.* 58, 950–5.
Piccolo, A. and Spiteller, M. (2003), 'Electrospray ionization mass spectrometry of terrestrial humic substances and their size fractions', *Anal. Bioanal. Chem.* 377, 1047–59.
Piccolo, A., Nardi, S. and Concheri, G. (1992), 'Structural characteristics of humus and biological activities', *Soil Biol. Biochem.* 24, 273-380.
Piccolo, A., Pietramellara, G. and Mbagwub, J. S. C. (1997), 'Use of humic substances as soil conditioners to increase aggregate stability', *Geoderma* 75, 267-77.

Pinton, R., Cesco, S., Santi, S. and Varanini, Z. (1997), 'Soil humic substances stimulate proton release by intact oat seedling roots', *J. Plant Nutr.* 20, 857–69.

Pinton, R., Cesco, S., Iacoletti, G., Astolfi, S. and Varanini, Z. (1999), 'Modulation of NO_3 uptake by water-extractable humic substances: involvement of root plasma membrane H^+ATPase', *Plant Soil* 215, 155–61.

Pizzeghello, D., Nicolini, G. and Nardi, S. (2001), 'Hormone-like activity of humic substances in *Fagus sylvaticae* L. forests', *New Phytol.* 151, 647–57.

Pizzeghello, D., Nicolini, G. and Nardi, S. (2002), 'Hormone-like activities of humic substances in different forest ecosystems', *New Phytol.* 155, 393–402.

Pizzeghello D., Francioso O., Ertani A., Muscolo A. and Nardi S. (2013), 'Isopentenyladenosine and cytokinin-like activity of four humic substances', *J. Geochem. Exp.* 129, 70–5.

Poapst, P.A. and Schnitzer, M. (1971), 'Fulvic acid and adventitious root formation', *Soil Biol. Biochem.* 3, 215–19.

Rauthan, B. S. and Schnitzer, M. (1981), 'Effects of a soil fulvic acid on the growth and nutrient content of cucumer (*Cucumis sativus*) plants', *Plant Soil* 63, 491–5.

Russell L., Stokes A. R., Macdonald H., Muscolo A. and Nardi S. (2006), 'Stomatal responses to humic substances and auxin are sensitive to inhibitors of phospholipase A_2', *Plant Soil* 283, 175–85.

Sadiq, S. A., Baloch, D. M., Ahmed, N. and Hidayatullah, P. (2014), 'Role of coal-derived humic acid in the availability of nutrients and growth of sunflower under calcareous soil', *J. Anim. Plant Sci.* 24, 1737–42.

Sanchez-Sanchez, A., Sanchez-Andreu, J., Juarez, M., Jorda, J. and Bermudez D. (2002), 'Humic substances and amino acids improve effectiveness of chelate FeEDDHA in lemon trees', *J. Plant Nutr.* 25, 2433–42.

Sanchez-Sanchez, A., Sanchez-Andreu, J., Juarez, M., Jorda, J. and Bermudez, D. (2006), 'Improvement of iron uptake in table grape by addition of humic substances', *J. Plant Nutr.* 29, 259–72.

Sauchelli, V. (1944), 'Humus: the working partner of chemical plant food', *Amer. Fertil.* 101, 11–12, 26–8.

Schaumann, G. E. and Bertmer, M. (2008), 'Do water molecules bridge soil organic matter molecule segments?' *Eur. J. Soil Sci.* 59 (3), 423–9.

Schiavon, M., Pizzeghello, D., Muscolo, A., Vaccaro, S., Francioso, O. and Nardi, S. (2010), 'High molecular size humic substances enhance phenlylpropanoid metabolism in Maize (*Zea mays* L.)', *J. Chem. Ecol.* 36, 662–9.

Schmidt, W., Cesco, S., Santi, S., Pinton, R. and Varanini, Z. (2005). 'Water-extractable humic substances as nutrient acquisition signals for root hairs development in *Arabidopsis*', *in* Hartmann, A., Schmid, M., Wenzel, W. and Hinnsinger, P. (Eds), *Rizosphere 2004, Perspectives and Challenges*, GSF-Berich, Neuherberg, p. 71.

Selim, E. M., Shaymaa, I. S., Asaad, F. F. and El-Neklawy, A. S. (2012), 'Interactive effects of humic acid and water stress on chlorophyll and mineral nutrient contents of potato plants', *J. Appl. Sci. Res.* 8, 531–7.

Sessi, E., Pizzeghello, D., Reniero, F., Tomasi, M., De Siena, C., Nicolini, G. and Nardi, S. (2001), 'Effect of humic substances extracted from pasture and forested soils on plant metabolism', *Fresen. Environ. Bull.* 10, 217–20.

Sitzia, T., Pizzeghello, D., Dainese, M., Ertani, A., Carletti, P., Semenzato, P., Nardi, S. and Cattaneo, D. (2014), 'Topsoil organic matter properties in contrasted hedgerow vegetation types', *Plant Soil* 383, 337–48.

Sitzia, T., Campagnaro, T., Kotze, D. J., Nardi, S. and Ertani, A. (2018), 'The invasion of abandoned fields by a major alien tree filters understory plant traits in novel forest ecosystems', *Sci. Rep.* 8, 8410.

Smejkalova, D. and Piccolo, A. (2008), 'Aggregation and disaggregation of humic supramolecular assemblies by NMR diffusion ordered spectroscopy (DOSY-NMR)', *Environ. Sci. Technol.* 42, 699–706.

Stevenson, F. J. (1982), *Humus Chemistry: Genesis, Composition, Reactions*, John Wiley & Sons, New York, p. 443.

Stevenson, F. J. (1994), *Humus Chemistry: Genesis, Composition, Reactions* (2nd edn.), Wiley, New York.

Swarup, R., Parry, G., Graham, N., Allen, T. and Bennett, M. J. (2002), 'Auxin crosstalk: integration of signalling pathways to control plant development', *Plant Mol. Biol.* 49, 411–26.

Tahir, M. M., Khurshid, M., Khan, M. Z., Abbasi, M. K. and Hazmi, M. H. (2011), 'Lignite-derived humic acid effect on growth of wheat plants in different soils', *Pedosphere* 2, 124–31.

Tan, K. H. (2003), *Humic Matter in Soil and the Environment*, Marcel Dekker, New York, p. 495.

Vaccaro, S., Ertani, A., Nebbioso, A., Muscolo, A., Quaggiotti, S., Piccolo, A. and Nardi, S. (2015), 'Humic substances stimulate maize nitrogen assimilation and amino acid metabolism at physiological and molecular level', *Chem. Biol. Tech. Agr.* 2, 5

Vaughan, D. and Malcolm, R. E. (1985), 'Influence of humic substances on growth and physiological processes', *in* Vaughan, D. and Malcolm, R. E. (Eds), *Soil Organic Matter and Biological Activity*, Martinus Nijhoff/Junk, Dordrecht, The Netherlands, pp. 37–76.

Vidal, K., Guermache, F. and Widmer, T. (2004), '*In vitro* culturing of yellow starthistle (*Centaurea solstitialis*) for screening biological control agents', *Biol. Control* 30, 330–5.

Visser, S. A. (1985), 'Physiologic action of humic substances on microbial cells', *Soil Biol. Biochem.* 17, 457–62.

Xudan, X. (1986), 'The effect of foliar application of fulvic acid on water use, nutrient uptake and yield in wheat', *Aust. J. Agric. Res.* 37, 343–50.

Yildrim, E. (2007), 'Foliar and soil fertilization of humic acid affect productivity and quality of tomato', *Acta Agric. Scand. Sect. B.* 57, 182–6.

Young, C. C. and Chen, L. F. (1997), 'Polyamines in humic acid and their effect on radical growth of lettuce seedlings', *Plant Soil* 195, 143–9.

Zandonadi, D. B., Canellas, L. P. and Façanha, A. R. (2007), 'Indolacetic and humic acids induce lateral root development through a concerted plasmalemma and tonoplast H^+-pumps activation', *Planta* 225, 1583–95.

Chapter 4

Seaweed extracts as plant biostimulants in agriculture

Izabela Michalak, Wrocław University of Science and Technology, Poland; Katarzyna Tyśkiewicz, Marcin Konkol and Edward Rój, New Chemical Syntheses Institute, Poland; and Katarzyna Chojnacka, Wrocław University of Science and Technology, Poland

1 Introduction

Modern agriculture relies on chemical fertilizers and synthetic pesticides to enhance crop production. Side effects of these practices include reduced soil efficiency and negative impacts on human health (Vasantharaja et al., 2019). Extracts from various algae can be useful in agricultural production. They belong to a group of photosynthetic and structure-complex organisms that can be divided into microalgae (eukaryotic and prokaryotic cyanobacteria - blue-green algae) and macroalgae (seaweeds) depending on their size (microalgae 0.2–2.0 μm diameter; macroalgae <60 m length) (Lordan et al., 2011; Ronga et al., 2019). The main role of algae in a coastal ecosystem is to provide nutrition, as well as shelter for other marine organisms (Lalopua and Latumahina, 2015). Both, microalgae and macroalgae have a great potential to be used in the production of biostimulants of plant growth. In the case of microalgae their main advantages are simple unicellular structure, high photosynthetic efficiency, ability to grow in

http://dx.doi.org/10.19103/AS.2020.0068.05

different environments, as well as flexibility to industrial wastewater and possibility to produce valuable biologically active compounds. Nevertheless, their use in agriculture is still in its infancy (Chiaiese et al., 2018; Ronga et al., 2019).

The main advantage of seaweeds is their species diversity – they are divided into red (*Rhodophyta*), green (*Chlorophyta*) and brown algae (*Phaeophyta*), depending on the pigment (Shahnaz and Shameel, 2009). The green pigmentation of green algae is attributed to the presence of chlorophyll a and *b*. In the case of brown algae, xanthophylls and fucoxanthin are responsible for the characteristic color of these species. The major and dominant compounds in red algae are those from the group of red pigments, such as phycobiliproteins (Hamed et al., 2018).

Macroalgae are characterized by various activities, such as anti-inflammatory, antidiabetic, antihypertensive, antimutagenic (Ioannou and Roussis, 2009; Andrade et al., 2013), antibacterial (Singh and Chaudhary, 2010), antifungal (de Felício et al., 2010) and antiviral (Bouhlal et al., 2010). These beneficial properties of marine algae may be attributed to the presence of a number of bioactive compounds, including polysaccharides and derived oligosaccharides, lipids, phenolic compounds, pigments, lectins, alkaloids, terpenes as well as halogenated compounds (furanones) (Pérez et al., 2016). The content of the bioactive compounds may differ in certain algae species. However, some species are characterized by a high content of fatty acids, fibers, sterols, proteins, phytocolloids, amino acids and vitamins (Alassali et al., 2016). These compounds may also inhibit the growth of viruses, bacteria, fungi and other algae (Silkina et al., 2012).

Recently, two major promising applications of algae in the form of extracts have been reported, including the activity of the algae as plant growth promoters (biostimulants), as well as plant protection agents. Khan et al. (2009) distinguished two types of application methods of algal extract-based products, the foliar and soil application methods (Fig. 1). The aerial/foliar application methods are all kinds of treatment involving seed treatment, seeding dip and foliar spray, whereas the soil application is based on the addition of algal extracts to hydroponics (du Jardin, 2015; Digruber et al., 2018; Ronga et al., 2019). Other forms of application include granules and seaweed paste as soil manure (Vasantharaja et al., 2019).

Algal formulations were found to enhance organ growth, development of disease tolerance and climatic stress (drought or cold) (Digruber et al., 2018). Biostimulants based on seaweeds are environmentally friendly solutions for sustainable crop production. The formulations based on algal extracts improve growth yield when used together with conventional fertilizers (Sharma et al., 2017). Seaweed extracts fit the concept of organic and sustainable agriculture, because they are biodegradable, nonpolluting and nontoxic to humans and animals (Rathore et al., 2009).

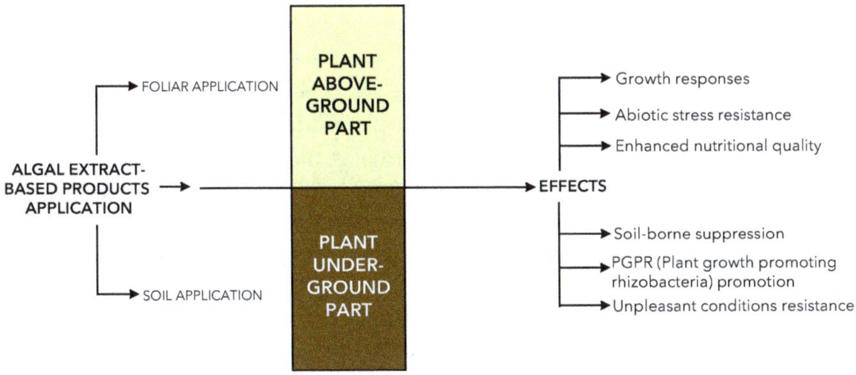

Figure 1 Application methods of algal extract-based products (on the basis of Khan et al., 2009).

Table 1 Benefits of seaweed extracts to soil (du Jardin, 2015)

Beneficial role of chemical compounds to plants
polysaccharides form a gel that retains water and aerates soil
polyanionic compounds
bind toxic metal ions and are useful in soil remediation
positive effect on microflora
fertilizer properties due to the presence of macronutrients and micronutrients
beneficial effect on germination of seeds - action of plant growth hormones - gibberellins, auxins, cytokinins, abscisic acid (ABA), polyamines, sterols

Seaweed extracts can play the role of both fertilizer (because they contain nutrients) and plant growth biostimulants due to enhanced germination of seeds and plant growth, root development, higher yield and quality of crops (e.g. vegetables: tomato, cucumber, spinach broccoli and bean), and they also increase postharvest shelf life. Seaweed formulations improve tolerance of plants to environmental (abiotic) stress (drought, increase nutrient uptake, antioxidant properties) (Vasantharaja et al., 2019).

The composition of algal extracts depends on the type of seaweed (green, brown or red), harvesting location and spatiotemporal and environmental changes that the alga experiences (Sharma et al., 2019) and the process used for manufacture and extraction (Goñi et al., 2018). Important components of algal extracts in terms of plant biostimulating effect is the presence of polysaccharides, laminarin, carrageenans and alginates, macronutrients, micronutrients, plant growth hormones, sterols and betaines (du Jardin, 2015). Compounds found in seaweed extracts show many beneficial effects on soil (Table 1).

Biologically active substances from seaweed extracts have been confirmed to enhance yield and quality of crops. The literature describes a range of application: vegetables, flowers, grasses or cereals (Rathore et al., 2009). Formulations are useful for the following plants: tomato, bean, wheat, maize and grasses (Digruber et al., 2018). Sharma et al. (2017) used extracts from seaweeds *Kappaphycus alvarezii* and *Gracilaria edulis* that contain plant growth regulators, macronutrients, micronutrients and quaternary ammonium compounds. It was found that the extracts from those seaweeds stimulate the growth of several plant species: soy, wheat, rice, the potato-rice bar-cultivation system and maize (Sharma et al., 2017). There are still not many studies on applications in orchards. Responses of woody fruit species to seaweed treatment are limited (Digruber et al., 2018).

2 Classification, legal requirements for biostimulant registration and the biostimulant market

In the work of du Jardin (2015), biostimulants of plant growth were classified into eight categories: (1) seaweed extracts, (2) humic substances, (3) complex organic materials, (4) beneficial chemical elements, (5) inorganic salts (including phosphite), (6) chitin and chitosan derivatives, (7) free amino acids and other N-containing substances and (8) antitranspirants.

According to the proposal of the Regulation of the European Parliament and of the Council laying down rules on making fertilizing products available in the EU market and amending Regulations (EC) No 1069/2009 and (EC) No 1107/2009 and repealing Regulation (EC) No 2003/2003 – 'plant biostimulant means a product stimulating plant nutrition processes independently of the product's nutrient content with the sole aim of improving one or more of the following characteristics of the plant or the plant rhizosphere – (a) nutrient use efficiency; (b) tolerance to abiotic stress; (c) quality traits and (d) availability of confined nutrients in soil or rhizosphere'. Moreover, 'plant biostimulants are by nature more similar to fertilising products than to most categories of plant protection products. They act in addition to fertilisers, with the aim of optimising the efficiency of those fertilisers and reducing the nutrient application rates'.

Seaweed extracts are classified within the group of biostimulants (Colla and Rouphael, 2015), because they contain plant growth hormones that are identifiable, for example, auxins, cytokinins (du Jardin, 2015). These extracts are important on the biostimulants market, because they have been found to increase tolerance to drought (Goñi et al., 2018). Consequently, higher demand for biostimulants is being observed, especially in the EU, which is the largest market. It is estimated that the global market increase will reach $ 2.91 billion by the year 2021 (Scaglia et al., 2017).

3 Effects and mode of action of seaweed extracts on plant primary and secondary metabolism

3.1 Primary metabolites of seaweeds

Primary metabolites are low-molecular chemical compounds that fulfill the fundamental functions of the living organism, for example, energy building blocks. They occur in all organisms and are necessary for the proper functioning of life processes. The primary metabolites of plants include carbohydrates, fats, proteins, nucleic acids, organic acids and many others. Most of the published works concern the influence of algae extracts on the content of secondary metabolites in plants. Secondary metabolites are very important because of their versatile influence on plant development. This also applies to primary metabolites. *Ascophylum nodosum* (*A. nodosum*) extract as a biostimulant to promote growth and productivity in agricultural crops, such as barley (AC Sterling variety), was presented by Rayorath et al. (2008). The effect of an *A. nodosum* extract was also studied on rapeseed and revealed a significant effect on root (102%) and shoot (23%) growth (Patier et al., 1993).

Several reports demonstrated that the application of various extracts enhanced root vigor, plant growth and development, and leaf numbers, resulting in higher yields of vegetables, fruit and field crops under normal and a variety of stress conditions (Sangha et al., 2014). Commercially produced extracts of *A. nodosum* are effective on fruit trees. Trees treated with Actiwave® decreased variations in yield between their 'on' and 'off' years, and increased the average fruit weight (Spinelli et al., 2009).

One of the important targets in agricultural production is the introduction of value-added nutritional factors. Improved crop nutrition may have a direct effect on the consumer's health. Therefore, the use of seaweed extracts, on a variety of crops, has enormous potential (McGloughlin, 2010). Biofortification, a process for improving the bioavailability of health-promoting minerals in food crops, has been commonly targeted through fertilization, plant breeding or biotechnological approaches (White and Broadley, 2009). In another study on wheat, the yield and nutritional quality, for example, carbohydrates, protein and minerals of the grains were improved after spray treatment of the plants with 0.25%, 0.50% and 1.0% of seaweed *K. alvarezii* (Zodape et al., 2009).

3.2 Plant growth stimulation

Macroalgae extracts are considered as plant biostimulants due to their ability to catalyze different plant metabolic pathways, increase nutrient intake, improve crop quality and stress resistance of plants (Sangha et al., 2014; Chojnacka et al., 2015). The crop characteristics, such as growth, yield and quality, may be enhanced through the use of natural preparations on the basis of algae with strong biostimulating properties (Hong et al., 2007). The appropriate period for algae biostimulation is primarily after plant germination in the form of spraying. However, soil fertilization is also recommended (Dobrzański et al., 2008). Algae-based preparations are also useful during unpleasant weather conditions (freezing, drought) (Sharma et al., 2014).

In the studies performed by Dobrzański et al. (2008), the seeds of carrot and parsley were treated with a biostimulant via the addition of 15% extract of seaweed from the group of brown algae, *Sargassum*. As it was tested, the formulation influenced the germination of seeds as well as improved the growth of plants. Moreover, the increase in the content of nitrates and carotenoids was observed. Khan et al. (2009) put the attention to a wide group of biostimulating formulations based on algae used in agriculture. The most common preparations are those with the addition of the *A. nodosum* extract, which is a source of different compounds, including osmolytic betaines but also betaine-like compounds (Blunden and Gordon, 1986; Khan et al., 2009). Apart from *A. nodosum*, *Ecklonia maxima*, *Macrocystis pyrifera* and *Durvillea potatorum* should be taken into consideration as the most used in different industries (Khan et al., 2009). According to Blunden et al. (2010), betaines are responsible for plant protection in terms of unpleasant conditions (drought, extreme temperature, stress). For instance, the brown macroalga *A. nodosum* extracts were investigated on different plants, such as spinach (*Spinacia oleracea* L.) (Fan et al., 2011), tomato (Ali et al., 2016; Goñi et al., 2018) and maize (*Zea mays*) (Ertani et al., 2018). Furthermore, one of the major components of seaweed extracts are the phytohormones (cytokinin, gibberellins, abscisic acids, auxins), which are said to regulate phytohormone metabolic genes as it was mentioned by Raj et al. (2018) in the example on *A. nodosum* extract.

Another formulation is based on *Ecklonia maxima* extracts as studied by Crouch and van Staden (1991), as well as by Rengasamy et al. (2015). It was assumed that compounds present in tested brown algae in the form of Kelpak (based on *E. maxima* macroalga) have an influence on a mung bean root promotion. The results provided by Matysiak et al. (2010) confirmed the positive activity of Kelpak on the winter wheat stimulation. It was also observed that lower concentration of the product was required to obtain such an activity. A similar observation was made by Kumar and Sahoo (2011). Among all tested concentrations (5, 10, 20, 30, 40, 50 and 100% weight) of the *Sargassum wightii* extract on common wheat (*Triticum aestivum*) growth and promotion, the optimal concentration was 20%. The extract guaranteed improved germination as well as root and stem length. The influence of algal extract on winter wheat was also studied by Michalak et al. (2016a) in an analogous manner to Matysiak et al. (2010). However, the extracts of *A. nodosum*, *Spirulina plantensis* and Baltic Sea green macroalgae were obtained with the use of supercritical fluid extraction (SFE) (Michalak et al., 2016a). The SFE was further used to obtain extracts of Baltic Marine macroalgae (*Polysiphonia fucoides*, *Ulva flexuosa*, *Cladophora glomerata*) (Michalak et al., 2016b). SFE extracts were found to be effective in both chlorophyll and carotenoid content increase in plant shoots.

The literature data provide also the application of algal extracts on different fruits (olives, grapes, apple trees) (El-Migeed et al., 2004; El-Moniem and Abd-Allah 2008; Spinelli et al., 2009). As it was pointed out by the authors, algal extracts contain elements from the group of macronutrients and micronutrients but also polyamines, vitamins and growth-enhancing regulators, which are necessary for growth, yield and fruit quality improvement. Al-Musawi (2018) treated oranges (*Citrus aurantium* L.) with two extracts from *Ecklonia* and *Fucox* at three levels of concentrations (1, 2 and 3 wt%). Two consecutive spraying of orange trees with algal extracts resulted in the fruits' enhancement in terms of length, width, size, fresh weight as well as quality and quantity. The best results were observed for *Fucox* (3%). An experiment with algal extract-based products was also conducted on mango trees by El-Sharony et al. (2015). The influence of the product was evaluated taking into consideration the leaves' macroelement content (N, P, K), physical and chemical characteristic of fruits, and growth increase. As it was highlighted by the authors, the 2% algal extract was able to improve all investigated parameters. However, the exact species of algae was not specified by the authors.

The studies carried out by Eyras et al. (1998) on the influence of two green macroalga (*Ulva* spp., *Codium* spp.) and one brown macroalga (*Dictyota* spp.) on tomatoes (*Licopersicum esculentum* var. Platense) growth showed positive effects in all experiments. The extracts were obtained in the form of the compost at concentrations of 35%, 18% and 17%, respectively, for *Ulva* spp., *Codium* spp., and *Dictyota* spp. Tomatoes were cultivated on a different medium, including sand, sandy-loam soil as well as sandy-loam soil with inorganic fertilizers. The growth of the plant proportionally increased with the dose of the extract.

In addition to green and brown algae, red alga, such as *K. alvarezii*, may also be used as a plant growth-promoting agent as it was shown in the surveys conducted by Rathore et al. (2009). Different concentrations of the extract (2.5, 5, 7.5, 10; 12.5, 15% v/v) were applied on soybean (*Glycine max*) leaves with a foliar spray method. The best results were obtained when the highest content of the extract (15%) was used. However, it is the reverse effect than that described by Matysiak et al. (2010) and Kumar and Sahoo (2011), who applied lower concentrations of algal extracts in order to obtain the best final results. Table 2 comprises the use of different macroalgae extracts as plant growth promoters.

3.3 Plant protecting agent

Marine macroalgae have also been used in the production of various metabolites against a number of fungicidal pathogens. The interest in marine algae in a

Table 2 Application of algae as plant growth-promoting agents

Alga species	Type of alga	Treated plant	Biostimulant product	Doses	Active substance(s)	Effect of a biostimulant	Reference
Ascophyllum nodosum	brown	Spinach (*Spinacia oleracea*)	Extract (70% aqueous methanol (*v/v*), 40°C, 2 h)	0.1, 1.0, 5.0 g/L; 50 mL per plant; 14 and 7 days prior to harvest	phenolic compounds	increase of total phenolic and flavonoid content	Fan et al. (2011)
Ascophyllum nodosum	brown	Winter wheat (*Triticum aestivum*)	Supercritical fluid extraction (50°C, 500 bar, 100 kg CO_2/kg)	1.0, 1.5, 1.8 L/ha, *Ascophyllum nodosum* extract (10 wt%)	not specified	similar effect as commercial products	Michalak et al. (2016a)
Ascophyllum nodosum	brown	Tomato (*Licopersicum esculentum*)	Commercial extract	foliar spray, 0.33 wt% extract, on 35-day-old tomato plants	sulfate (8.02-15.98 wt%), uronic acid (8.49-13.05 wt%), fucose (6.90-12.12 wt%), polyphenol (3.60-12.10 wt%), laminarin (2.30-3.57 wt%), mannitol (1.93-8.02 wt%)	enhance the tolerance to drought stress	Goñi et al. (2018)
Ascophyllum nodosum	brown	Tomato (*Licopersicum esculentum*)	Extract (KOH followed by neutralization)	foliar spray, soil drench, 0.2, 0.5% extract, once every 15 days	not specified	improved size, color, firmness, increased total soluble solids, ascorbic acid levels and mineral levels	Ali et al. (2016)
Ascophyllum nodosum	brown	Maize (*Zea mays*)	Extract (water + sulfuric acid to pH 3-3.5; neutralization by KOH)	0.5 mL/L, for 2 days	phenolic acids (gallic, protocatechuic, vanillic, caffeic, p-courmaric, syringic, p-hydroxybenzoic), hormones (IAA, IPA)	root promotion, improved plant nutrition, total phenolic content increase	Ertani et al. (2018)
Baltic green macroalgae	green	Winter wheat (*Triticum aestivum*)	SFE (50°C, 500 bar, 100 kg CO_2/kg)	1.0, 1.5, 1.8 L/ha, 10 wt% of Baltic Sea algal extract	not specified	similar effect as commercial products	Michalak et al. (2016a)

Cladophora glomerata	green	Garden cress (*Lepidium sativum*) Wheat (*Triticum aestivum*)	SFE (500 bar, CO_2 flow: 200 kg/h)	10 mL, 50 mL, concentration: 13.8 mg/L; 27.6 mg/L; on the 4th day (wheat), on the 6th day (cress)	polyphenols, chlorophyll, carotenoids	enhanced chlorophyll and carotenoid content in plant shoots, enhanced root thickness and above-ground biomass	Michalak et al. (2016b)
Codium spp.	green	Tomato (*Licopersicum esculentum* var. Platense)	Compost	10, 25, 50, 100% of compost; mixtures of compost, compost plus local soil, local soil plus inorganic fertilizers at different ratio, laboratory experiments	not specified	increase of the aerial biomass of tomato plants	Eyras et al. (1998)
Dictyota sp.	brown	Tomato (*Licopersicum esculentum* var. Platense)	Compost	10, 25, 50, 100% of compost; mixtures of compost, compost plus local soil, local soil plus inorganic fertilizers at different ratio, laboratory experiments	not specified	increase of the aerial biomass of tomato plants	Eyras et al. (1998)
Ecklonia maxima	brown	Mung bean (*Vigna mungo*) Winter wheat (*Triticum aestivum*)	Kelpak	10% extract solution	root promotory compounds	root promotion	Khan et al. (2009), Crouch and Van Staden (1991), Rengasamy et al. (2015), Matysiak et al. (2010)
Ecklonia maxima	brown	Orchard grass (*Dactylis glomerata*) Braun's festulolium (*Festulolium braunii*)	Kelpak	application at 2 dm^3/ha	not specified	reduction of cellulose, hemicellulose, lignin, increase of non-structural carbohydrates	Ciepiela et al. (2016)

(*Continued*)

Table 2 (*Continued*)

Alga species	Type of alga	Treated plant	Biostimulant product	Doses	Active substance(s)	Effect of a biostimulant	Reference
Ecklonia	brown	Orange (*Citrus aurantium*)	not specified	1, 2, 3%, spraying twice	auxin (11 mg/L), cytokinins (31 mg/L), amino acids (3%)	reduction of the acidity and carotene content, increase of vitamin C content, enhance of fruit length, width, size, fresh weight, peel thickness	Al-Musawi (2018)
Fucox	brown	Orange (*Citrus aurantium*)	not specified	1, 2, 3%, spraying twice	fucoxanthin (70 mg/L), auxin (20 mg/L), cytokinins (35 mg/L), vitamin C (9 mg/L), amino acids (6%)	reduction of the acidity and carotene content, increase of vitamin C content, enhance of fruit length, width, size, fresh weight, peel thickness	Al-Musawi (2018)
Kappaphycus alvarezii	red	Soybean (*Glycine max*)	Extract (homogenization by grinder, filtration)	foliar spray, 2.5, 5, 7.5, 12.5, 15 (*v/v*)	not specified	enhanced yield parameters, improved nutrient uptake	Rathore et al. (2009)
Laminaria	brown	Maize (*Zea mays*)	Extract (water + sulfuric acid to pH 3-3.5; neutralization by KOH)	0.5 mL/L, for 2 days	phenolic acids (gallic, protocatechuic, vanillic, caffeic, p-courmaric, syringic, p-hydroxybenzoic), hormones (IAA, IPA)	decrease of total phenolic content	Ertani et al. (2018)
Polysiphonia fucoides	red	Winter wheat (*Triticum aestivum*) Garden cress (*Lepidium sativum*)	SFE (500 bar, CO_2 flow: 200 kg/h)	10 mL, 50 mL, concentration: 13.8 mg/L; 27.6 mg/L; on the 4th day (wheat), on the 6th day (cress)	polyphenols, chlorophyll, carotenoids	enhanced chlorophyll and carotenoid content in plant shoots, enhanced root thickness and above-ground biomass	Michalak et al. (2016b)

Sargassum species	brown	Carrot seeds Parsley seeds	Algamino Plant	seeds soaked in 5–50% extract for 24 h	not specified	enhanced percentage of seed germination, growth, yield	Kumar and Sahoo (2011)
Sargassum wightii	brown	Common wheat (*Triticum aestivum* var. Pusa Gold)	Extract (boiling with distilled water, 1 h, filtration)	seeds soaked in 5–50% extract for 24 h	not specified	percentage of seed germination, growth, yield	Kumar and Sahoo (2011)
Ulva flexuosa	green	Winter wheat (*Triticum aestivum*) Garden cress (*Lepidium sativum*	SFE (500 bar, 200 kg/h)	10 mL, 50 mL, concentration: 13.8 mg/L; 27.6 mg/L; on the 4th day (wheat), on the 6th day (cress)	polyphenols, chlorophyll, carotenoids	enhanced chlorophyll and carotenoid content in plant shoots, enhanced root thickness and above-ground biomass	Michalak et al. (2016b)
Ulva sp.	green	Tomato (*Licopersicum esculentum* var. Platense)	Compost	10, 25, 50, 100% of compost; mixtures of compost, compost plus local soil, local soil plus inorganic fertilizers at different ratio	not specified	increase of the aerial biomass of tomato plants	Eyras et al. (1998)
not specified	not specified	Mango (*Mangifera indica*)	not specified	foliar spray, 2%	not specified	fruit set, retention, yield and fruit quality improved, leaf N, P and K content increased	El-Sharony et al. (2015)

commercial production as well as biotechnology is highlighted due to the fact that seaweed extracts may serve as plant protecting agents (Hamed et al., 2018). The attention should be paid particularly to a group of polysaccharides that are able to chelate metal ions. Moreover, it has been proved that structural groups of these compounds may take part in a reaction with microelements of various plants. The stimulation of plant defense responses may be assigned to particular polysaccharides from different types of macroalgae, including fucans, laminarin and alginates from brown algae, ulvans from green algae as well as carrageenans and porphyran from red algae (Vera et al., 2011). In addition, another group of compounds, belonging to mannose and mannitol alcohols may be responsible for osmotic process regulation, radical-scavenging properties and plant pathogen inhibition (Stoop et al., 1996; Bohnert and Jensen 1996; Prabhavathi and Rajam, 2007).

Nowadays, a number of different chemical plant protection products are used in order to support actions against pathogens and pests. All kinds of treatments on the basis of algal extracts have an influence on good crop quality. Moreover, such preparations are considered to cause small adverse effects on the natural environment (Chojnacka et al., 2012).

In the contribution by de Corato et al. (2017), the extracts of *Laminaria digitata*, *Undaria pinnatifida* and *Porphyra umbilicalis* have been demonstrated to decrease the level of infection caused by mold in strawberries, peaches and lemons. In this case, the undesirable impact of the fungi could have been reduced by the high content of fatty acids. The increase of the specific gene expression toward defense pathways of a plant was also mentioned (de Corato et al., 2017).

Several authors also found some antifungal effects induced by all kinds of algae (brown, green, red) on tomato diseases. For instance, as for brown algae, the most tested were the extracts of *Cystoseira myriophylloides* (aqueous extract), *Fucus spiralis* (aqueous), *Laminaria digitata* (aqueous) (Esserti et al., 2017), *Padina gymnospora* (crude, purified extract) (Hernández-Herrera et al., 2014a), but also *Spatoglossum variabile* (ethanolic, aqueous extract) (Sultana et al., 2011). Esserti et al. (2017) pointed out the high antioxidants content as a factor affecting antifungal properties of tested macroalgae. Hernández-Herrera et al. (2014a) also investigated green algal extract of *Ulva lactuca* and *Caulerpa sertularioides* and brown alga *Sargassum liebmannii* as eliciting plant protection responses against *Alternaria solani*. Apart from *Spatoglossum variabile*, *Melanothamnus afaqhusainii* and *Halimeda tuna* were also characterized by positive effects on pathogenic fungi infecting the roots of tomato and sunflower plants as indicated by Sultana et al. (2011).

A considerable number of studies showed that the algal preparations on the basis of the brown algae *Sargassum* as well as green *Ulva* are able to protect the plants against several pathogenic fungi (Jaulneau et al., 2011; Galal et al.,

2011; Borbón et al., 2012; Hernández-Herrera et al., 2014a; Raj et al., 2016; Ramkissoon et al., 2017; Ibraheem et al., 2017; Ammar et al., 2017). Tomato (*Solanum lycopersicum*) plants were treated with alkaline extracts of *Gelidium serrulatum* (red), *Sargassum filipendula* (brown) and *Ulva lactuca* (green) in order to inhibit *Alternaria solani*, one of the most often fungal pathogen-affecting tomatoes (Ramkissoon et al., 2017). A variety of solvents have been employed in screening seaweeds for antifungal activity. It was shown that both lipophilic and hydrophilic algal extracts show such properties. The seaweeds were extracted with the use of acetone, methanol, methanol/chloroform mixtures, ethanol, diethyl ether, alkaline solutions (Tuney et al., 2006; Galal et al., 2011), but also crude/purified preparations (Cluzet et al., 2004; Paulert et al., 2010) and supercritical fluid extracts (de Corato et al., 2017) were utilized. Similar to studies performed by Esserti et al. (2017), Raj et al. (2016) and Ammar et al. (2017) stated that the high level of phenolic compounds and flavonoids enhanced the overall plant resistance to fungi. The growth of fungi *Pythium aphanidermatum* treated with methanolic extract of *Sargassum vulgare* was inhibited by about 51%.

Horoszkiewicz-Janka and Jajor (2006) and Horoszkiewicz-Janka and Michalski (2006) applied two formulations, such as Kelpak (*Ecklonia maxima*) and Bio Algeen S 90 Plus 2 on the grains of the cereal species (spring rape, barley, oat) in order to evaluate their activity both on plant growth improvement as well as against plant pathogenic fungi. As it was observed, both plant protection products increased the resistance to pathogen attacks. The plants were characterized by the increased levels in terms of quality and quantity. The positive effect of preparations may also be used in grape, citrus fruits, agricultural crops and ornamental plants.

Okra (*Abelmoschus esculentus*) is often infected by pathogens such as *Macrophomina phaseolina*, *Rhizoctonia solani* and *Fusarium solani* (Sultana et al., 2005). The use of brown algae of *Stokeyia indica* and *Padina pavonia* resulted in a reduction of the okra roots infection. Other *Phaeophyta* species (*Stoechospermum marginatum* and *Sargassum tenerrimum*) were also positive in *Meloidogyne javanica* as well as root-infecting pathogen neutralization as it was investigated by Ehteshamul-Haque et al. (1996). The authors observed better activity of *Solieria robusta* on *Fusarium solani* colonizing the roots of soybean (*Glycine max*) in a comparison with Topsin-M, which is a commercial plant protection product.

One of the most serious diseases of fruits and vegetables are powdery mildews infecting, for instance, bean, grapevine and cucumber. However, the fungal activity of such diseases may be limited by algal extracts from *Ulva armoricana* as proved by Jaulneau et al. (2011). Similar studies were performed by Lizzi et al. (1998) and Jayaraman et al. (2011) on pepper, carrots and cucumber resistance to fungi from the group of *Phytophthora capsici*, *Alternaria radicina*,

Botrytis cinerea, *Alternaria cucumerinum*, *Didymella applanata* and *Fusarium oxysporum* caused by the plants' treatment with *Ascophyllum nodosum* extracts. A broad group of macroalgal extracts with antifungal properties has been provided in Table 3.

3.4 Antibacterial properties of algal extracts

According to numerous studies, bioactive compounds from marine seaweeds extracts may be protective against various bacteria (Rao and Parekh, 1981; Pesando and Caram, 1984; Arunkumar et al., 2005; Liu et al., 2017; Kumar et al., 2008; Bazes et al., 2009). The growth of *Pseudomonas syringae*, which causes *Gymnema sylvestre* diseases, was suppressed in the highest degree as was shown by Kumar et al. (2008). Moreover, ethyl acetate extracts of different types of macroalgae (red, green, brown) were found to be effective in *Pseudomonas syringae* inhibition (Kumar et al., 2008). The antibacterial activity of algal extracts is related to the presence of phenolic compounds as well as tannins and saponins (Jeyaseelan et al., 2012). A broad range of macroalgal extracts (10 *Rhodophyta*, 6 *Chlorophyta*, 3 *Phaeophyta*) obtained with diethyl ether, chloroform, ethanol and acetone inhibited the growth of gram-positive bacteria, including *Bacillus megatherium* and *Staphylococcus aureus* (Rao and Parekh, 1981). The negative effect of *Xanthomonas oryzae pv. oryzae*, a bacterium which causes a serious blight of rice, may be decreased by the plant treatment with a methanolic extract of *Sargassum swartzii* as was indicated by Arunkumar et al. (2005). Table 4 lists macroalgal extracts with antibacterial activity.

4 Effect of seaweed extracts on plant physiology

The effect of seaweed extracts on plant physiology depends not only on the extract itself, but also on the variety of plants - individual cultivars of the same species may respond differently to the treatments with seaweed extracts (Masny et al., 2004; Basak, 2008). Moreover, seed germination, root and shoot growth, yield of vegetables, fruit and field crops depend on multiple factors such as the genetic potential of the plant, the growth environment, agricultural inputs and practices (Sangha et al., 2014).

In the experiments aimed at determining the effect of seaweed extracts on plant physiology, extracts produced from brown seaweeds - *Ecklonia maxima* and *Ascophyllum nodosum* - are the most often used. Among the commercial products we can distinguish Kelpak® - an extract produced from *Ecklonia maxima*, which is known to contain natural auxins (e.g. indole-3-carboxylic acid, indole-3-acetic acid (IAA), indole-3-aldehyde, N-hydroxyethylphalimide, N,N-dimethyltryptryptamine) and cytokinins (e.g. trans-zeatin, cis-zeatin, transribosyizeatin, dihydrozeatin, isopentenyladenine, isopentenyladenosine)

Table 3 Antifungal properties of seaweed extracts

Alga	Type of macroalga	Type of extract	Pathogen	Plant	Doses/Active substance(s)	Effect of a product	Reference
Ascophyllum nodosum	brown	not specified	*Phytophthora capsici*	Pepper	foliar spray	reduced effect of phytopathogens	Lizzi et al. (1998)
Ascophyllum nodosum	brown	commercial (Stimplex™)	*Alternaria radicina, Botrytis cinerea*	Blights, carrots	0.5, 1.0% concentration, twice at 10-day intervals/ oligosaccharides, polysaccharides	enhanced activities of various defenserelated enzymes including chitinase, β-1,3-glucanase, peroxidase, polyphenol oxidase, phenylalanine ammonia lyase, and lipoxygenase, increased level of phenolics	Jayaraman et al. (2011)
Ascophyllum nodosum	brown	commercial (Stimplex™)	*Alternaria cucumerinum, Didymella applanata, Fusarium oxysporum, Botrytis cinerea*	Cucumber	0.5, 1.0% concentration, twice at 10-day intervals/ oligosaccharides, polysaccharides	enhanced activities of various defenserelated enzymes including chitinase, β-1,3-glucanase, peroxidase, polyphenol oxidase, phenylalanine ammonia lyase, and lipoxygenase, increased level of phenolics	Jayaraman et al. (2011)
Caulerpa sertularioides	green	crude, purified	*Alternaria solani*	Tomato (*Solanum lycopersicum*)	0.1 mg/mL/ carbohydrate and protein composition	increased antifungal properties	Hernández-Herrera et al. (2014a)
Codium iyengarii	green	powder	*Fusarium solani*	Okra (*Abelmoschus esculentus*)	0.5, 1.0 wt% extract	reduced effect of phytopathogens	Sultana et al. (2005)

(*Continued*)

Table 3 *(Continued)*

Alga	Type of macroalga	Type of extract	Pathogen	Plant	Doses/Active substance(s)	Effect of a product	Reference
Codium iyengarii	green	ethanol	*Fusarium solani*	not specified	200 mg/mL in ethanol	growth inhibition of pathogens	Ara et al. (1998)
Cystoseira mediterranea	brown	methanol, acetone diethyl, ether, ethanol	*Alternaria alternate, Alternaria brassicicola, Fusarium oxysporium, Ulocladium botrytis, Botryotricum piluliferum*	not tested	extract concentration of 50 mg/mL	pectinase and cellulose activities inhibition	Galal et al. (2011)
Cystoseira. myriophylloides	brown	aqueous	*Verticillium dahlia*	Tomato (*Solanum lycopersicum*)	5 mL of 0.5 and 1.5% extract for 24 h, application for 6 days	reduction of fungal diseases	Esserti et al. (2017)
Fucus spiralis	brown	aqueous	*Verticillium dahlia*	Tomato (*Solanum lycopersicum*)	5 mL of 0.5 and 1.5% extract for 24 h, application for 6 days	reduction of fungal diseases	Esserti et al. (2017)
Gelidium serrulatum	red	alkaline	*Alternaria solani*	Tomato (*Solanum lycopersicum*)	5% (w/v); 0.5%, 0–96 h treatment/ defense enzymes and total phenols	increased level of defense enzymes and total phenols	Ramkissoon et al. (2017)
Halimeda tuna	green	ethanol aqueous	*Macrophomina phaseolina, Rhizoctonia solani, Fusarium solani, Meloidogyne javanica*	Tomato (*Lycopersicon esculentum*), sunflower (*Helianthus annuus*)	1.0 wt%	improved weight, root length, yield	Sultana et al. (2011)

Hydroclathrus clathratus	brown	methanol, crude	*Fusarium solani*, *Rhizoctonia solani*	Eggplant (*Solanum melongena*)	1.25, 3.0, 5.0 per disk/fatty acids, phenolic compounds, polysaccharides	fungal growth inhibition	Ibraheem et al. (2017)
Laminaria digitata	brown	crude, purified	*Botrytis cinerea*, *Plasmopara viticola*	Grapevine	laminarin extracted and purified	laminarin-induced protection	Aziz et al. (2003)
Laminaria digitata	brown	aqueous	*Verticillium dahlia*	Tomato (*Solanum lycopersicum*)	5 mL of 0.5 and 1.5% extract for 24 h, application for 6 days	reduction of fungal diseases	Esserti et al. (2017)
Laminaria digitata	brown	SFE	*Bortrytis cinerea*, *Monilinia laxa*, *Penicillium digitatum*	not tested	10, 20, 30 g/L of extract/fatty acids, phenolic compounds, phlorotannins, polysaccharides	inhibitory effect against pathogens	de Corato et al. (2017)
Melanothamnus afaqhusainii	red	powder	*Fusarium solani*, *Macrophomina phaseolina*	Eggplant (*Solanum melongena*), watermelon (*Citrullus lanatus*)	dose not specified	number of fruits and weight increased	Baloch et al. (2013)
Melanothamnus afaqhusainii	red	ethanol, aqueous	*Macrophomina phaseolina*, *Rhizoctonia solani*, *Fusarium solani*, *Meloidogyne javanica*	Tomato (*Lycopersicon esculentum*), sunflower (*Helianthus annuus*)	1.0 wt%	improved weight, root length, yield	Sultana et al. (2011)

(*Continued*)

Table 3 (*Continued*)

Alga	Type of macroalga	Type of extract	Pathogen	Plant	Doses/Active substance(s)	Effect of a product	Reference
Padina gymnospora	brown	methanol, crude	*Fusarium solani, Rhizoctonia solani*	Eggplant (Solanum melongena)	1.25, 3.0, 5.0 per disk/fatty acids, phenolic compounds, polysaccharides	fungal growth inhibition	Ibraheem et al. (2017)
Padina gymnospora	brown	crude, purified	*Alternaria solani*	Tomato (*Solanum lycopersicum*)	0.1 mg/mL/ carbohydrate and protein composition	increased antifungal properties	Hernández-Herrera et al. (2014a)
Padina pavonia	brown	powder	*Macrophomina phaseolina, Rhizoctonia solani, Fusarium solani*	Okra (*Abelmoschus esculentus*)	1.0 wt% extract	enhanced plant growth	Sultana et al. (2005)
Polycladia indica	brown	powder	*Fusarium solani, Macrophomina phaseolina*	Eggplant (*Solanum melongena*), watermelon (*Citrullus lanatus*)	dose not specified	number of fruits and weight increased	Baloch et al. (2013)
Porphyra umbilicalis	red	SFE	*Bortrytis cinerea, Monilinia laxa Penicillium digitatum*	not tested	10, 20, 30 g/L of extract/fatty acids, phenolic compounds, phlorotannins, polysaccharides	inhibitory effect against pathogens	de Corato et al. (2017)
Sargassum polyceratium	brown	acetone	*Geotrichum candidum*	not tested	5 g powdered algae (Soxhlet, acetone, ethanol)	growth inhibition of phytopathogens	Borbón et al. (2012)

Sargassum filipendula	brown	alkaline	*Alternaria solani*	Tomato (*Solanum lycopersicum*)	5% (w/v); 0.5%, 0–96 h treatment/ defense enzymes and total phenols	increased level of defense enzymes and total phenols	Ramkissoon et al. (2017)
Sargassum latifolium	brown	methanol, crude	*Fusarium solani, Rhizoctonia solani*	Eggplant (*Solanum melongena*)	1.25, 3.0, 5.0 per disk/fatty acids, phenolic compounds, polysaccharides	fungal growth inhibition	Ibraheem et al. (2017)
Sargassum liebmannii	brown	crude, purified	*Alternaria solani*	Tomato (*Solanum lycopersicum*)	0.1 mg/mL/ carbohydrate and protein composition	increased antifungal properties	Hernández-Herrera et al. (2014a)
Sargassum wightii	brown	methanol/ chloroform (1:1)	*Rhizoctonia solani*	Rice seeds	20 wt%	spore germination reduced	Raj et al. (2016)
Sargassum vulgare	brown	methanol, aqueous	*Pythium aphanidermatum*	Potato	40, 50 g/L/phenolic acids, flavonoids	antifungal properties increased	Ammar et al. (2017)
Solieria robusta	red	powder	*Macrophomina phaseolina, Rhizoctonia solani, Fusarium solani*	Okra (*Abelmoschus esculentus*)	1.0 wt% extract	enhanced plant growth	Sultana et al. (2005)
Solieria robusta	red	powder	*Macrophomina phaseolina, Rhizoctonia solani, Fusarium solani*	Soybean (*Glycine max*)	1.0 wt%	improved weight, root length, yield	Sultana et al. (2011)
Spatoglossum variabile	brown	powder	*Fusarium solani, Macrophomina phaseolina*	Eggplant (*Solanum melongena*), watermelon (*Citrullus lanatus*)	dose not specified	number of fruits and weight increased	Baloch et al. (2013)

(*Continued*)

Table 3 (*Continued*)

Alga	Type of macroalga	Type of extract	Pathogen	Plant	Doses/Active substance(s)	Effect of a product	Reference
Spatoglossum variabile	brown	ethanol, aqueous	*Macrophomina phaseolina, Rhizoctonia solani, Fusarium solani, Meloidogyne javanica*	Tomato (*Lycopersicon esculentum*), sunflower (*Helianthus annuus*)	1.0 wt%	improved weight, root length, yield	Sultana et al. (2011)
Stoechospermum marginatum	brown	not specified	*Meloidogyne javanica, Macropholina phaseolina, Fusarium solani, Rhizoctonia solani*	not tested	not specified	root infecting fungi inhibited	Ehteshamul-Haque et al. (1996)
Stoechospormum marginatum	brown	ethanol	*Fusarium solani*	not specified	200 mg/mL in ethanol	growth inhibition of pathogens	Ara et al. (1998)
Stokeyia indica	brown	powder	*Macrophomina phaseolina, Rhizoctonia solani, Fusarium solani*	Okra (*Abelmoschus esculentus*)	1.0 wt% extract	enhanced plant growth	Sultana et al. (2005)
Ulva armoricana	green	not specified	*Erysiphe polygoni, Erysiphe necator, Sphareotheca fuliginea*	Bean, grapevine, cucumber	extract diluted in water (1/18; 1/9; 1/2)/polysaccharides	powdery mildew reduced	Jaulneau et al. (2011)
Ulva lactuca	green	alkaline	*Alternaria solani*	Tomato (*Solanum lycopersicum*)	5% (w/v); 0.5%, 0–96 h treatment/ defense enzymes and total phenols	increased level of defense enzymes and total phenols	Ramkissoon et al. (2017)

Ulva lactuca	green	crude, purified	*Alternaria solani*	Tomato (*Solanum lycopersicum*)	0.1 mg/mL/ carbohydrate and protein composition	increased antifungal properties	Hernández-Herrera et al. (2014a)
Ulva rigida	green	methanol, acetone, diethyl ether, ethanol	*Alternaria alternate, Alternaria brassicicola, Fusarium oxysporium, Ulocladium botrytis, Botryotricum piluliferum*	not tested	extract concentration of 50 mg/mL	pectinase and cellulose activities inhibition	Galal et al. (2011)
Undaria pinnatifida	green	SFE	*Bortrytis cinerea, Monilinia laxa, Penicillium digitatum*	not tested	10, 20, 30 g/L of extract/fatty acids, phenolic compounds, phlorotannins, polysaccharides	inhibitory effect against pathogens	de Corato et al. (2017)

Table 4 Antibacterial properties of seaweed extracts

Alga	Type of alga	Type of extract	Bacterium	Plant	Doses/active substance(s)	Effect of a product	Reference
Acanthophora delile	red	diethyl ether, chloroform, acetone, ethanol	*Bacillus megatherium, Staphylococcus aureus*	-	not specified	antibacterial activity	Rao and Parekh (1981)
Caulerpa racemosa	green	ethanol	*Staphylococcus aureus*	-	5 g powdered algae (Soxhlet, acetone, ethanol)	growth inhibition of phytopathogens	Borbón et al. (2012)
Caulerpa scalpelliformis	green	diethyl ether, acetone, ethanol	*Bacillus megatherium, Staphylococcus aureus*	-	not specified	antibacterial activity	Rao and Parekh (1981)
Caulerpa sertularioides	green	acetone	*Bacillus megatherium, Staphylococcus aureus*	-	not specified	antibacterial activity	Rao and Parekh (1981)
Caulerpa taxifolia	green	diethyl ether, acetone	*Bacillus megatherium, Staphylococcus aureus*	-	not specified	antibacterial activity	Rao and Parekh (1981)
Chaetomorpha antennina	green	ethyl acetate	*Pseudomonas syringae*	*Gymnema sylvestre*	50 g of seaweed extracted in 200 mL petroleum ether, ethyl acetate, chloroform, acetone and methanol (1:4 w/v)	growth inhibition of pathogens	Kumar et al. (2008)
Chondria armata	red	diethyl ether, chloroform, acetone, ethanol	*Bacillus megatherium, Staphylococcus aureus*	-	not specified	antibacterial activity	Rao and Parekh (1981)
Cladophora monumentalis	green	diethyl ether	*Bacillus megatherium, Staphylococcus aureus*	-	not specified	antibacterial activity	Rao and Parekh (1981)
Corallina officinalis	red	diethyl ether, ethanol, chloroform, acetone,	*Bacillus megatherium, Staphylococcus aureus*	-	not specified	antibacterial activity	Rao and Parekh (1981)
Cystoseira myriophylloides	brown	aqueous	*Agrobacterium tumefaciens*	Tomato (*Solanum lycopersicum*)	5 mL of 0.5 and 1.5% extract for 24 h, application for 6 days	bacterial growth inhibition	Esserti et al. (2017)

Cystoseira stricta	brown	ethanol	Various bacteria	-	not specified	antimicrobial properties	Pesando and Caram (1984)
Dictyota sp.	brown	diethyl ether, chloroform, acetone, ethanol	*Bacillus megatherium, Staphylococcus aureus*	-	not specified	antibacterial activity	Rao and Parekh (1981)
Dictyota dichotoma	brown	diethyl ether, chloroform, acetone, ethanol	*Bacillus megatherium, Staphylococcus aureus*	-	not specified	antibacterial activity	Rao and Parekh (1981)
Enteromorpha intestinalis	green	diethyl ether, chloroform, acetone, ethanol	*Bacillus megatherium, Staphylococcus aureus*	-	not specified	antibacterial activity	Rao and Parekh (1981)
Fucus spiralis	brown	aqueous	*Agrobacterium tumefaciens*	Tomato (*Solanum lycopersicum*)	5 mL of 0.5 and 1.5% extract for 24 h, application for 6 days	bacterial growth inhibition	Esserti et al. (2017)
Gelidiella acerosa	red	diethyl ether, acetone, ethanol, chloroform	*Bacillus megatherium, Staphylococcus aureus*	-	not specified	antibacterial activity	Rao and Parekh (1981)
Gracilaria cervicornis	red	ethanol	*Staphylococcus aureus*	-	5 g powdered algae (Soxhlet, acetone, ethanol)	growth inhibition of phytopathogens	Borbón et al. (2012)
Gracilaria corticata	red	ethyl acetate	*Pseudomonas syringae*	Medicinal plant (*Gymnema sylvestre*)	50 g of seaweed extracted in 200 mL petroleum ether, ethyl acetate, chloroform, acetone and methanol (1:4 w/v)	growth inhibition of pathogens	Kumar et al. (2008)
Gradlaria corticata	red	diethyl ether, acetone, ethanol, chloroform	*Bacillus megatherium, Staphylococcus aureus*	-	not specified	antibacterial activity	Rao and Parekh (1981)

(Continued)

Table 4 (*Continued*)

Alga	Type of alga	Type of extract	Bacterium	Plant	Doses/active substance(s)	Effect of a product	Reference
Gracilariopsis longissima	red	ethyl acetate	*Pseudomonas syringae*	Medicinal plant (*Gymnema sylvestre*)	50 g of seaweed extracted in 200 mL petroleum ether, ethyl acetate, chloroform, acetone and methanol (1:4 w/v)	growth inhibition of pathogens	Kumar et al. (2008)
Grateloupia lithophila	red	ethyl acetate	*Pseudomonas syringae*	Medicinal plant (*Gymnema sylvestre*)	50 g of seaweed extracted in 200 mL petroleum ether, ethyl acetate, chloroform, acetone and methanol (1:4 w/v)	growth inhibition of pathogens	Kumar et al. (2008)
Halimeda tuna	green	ethyl acetate	*Pseudomonas syringae*	Medicinal plant (*Gymnema sylvestre*)	50 g of seaweed extracted in 200 mL petroleum ether, ethyl acetate, chloroform, acetone and methanol (1:4 w/v)	growth inhibition of pathogens	Kumar et al. (2008)
Halimeda tuna	green	diethyl ether, acetone, ethanol	*Bacillus megatherium, Staphylococcus aureus*	-	not specified	antibacterial activity	Rao and Parekh (1981)
Hypnea musciformis	red	diethyl ether, acetone, ethanol, chloroform	*Bacillus megatherium, Staphylococcus aureus*	-	not specified	antibacterial activity	Rao and Parekh (1981)
Laminaria japonica	brown	crude fucoidans fraction	*Escherichia coli, Staphylococcus aureus*	-	isolated fucoidans	antibacterial activity of depolymerized products of fucoidans	Liu et al. (2017)

Laurencia obtusa	red	ethyl acetate	*Pseudomonas syringae*	Medicinal plant (*Gymnema sylvestre*)	50 g of seaweed extracted in 200 mL petroleum ether, ethyl acetate, chloroform, acetone and methanol (1:4 w/v)	growth inhibition of pathogens	Kumar et al. (2008)
Laurencia papillosa	red	diethyl ether, acetone, ethanol, chloroform	*Bacillus megatherium, Staphylococcus aureus*	-	not specified	growth inhibition of pathogens	Arunkumar et al. (2005)
Padina boergesenii	brown	ethyl acetate	*Pseudomonas syringae*	Medicinal plant (*Gymnema sylvestre*)	50 g of seaweed extracted in 200 mL petroleum ether, ethyl acetate, chloroform, acetone and methanol (1:4 w/v)	growth inhibition of pathogens	Kumar et al. (2008)
Padina gymnospora	brown	diethyl ether, acetone	*Bacillus megatherium, Staphylococcus aureus*	-	not specified	antibacterial activity	Rao and Parekh (1981)
Polysiphonia kappannae	red	diethyl ether, acetone, ethanol, chloroform	*Bacillus megatherium, Staphylococcus aureus*	-	not specified	antibacterial activity	Rao and Parekh (1981)
Rhodymenia palmata	red	diethyl ether, acetone, ethanol, chloroform	*Bacillus megatherium, Staphylococcus aureus*	-	not specified	antibacterial activity	Rao and Parekh (1981)
Sarconema furcettatum	red	diethyl ether, acetone, ethanol, chloroform	*Bacillus megatherium, Staphylococcus aureus*	-	not specified	antibacterial activity	Rao and Parekh (1981)
Sargassum muticum	brown	dichloromethane	*Cylindrotheca closterium*	-	2 g of extract suspended in 12 L of ethanol/fatty acids	antibacterial activity of fatty acids against bacteria	Bazes et al. (2009)

(*Continued*)

Table 4 (*Continued*)

Alga	Type of alga	Type of extract	Bacterium	Plant	Doses/active substance(s)	Effect of a product	Reference
Sargassum polyceratium	brown	acetone	*Staphylococcus aureus*, *Erwinia carotovora*	Medicinal plant (*Gymnema sylvestre*)	50 g of seaweed extracted in 200 mL petroleum ether, ethyl acetate, chloroform, acetone and methanol (1:4 w/v)	growth inhibition of pathogens	Kumar et al. (2008)
Sargassum polyceratium	brown	ethanol, acetone	*Staphylococcus aureus*, *Erwinia carotovora*	-	5 g powdered algae (Soxhlet, acetone, ethanol)	growth inhibition of phytopathogens	Borbón et al. (2012)
Sargassum swartzii	brown	methanol	*Xanthomonas oryzae* pv. *oryzae*	Rice	not specified	growth inhibition of pathogens	Arunkumar et al. (2005)
Sargassum wightii	brown	methanol	*Pseudomonas syringae*	Medicinal plant (*Gymnema sylvestre*)	50 g of seaweed extracted in 200 mL petroleum ether, ethyl acetate, chloroform, acetone and methanol (1:4 w/v)	growth inhibition of pathogens	Kumar et al. (2008)
Turbinaria conoides	brown	ethyl acetate	*Pseudomonas syringae*	Medicinal plant (*Gymnema sylvestre*)	50 g of seaweed extracted in 200 mL petroleum ether, ethyl acetate, chloroform, acetone and methanol (1:4 w/v)	growth inhibition of pathogens	Kumar et al. (2008)
Ulva lactuca	green	ethyl acetate	*Pseudomonas syringae*	Medicinal plant (*Gymnema sylvestre*)	50 g of seaweed extracted in 200 mL petroleum ether, ethyl acetate, chloroform, acetone and methanol (1:4 w/v)	growth inhibition of pathogens	Kumar et al. (2008)

(Basak 2008; Colla et al., 2017a). The second commercial product - Goëmar BM 86®, produced from *Ascophyllum nodosum* is rich in vitamins, amino acids, oligosaccharides, phytohormones and mineral nutrients such as N, Mg and B (Basak, 2008). The stimulating properties of the seaweed extracts are attributed to the presence of these bioactive compounds. Some of the examples concerning the effect of seaweed extracts on plant physiology are presented in Table 5.

4.1 Seed germination

In the case of the effect of seaweed extracts on seeds germination, several germination parameters such as germination percentage, germination index, mean germination time, germination energy and seedling vigor index should be taken into account (Hernández-Herrera et al., 2014b; Castellanos-Barriga et al., 2017). Seaweed extract can also positively influence seed imbibition which can increase seeds germination (Castellanos-Barriga et al., 2017).

The increase of seed germination can result from the presence of plant hormones (indole 3-acetic acid and indole-3-butyric acid (IBA), gibberellins, cytokinins), vitamins, amino acids and micronutrients such as Co, Cu, Fe, Mo, Mn, Ni and Zn (Crouch and van Staden, 1993; Sivasankari et al., 2006; Hernández-Herrera et al., 2014b). Mzibra et al. (2018) showed that seaweed polysaccharides can also act as biostimulants of seed germination and plant growth. Their polymeric structure allows them to act as water holders. Moreover, polysaccharides can increase the activity of germination enzymes and accelerate the seed metabolic activity (Hu et al., 2004).

Seaweed extracts are known to be bioactive at low concentrations - diluted as 1:1000 or more (Crouch and van Staden, 1993). Usually, lower concentrations of seaweed extracts show higher rates of seeds germination, while the higher concentrations can suppress germination (Sivasankari et al., 2006; Hernández-Herrera et al., 2014b; Castellanos-Barriga et al., 2017). This negative effect can result from the inhibition of seeds' ability to imbibe water (Hernández-Herrera et al., 2014b). Hernández-Herrera et al. (2014b) showed that extracts produced from *Ulva lactuca* and *Padina gymnospora* and applied at lower concentrations (0.2 among 0.4 and 1.0%) enhanced germination of tomato seeds. The same concentration of *Ulva lactuca* extract (0.2%), obtained by hydrolysis of seaweeds with H_2SO_4, was optimal for the germination of mung bean (*Vigna radiata*) (Castellanos-Barriga et al., 2017). The inhibition of seed germination may also result from the solvent used for the production of seaweed extract. The preliminary results of Choi and Choi (2016) showed that methanol extracts of *Capsosiphon fulvescens* and *Monostroma nitidum* completely inhibited germination of lettuce (*Lactuca sativa* L.). In contrast to methanol extracts, water extracts of these seaweeds slightly inhibited

Table 5 Examples of the effect of seaweed extracts on plant physiology

Alga	Type of alga	Type of extract	Plant	Doses/active substance(s)	Effect of a product	Reference
Ascophyllum nodosum - Actiwave® product	brown	not specified	strawberry (*Fragaria ananassa* cv. Queen Elisa)	500 mL/L of water/ kahydrin, alginic acid and betaines	• increase of the mineral nutrient uptake; • increase of the vegetative growth, photosynthetic rate, leaf chlorophyll content, stomata density, the fruit production; • increase of the plant biomass: the shoot and root dry matter; • positively influence on the root-associated microbial biocoenosis	Spinelli et al. (2010)
Ascophyllum nodosum - Acadian™ product	brown	water soluble, alkaline extract	strawberries - varieties Albion, Camarosa, Chandler, Festival	1, 2 or 4 g/L/not specified	• increase of root and shoot growth, berry yield; • increase/stimulation of rhizosphere microbial diversity (increase of colony counts in soil samples) and physiological activity (increase of metabolic activities of soil microbes)	Alam et al. (2013)
(1) *Ulva lactuca*, (2) *Caulerpa sertularioides*, (3) *Padina gymnospora*, (4) *Sargassum liebmannii*	(1) and (2) - green; (3) and (4) - brown	autoclaving of seaweeds mixed with water at 121°C for 1 h	tomato (*Solanum lycopersicum* L.)	0.2, 0.4, 1.0%/macroelements	• *U. lactuca* and *P. gymnospora* (0.2%) enhanced germination (lower mean germination time, high germination index and germination energy, greater seedling vigor, greater plumule and radicle length) and shoot length, root length and weight; • height of the plant was higher for soil drench than the foliar spray application	Hernández-Herrera et al. (2014)

(1) *Sargassum wightii*, (2) *Caulerpa chemnitzia*	(1) brown; (2) green	boiling of seaweeds in distilled water	cowpea (*Vigna sinensis*)	5, 10, 20, 30, 40, 50, 100%/microelements and macroelements	• increase of shoot and root length, fresh and dry weight of seedlings; • increase of chlorophyll, carotenoids, protein, amino acid, reducing sugar and root sugar content of shoot and root; • increase of α and β-amylase activities in *V. sinensis*	Sivasankari et al. (2006)
Ulva lactuca	green	hydrolysis with H_2SO_4 at 2, 4, 6, 8 and 10% concentrations	mung bean (*Vigna radiata*)	0.2, 0.4, 0.6, 0.8 and 1.0%/carbohydrates, minerals and proteins	• increase at low concentrations (0.2%) of seed germination rates (lower mean germination time, high germination index, greater seedling vigour); • increase of plumule, radicle, shoot and root length; • increase of fresh and dry weight of mung bean; • increase of protein, chlorophyll, total and reducing sugars content	Castellanos-Barriga et al. (2017)
Ascophyllum nodosum	brown	aqueous alkaline extract	tomato, dwarf French bean, wheat, barley, maize	8 mL/L of water/betaines (γ-aminobutyric acid betaine, δ-aminovaleric acid betaine, glycinebetaine)	• increase of chlorophyll content in the leaves	Blunden et al. (1996)

(*Continued*)

Table 5 (*Continued*)

Alga	Type of alga	Type of extract	Plant	Doses/active substance(s)	Effect of a product	Reference
Ecklonia maxima – Kelpak product	brown	process called 'cold cell burst' (high pressure, low temperatures)	tomato (*Solanum lycopersicum* L.)	3 mL/L/phytohormones: auxins, cytokinins, carbohydrates, amino acids, vitamins: B1, B2, C, E, elements: N, P, K, Ca, Mg, Fe, Mn, B, Zn, Cu	• increase of the early and total marketable yield of fresh tomato; • increase of the Ca concentration in the fruit tissue	Colla et al. (2017a,b)
(1) *Ulva rigida*, (2) *Codium decorticatum*, (3) *Fucus spiralis*, (4) *Bifurcaria bifurcata*, (5) *Gigartina* sp., (6) *Chondracanthus acicularis*	(1) and (2) - green; (3) and (4) - brown; (5) and (6) - red	hot water method under neutral conditions	tomato	in vitro conditions: 0.2 mg/mL; greenhouse conditions: 0.1 mg/mL/ polysaccharides	• increase of the seeds germination; • increase of the plant biomass (shoots and roots); • increase of the chlorophyll content	Mzibra et al. (2018)
A commercial seaweed extract from *Ascophyllum nodosum*	brown	not specified	strawberry plants cv. Camarosa	0.5 g/L of seaweed extract in combination with a commercial product of nitrophenolates/possible action due to the content of cytokinins, auxins and betaines	• increase of the marketable yield and fruit size, • no significant impact on fruit juice pH, titratable acidity, total soluble solids content; fruit organic acid and carbohydrate concentration and on fruit color; • enhancement of the total anthocyanin concentration	Roussos et al. (2009)

Ascophyllum nodosum – Goemar GA 14 product	brown	not specified	maize (*Zea mays* L. cv. DEA)	3.3 and 6.6 g/L of water/ phenolic compounds, mannitol, laminaran, fucoidan, alginates, auxin, gibberellins, substances with a cytokinin-like activity, cytokinins, abscisic acid, 1-amino-cyclopropane-l-carboxylicacid, betaines	• increase of the total fresh matter production of maize seedlings (increase of root and stem mass per plant) for 3.3 g/L; • increase in stomatal resistance and a decrease in transpiration rate (for a dose 3.3 g/L); • no significant effects on leaf number and plant height	Jeannin et al. (1991)
Ascophyllum nodosum – Goemar Spring Cítricos®	brown	not specified	trees of 'de Nules' clementine mandarin (*Citrus clementina* Hort. *ex* Tanaka) and Navelina orange (*Citrus sinensis* L. Osbeck)	concentrations from 0.15% to 0.30% (full cover spray)	• increase of the yield of 'de Nules' clementine mandarin; • increase of the marketable fruit size	Fornes et al. (2002)
Ecklonia maxima – Kelpak® product, *Ascophyllum nodosum* – Goëmar BM 86®	brown	not specified	apple trees cv. Gala Must, Golden Delicious, Jonagold Decosta, Elstar	200 and 250 mL/100 L of water (Kelpak and Goëmar, respectively)/Kelpak – auxins and cytokinins; Goëmar – amino acids, phytohormones, vitamins, poligosaccharides, minerals: N, Mg, B	• improvement in fruit set and the size of apples; • improvement in the distribution of apples in size classes; • reduction of russeting; • increase of the total nitrogen level in fruit; • stimulation of the growth of shoots and leaves; • improvement in flower quality and prolonged blooming time; • no effect on the storage quality of apples	Basak (2008)

(Continued)

Table 5 (*Continued*)

Alga	Type of alga	Type of extract	Plant	Doses/active substance(s)	Effect of a product	Reference
Seaweeds extract	not specified	not specified	strawberry cv. Honeoye	0.5% solution	• enhancement of root growth characteristics (root weight, total root length, diameter, surface area, volume and total number of root tips); • had a positive influence on the colonization of roots by arbuscular mycorrhizal fungi (*Claroideoglomus claroideum* and *Funneliformis mosseae*)	Sas Paszt et al. (2015)
(1) *Codium iyengarii*, (2) *Jania capillacea*, (3) *Solieria robusta*, (4) *Stokeyia indica*	(1) green; (2) and (3) red; (4) brown	ethanol extracts	chili	50% mortality of root knot nematode within 24 h at 10 mg/mL	• extracts from all seaweeds caused mortality of the root knot nematode – *Meloidogyne javanica* juveniles • *S. indica* and *S. robusta* extracts applied as soil amendments alone or with *Pseudomonas aeruginosa*, significantly suppressed infection of chili roots by root-infecting fungi *Fusarium solani*, *Rhizoctonia solani* and *Macrophomina phaseolina*	Sultana et al. (2008)

germination. It is supposed that the lipid and phenolic fraction of extracts may be the reason of germination hindering. The application of seaweed extracts in a greenhouse (pot experiments) or in the field trials should be preceded by the detailed germination tests aimed at the selection of the best algal extract produced from different seaweed species by different extraction methods with the use of a wide range of solvents. These tests also allow in choosing the optimal concentration of seaweed extract conditioning the largest shoot, root growth, fruit set and, finally, the product quality.

4.2 Shoot growth

The effect of seaweed extracts on the shoot growth was investigated by many authors. The complexity of numerous bioactive compounds in algal extracts may lead to their biostimulant activities of vegetables, fruits, flowers, crops and so on. Among them, plant hormones play an imperative role in the increase of cell size and cell division. For example, cytokinins are effective in shoot formation (Castellanos-Barriga et al., 2017). Another compound - betaines are organic osmolites that present a cytokinin-like activity. Their exogenous application in the form of algal extracts can increase not only the growth of shoots, but also roots and the chlorophyll content in leaves (Blunden et al., 1996). The plant growth can be also stimulated by seaweeds' polysaccharides and oligosaccharides through the enhancement of carbon and nitrogen assimilation and plant cell division (Castro et al., 2012).

Several commercial seaweed extracts, as well as those produced in laboratory conditions, exhibited a positive effect on the shoot growth. For example, Acadian™, a commercial product (water-soluble alkaline extract) from a brown seaweed *Ascophyllum nodosum*, applied in the field study increased strawberry shoot dry weight and also the leaf area (Alam et al., 2013). Tested by Ji et al. (2017) a commercial seaweed extract containing polysaccharide, alginate, increased the chrysanthemum shoot height, weight and leaf growth (length, width, area and weight) in a greenhouse. Another commercial product - seaweed extract from *Ascophyllum nodosum* (Goemar GA 14) applied foliarly in a greenhouse - increased maize shoot height and weight (Jeannin et al., 1991).

4.3 Root growth

The root system of plants is of great importance, taking into account the nutrients uptake and their transport in the plant (Ji et al., 2017). Seaweed extracts are known to enhance root development by the improvement of lateral formation and by the increase of the total volume of the root system and root length. Then, the vigorous root system can provide a larger surface area for nutrient and water uptake (Alam et al., 2013; Polo and Mata, 2017). The induction of the root growth can be due to phytohormones activity (auxin and gibberellin-like activities) which is involved in the regulation and improvement of the vegetal growth (Khan et al., 2009; Castellanos-Barriga et al., 2017; Polo and Mata, 2017). The application of seaweed extracts can also increase significantly the electrical conductivity of rhizospheric soil, which could be related to the mineralization of organic matter after the application of seaweed extract and consequently with the increase of the mineral nitrogen content in soil and increase of plant growth (Ji et al., 2017).

Literature data confirm that seaweed extracts increase not only the root length, but also the total root surface area, total root volume and number of roots as it was presented by Alam et al. (2013) who tested a commercial product - Acadian™ (*Ascophyllum nodosum*) in a strawberry cultivation. Similar effect showed seaweed extract containing alginate applied in a greenhouse on chrysanthemum (increase of the root dry weight, total root length, root surface area, root volume and root tip number) (Ji et al., 2017). Polo and Mata (2017) observed the increased root length of tomato at harvest after the application of Acadian, whereas foliar application of a commercial seaweed extract (Goemar GA 14) increased maize root (Jeannin et al., 1991). The effect of seaweed extract on the root growth can also depend on the time of its application. Jeannin et al. (1991) suggested that the application of seaweed extract from *Ascophyllum nodosum* was generally most effective on maize root growth when applied during early vegetative growth.

4.4 Fruit set

The effect of seaweed extracts on fruit formation remains unclear, but it is supposed that it can result from the content of growth regulators (1,3-1,6 D-glucane, oligosaccharides known as laminarin) that can elicit the polyamine synthesis and stimulate cell division (Patier et al., 1993; Castro et al., 2012). The cell division during the early stages of growth along with the induction of flower formation can be influenced also by the seaweeds' phytohormones (cytokinins, auxins) and betaines (Roussos et al., 2009). Cytokinins and betaines were suggested as active compounds present in *Ascophyllum nodosum* extract that can be responsible for the increasing of fruit set - clementine mandarin and orange (Fornes et al., 2002). Colavita et al. (2011) found that the commercial extract from *Ascophyllum nodosum* (Göemar BM®) improved pears' fruit set, size and total yield by increasing the fruits' cell number early in the growth season. Gutierrez-Gamboa et al. (2018) examined the effect of seaweed extract on fruit set in a grapevine (*Vitis vinifera* L.) variety 'Carménère'. The foliar application of *Ascophyllum nodosum* extract - especially in combination with micronutrients such as boron and zinc, before the beginning of flowering - improved fruit set, grapevine production and the content of phenolic compounds in fruits, which are classified as an elicitor

(Khan et al., 2012, Gutierrez-Gamboa et al., 2018). In the case of fruit set, the application time of seaweed extracts is of particular importance. Basak (2008) found that extracts from brown seaweeds, *Ecklonia maxima* and *Ascophyllum nodosum*, when applied to apples from the end of bloom until up to 4 weeks before harvest, improved fruit set and the size of apples. Grapevines ('Perlette') sprayed with the mixture of amino acids and seaweed extract (from *Ascophylum nodosum*) exhibited a significant increase in the berry setting, especially when applied at flowering, fruit setting and one month after fruit stages (Khan et al., 2012).

4.5 Product quality

The production of high-quality fruits by growers is forced by the increasing market competition (Colavita et al., 2011). This demand can be achieved by the reduction of the use of synthetic chemical products and the application of biological formulations, for example, seaweed extracts. Fruit quality parameters can be considered in terms of storability, ripening, firmness, sugar, nutrients and acid content, as well as external fruit quality parameters such as form, size, weight, length, diameter, thickness, color and so on (Nicola et al., 2009). Seaweed extracts, when applied as a foliar spray, can contribute to the supply of plant nutrients (Jeannin et al., 1991). Mineral content (e.g. calcium) is one of the major factors determining the quality and storability of fruits (Błaszczyk, 2008). For example, in the work of Colla et al. (2017a) it was shown that seaweed extract (Kelpak from *Ecklonia maxima*) enhanced the calcium, nitrogen and ascorbic acid content in tomato tissue. Grapevines treated with multiple spray applications of mixture of amino acids and seaweed extract at flowering and fruit setting stages showed higher levels of leaf N, P, K, B, Fe and Zn contents (Khan et al., 2012). The presence of bioactive compounds (e.g. amino acids, carbohydrates, peptides) in biostimulants of plant growth can increase the content of minerals in the fruit due to increased sink strength that influences the movement of substrates, including minerals, within the plant (Calvo et al., 2014).

Several authors reported a beneficial effect of seaweed extracts on the internal product quality. Błaszczyk (2008) showed that Goëmar BM 86 (from *Ascophyllum nodosum*) had a beneficial effect on the internal pears quality traits such as fruit firmness and titratable acidity during storage and maturation. Another trait of product quality is the content of soluble solids (the measure of sweetness). Grapevines treated with the mixture of amino acids and *Ascophyllum nodosum* seaweed extract exhibited significantly higher soluble solid concentrations, titrable acidity, total and reducing sugars than in the control group (Khan et al., 2012). Basak (2008) showed that the apples from the trees treated with seaweed extracts from *Ecklonia maxima* and *Ascophyllum nodosum* (Kelpak and Goëmar) had a little lower soluble solids content than the control group. But, after the storage, apples from the groups with seaweed extracts had better internal quality (a slightly higher firmness and significantly higher soluble solids content), but unfortunately they were more inclined to rot (Basak, 2008). Contradictory results were presented by Masny et al. (2004), who tested Kelpak and Goëmar on strawberry. Both seaweed extracts reduced strawberry firmness (varieties 'Elkat' and 'Salut'). Also in the case of strawberry quality, Spinelli et al. (2010) did not find a significant increase of the studied parameters such as firmness, sugar content, titratable acidity and pH after the use of the biostimulant Actiwave® (from brown seaweeds). The same result was observed in the study by Roussos et al. (2009) – a commercial seaweed extract from *Ascophyllum nodosum* did not have a significant impact on strawberry juice pH, titratable acidity, total soluble solids, organic acid and carbohydrate concentration but they enhanced total anthocyanin concentration. In the case of vegetables, for example, cowpea (*Vigna sinensis*), seaweed extracts from *Sargassum wightii* and *Caulerpa chemnitzia* applied at low doses (till 20%) increased the chlorophyll, carotenoids, protein and amino acids, reducing sugar and total sugar content of shoot and root (Sivasankari et al., 2006).

In terms of external fruits and vegetables quality parameters, positive effects of seaweed extracts are reported. In order to achieve a larger fruit size, many plant growth-regulating compounds such as phytohormones (auxins, cytokinins and gibberellins) have been used in various crops (Guardiola and Garcia-Luis, 2000). Basak (2008) found that extracts from brown seaweeds *Ecklonia maxima* and *Ascophyllum nodosum*, applied before harvest, increased the average size of apples, depending on the fruit set, ameliorated the distribution of apples in size classes and reduced russeting. Moreover, these natural products increased the yield of marketable apples (diameter > 70 mm). Grapevines treated with the mixture of amino acids and *Ascophyllum nodosum* seaweed extract exhibited significantly higher berry size, weight, than in the control group (Khan et al., 2012). In the case of strawberry quality, Spinelli et al. (2010) observed increased yield and berry size after the application of biostimulant Actiwave® (from brown seaweeds). The commercial seaweed extract from *Ascophyllum nodosum* increased the marketable yield, fruit size, as well as strawberry color (Roussos et al., 2009). Seaweed biostimulants also have a positive effect on vegetables – for example, Acadian produced from seaweed *Ascophyllum nodosum* increased tomato fruit diameter (Polo and Mata, 2017); Kelpak, applied foliarly, improved the early and total marketable yield of fresh tomato (Colla et al., 2017a); extracts from *Ulva lactuca*, *Padina gymnospora* showed increased tomato weight (Hernández-Herrera et al., 2014b); and so on.

5 Effects and mode of action of seaweed extracts on abiotic stress tolerance of horticultural and agronomic crops

Biostimulants modulate the stress of plants by increasing productivity. Using plant growth biostimulants is beneficial in terms of environmental impact and life cycle assessment (Sharma et al., 2017). Biostimulants are natural substances that support vegetative growth, uptake of nutrients, tolerance to abiotic stress by roots and plant tolerance to drought and salt stress (Spinelli et al., 2009). Plant growth biostimulants were defined by European Biostimulants Industry Council (EBIC) as stimulating the natural processes of nutrients uptake, efficiency, tolerance to abiotic stress and improving crop quality in rhizosphere (Colla and Rouphael, 2015). In some cases it is possible to reduce the application rates of chemical fertilizers sustaining the crop production yield (Sharma et al., 2017).

Various compounds present in seaweed extract work synergistically, but their mechanism of action still remains unknown. They are thought to enhance absorption of nutrients and thus the reduction of fertilizer application rates is lower (Rathore et al., 2009).

Significant components of seaweed extracts are phytohormones that regulate growth, development and stress adaptation. An example is abscisic acid, a stress hormone synthesized during abiotic stress. It causes closing of stomata, affects development of root architecture, regulation of growth, as well as expression of genes important for coping with drought, resulting in better performance during dehydration stress (Sharma et al., 2019).

Useful algae for the production of biostimulant formulations are brown seaweeds (*Laminaria, Fucus Ascophyllum*), but red seaweeds are also interesting f given the presence of carrageenans (du Jardin, 2015). Extract from brown algae *A. nodosum* contains plant growth regulators, amino acids, organic osmolites (e.g. betaines), minerals, vitamins (vitamin K kahydrin) and alginic acid (Spinelli et al., 2009). *Ascophyllum nodosum* extract increases drought stress tolerance of crops and grasses through dysregulation of abiotic stress genes (Goñi et al., 2018). Fei et al. (2017) applied extract from seaweed (*A. nodosum*) in the cultivation of poplars and switch grass grown in greenhouse and in the field and confirmed that the productivity of plants was enhanced. A red seaweed, *Gracilaria, dura* contains biostimulating compounds: indole 3-acetic acid (IAA), zeatin, gibberlin (GA3), choline, glycine, betaine, macronutrients and micronutrients (Sharma et al., 2019). Kelpak® - extract from *Ecklonia maxima* - contains auxins, gibberellins, cytokinins, abscisic acid, polyamines, brassinolide and castasterone (Digruber et al., 2018).

6 Effect of seaweed extracts on modulation of the rhizosphere microbial population

Agricultural treatments such as soil management, irrigation and fertilization on the one hand increase the production yield, but at the same time they can negatively influence the root biocoenosis by reducing mycorrhizal fungi and beneficial microorganisms (Spinelli et al., 2010). Destruction of soil mycorrhizal communities can be suppressed by the application of new products obtained from natural resources (e.g. seaweed extracts), which can be an alternative or supplementation of the traditional mineral fertilizers (Sas Paszt et al., 2015). Natural products, such as seaweed extracts can have a positive effect on the inoculation, sporulation, diversity of soil bacterial and fungal communities (e.g. symbiotic arbuscular mycorrhizal fungi from phylum *Glomeromycota*) present in the rhizospheric soil, colonization of roots by microorganisms and root growth (Sas Paszt et al., 2015; Kuwada et al., 2006). The presence of easily degradable organic matter in seaweed extracts can provide the nutrient base for the multiplication of bacteria, which increase their population in the rhizosphere of higher plants (Sultana et al., 2008). Seaweeds are known to contain microorganism-stimulatory substances (Kuwada et al., 2006) and can also stimulate the growth of bacteria and fungi to produce molecules that will be beneficial for plants – increased plant health and productivity (Polo and Mata, 2017). The stimulated microorganisms in the rhizosphere are able to improve the utilization of soil nutrients by plants, increase their availability for plants, enhance the nutrient cycle in the soil (Sas Paszt et al., 2015; Ji et al., 2017) as well as the growth of terrestrial plants by forming a symbiotic relationship with the roots of its host plant (Kuwada et al., 2006).

Moreover, the chemical composition of seaweed extract can have a positive effect on the rhizosphere. An example is Actiwave – a seaweed extract product derived from *A. nodosum* which contains kahydrin, a derivate of the K vitamin which can lead to the acidification of the rhizosphere. Actiwave also contains alginic acid which can act as a soil conditioning agent that increases water-holding characteristics of the rhizosphere and organic osmolites (e.g., betaines) that protect plants from biotic and abiotic stress, for example, drought, salinity, extreme temperatures and chilling (Spinelli et al., 2010). The increase of the soil moisture (rhizosphere) due to the presence of algal alginic acid can contribute to creating the environment which will be suitable for the growth of roots and root-associated beneficial microorganisms (Chen et al., 2003).

Literature data confirm the positive effect of seaweed extracts on modulation of the rhizosphere and phyllosphere microbial population. In the work of Alam et al. (2013), a commercially available alkaline extract derived from the brown seaweed *A. nodosum* (Acadian™) was examined in terms of soil microbial communities in the cultivation of strawberry. Application of extract in a greenhouse and field study increased the total soil microbial colony counts and the total microbial physiological activity in soil. Spinelli et al. (2010) reported that Actiwave (a seaweed extract derived from *A. nodosum*) positively influenced root-associated microbial biocoenosis composed of *Pseudomonas* spp., *Bacillus subtilis* and related species and *Streptomyces* spp. The population of *Pseudomonas* associated with roots was significantly increased. These bacteria can also act as biological control agents of plant pathogens and plant growth-promoting rhizobacteria, which can enhance the solubilization of insoluble phosphorus sources (Avis et al., 2008; Sultana et al., 2008). Kuwada et al. (2006) confirmed the stimulation of root colonization of papaya and passion fruit by arbuscular mycorrhizal fungi (*Gigaspora margarita* and *Glomus caledonium*) after the application of 25% methanol extracts from red seaweeds (*Gracilaria verrucosa*, *Gelidium amansii* and *Eucheuma cottonii*). The colonization of roots by microorganisms does not occur in every case – Sas Paszt et al. (2015) showed that seaweed extract did not give a positive effect on the strawberry root colonization of arbuscular mycorrhizal fungi.

It is also worth mentioning that seaweed extracts show bioactivity against soilborne plant pathogens and can suppress infection of roots by theses pathogens. For example, Sultana et al. (2008) showed that extracts from seaweeds *Stokeyia indica* and *Solieria robusta* applied as soil amendments significantly suppressed infection of chili roots by root-infecting fungi *Rhizoctonia solani*, *Macrophomina phaseolina*, *Fusarium solani* and the root knot nematode *Meloidogyne javanica*. Seaweeds compounds with antimicrobial activity, such as 1-aminocyclopropane-1-carboxylic acid or betaines, can be responsible for this effect (Nelson and Van Standen, 1985). Seaweed extracts can be an alternative to the prevalent use of synthetic pesticides.

Biostimulants of plant growth generally affect plant microbiomes residing in both rhizosphere and phyllosphere and thus improve plant performance (higher water and nutrient uptake, enhanced resilience against biotic and abiotic stress) (Colla et al., 2017b).

7 Conclusion

Seaweed extracts are considered as one of the categories of the group of biostimulants of plant growth. Biostimulants improve the growth of plants through enhancing nutrient uptake efficiency, by improving tolerance to abiotic stress and increasing the quality of crops. Already, it has been well known that the mode of action of biostimulants is related with primary and secondary metabolites (e.g., carbohydrates, lipids, proteins, nucleic acids, organic acids and many others) that protect crops from abiotic and biotic stress. Seaweed extracts are also active against a number of fungicidal pathogens, as well as various bacteria. Therefore they have the characteristics of the plant protection products.

Algal-based biostimulants act on the level of plant physiology: germination of seeds, growth of shoots and roots, set of fruits. Growth enhancement is also achieved by the positive effect on microbial populations of rhizosphere and phyllosphere that are an important part of plant growth. They also beneficially improve the quality of crop products – that is, shelf life. The number of publications cited in this chapter proves that there is scientific evidence that supports the use of seaweed extracts as biostimulants of diverse plant species.

The market of biostimulants is still in the stage of development, but different formulations with seaweed extracts are now available commercially. More and more agrochemical companies include algal extracts in the formulations that stimulate plant growth and development. It is necessary to carry out further research to prove the effectiveness of biostimulants on plants, as well as show the mechanisms of action of these products.

8 References

Alam, M. Z., Braun, G., Norrie, J. and Hodges, D. M. 2013. Effect of *Ascophyllum* extract application on plant growth, fruit yield and soil microbial communities of strawberry. *Can. J. Plant Sci.* 93(1), 23–36. doi:10.4141/cjps2011-260.

Alassali, A., Cybulska, I., Brudecki, G. P., Farzanah, R. and Thomsen, M. H. 2016. Methods for upstream extraction and chemical characterization of secondary metabolites from algae biomass. *Adv. Tech. Biol. Med.* 4(1), 163. doi:10.4172/2379-1764.1000163.

Ali, N., Farrell, A., Ramsubhag, A. and Jayaraman, J. 2016. The effect of *Ascophyllum nodosum* extract on the growth, yield and fruit quality of tomato grown under tropical conditions. *J. Appl. Phycol.* 28(2), 1353–62. doi:10.1007/s10811-015-0608-3.

Al-Musawi, Md. A. H. M. 2018. Effect of foliar application with algae extracts on fruit quality of sour orange, *Citrus aurantium*, L. *J. Environ. Sci. Pollut. Res.* 4, 250–2.

Ammar, N., Jabnoun-Khiareddine, H., Mejdoub-Trabelsibi, B., Nefzi, A., Mahjoub, M. A. and Daami-Remadi, M. 2017. Pythium leak control in potato using aqueous and organic extracts from the brown alga *Sargassum vulgare* (C. Agardh, 1820). *Postharvest Biol. Technol.* 130, 81–93. doi:10.1016/j.postharvbio.2017.04.010.

Andrade, P. B., Barbosa, M., Matos, R. P., Lopes, G., Vinholes, J., Mouga, T. and Valentão, P. 2013. Valuable compounds in macroalgae extracts. *Food Chem.* 138(2–3), 1819–28. doi:10.1016/j.foodchem.2012.11.081.

Ara, J., Sultana, V., Ehteshamul-Haque, S., Qureshi, S. A. and Ahmed, V. U. 1998. Bioactivity of seaweeds against soil borne plant pathogens. *Phytologia* 85, 292–9.

Arunkumar, K., Selvapalam, N. and Rengasamy, R. 2005. The antibacterial compound sulphoglycerolipid 1-0 palmitoyl-3-0(6-sulpho-aquinovopyranosyl)-glycerol from *Sargassum wightii* Greville (*Phaeophyceae*). *Bot. Mar.* 40, 441–5.

Avis, T. J., Gravel, V., Antoun, H. and Tweddell, R. J. 2008. Multifaceted beneficial effects of rhizosphere microorganisms on plant health and productivity. *Soil Biol. Biochem.* 40(7), 1733–40. doi:10.1016/j.soilbio.2008.02.013.

Aziz, A., Poinssot, B., Daire, X., Adrian, M., Bézier, A., Lambert, B., Joubert, J. M. and Pugin, A. 2003. Laminarin elicits defense responses in grapevine and induces protection against *Botrytis cinerea* and *Plasmopara viticola*. *Mol. Plant Microbe Interact.* 16(12), 1118–28. doi:10.1094/MPMI.2003.16.12.1118.

Baloch, G. N., Tariq, S., Ehteshamul-Haque, S., Athar, M., Sultana, V. and Ara, J. 2013. Management of root diseases of eggplant and watermelon with the application of asafoetida and seaweeds. *J. Appl. Bot. Food Qual.* 86, 138–42.

Basak, A. 2008. Effect of preharvest treatment with seaweed products, Kelpak® and Goëmar BM 86®, on fruit quality in apple. *Int. J. Fruit Sci.* 8(1–2), 1–14. doi:10.1080/15538360802365251.

Bazes, A., Silkina, A., Douzenel, P., Faÿ, F., Kervarec, N., Morin, D., Berge, J. P. and Bourgougnon, N. 2009. Investigation of the antifouling constituents from the brown alga *Sargassum muticum* (Yendo) Fensholt. *J. Appl. Phycol.* 21(4), 395–403. doi:10.1007/s10811-008-9382-9.

Błaszczyk, J. 2008. Quality of 'Conference' pears as affected by GOËMAR BM 86 and FRUTON. In: Sadowski, A. (Ed.), *Biostimulators in Modern Agriculture: Fruit Crops*. Editorial House Wieś Jutra, Ltd., Warsaw, Poland, pp. 18–24.

Blunden, G. and Gordon, S. M. 1986. Betaines and their sulphono analogues in marine algae. In: Round, F. E. and Chapman, D. J. (Eds), *Progress in Phycological Research*. Biopress Ltd, Bristol, UK.

Blunden, G., Jenkins, T. and Liu, Y.-W. 1996. Enhanced leaf chlorophyll levels in plants treated with seaweed extract. *J. Appl. Phycol.* 8(6), 535–43. doi:10.1007/BF02186333.

Blunden, G., Morse, P. F., Mathe, I., Hohmann, J., Critchleye, A. T. and Morrell, S. 2010. Betaine yields from marine algal species utilized in the preparation of seaweed extracts used in agriculture. *Nat. Prod. Commun.* 5(4), 581–5. doi:10.1177/1934578X1000500418.

Bohnert, H. J. and Jensen, R. G. 1996. Strategies for engineering water stress tolerance in plants. *Trends Biotechnol.* 14(3), 89-97. doi:10.1016/0167-7799(96)80929-2.

Borbón, H., Herrera, J. M., Calvo, M., Trimino, H., Sierra, L., Soto, R. and Vega, I. 2012. Antimicrobial activity of most abundant marine macroalgae of the Caribbean coast of Costa Rica. *J. Asian Sci. Res.* 2, 292–9.

Bouhlal, R., Riadi, H. and Bourgougnon, N. 2010. Anti-viral activity of the extracts of *Rhodophyceae* from Morocco. *Afr. J. Biotechnol.* 9(46), 7968–75. doi:10.5897/AJB09.2023.

Calvo, P., Nelson, L. and Kloepper, J. W. 2014. Agricultural uses of plant biostimulants. *Plant Soil* 383(1–2), 3–41. doi:10.1007/s11104-014-2131-8.

Castellanos-Barriga, L. G., Santacruz-Ruvalcaba, F., Hernandez-Carmona, G., Ramirez-Briones, E. and Hernandez-Herrera, R. M. 2017. Effect of seaweed liquid extracts from *Ulva lactuca* on seedling growth of mung bean (*Vigna radiata*). *J. Appl. Phycol.* 29(5), 2479–88. doi:10.1007/s10811-017-1082-x.

Castro, J., Vera, J., Gonzalez, A. and Moenne, A. 2012. Oligo-carrageenans stimulate growth by enhancing photosynthesis, basal metabolism, and cell cycle in tobacco plants (var. Burley). *J. Plant Growth Regul.* 31(2), 173–85. doi:10.1007/s00344-011-9229-5.

Chen, S. K., Edwards, C. A. and Subler, S. 2003. The influence of two agricultural biostimulants on nitrogen transformations, microbial activity, and plant growth in soil microcosms. *Soil Biol. Biochem.* 35(1), 9–19. doi:10.1016/S0038-0717(02)00209-2.

Chiaiese, P., Corrado, G., Colla, G., Kyriacou, M. C. and Rouphael, Y. 2018. Renewable sources of plant biostimulation: microalgae as a sustainable means to improve crop performance. *Front. Plant Sci.* 9, 1782. doi:10.3389/fpls.2018.01782.

Choi, J. S. and Choi, I. S. 2016. Inhibitory effect of marine green algal extracts on germination of *Lactuca sativa* seeds. *J. Environ. Biol.* 37(2), 207–13.

Chojnacka, K., Saeid, A. and Michalak, I. 2012. Możliwości zastosowania biomasy alg w rolnictwie. *Chemik* 66, 1235–48.

Chojnacka, K., Michalak, I., Dmytryk, A., Gramza, M., Słowiński, A. and Górecki, H. 2015. Algal extracts as plant growth biostimulants. In: Chojnacka, K. and Se-Kwon, K. (Eds), *Marine Algal Extracts: Process, Products and Applications*. Wiley.

Ciepiela, G. A., Godlewska, A. and Jankowska, J. 2016. The effect of seaweed *Ecklonia maxima* extract and mineral nitrogen on fodder grass chemical composition. *Environ. Sci. Pollut. Res.* 23(3), 2301–7. doi:10.1007/s11356-015-5417-3.

Cluzet, S., Torregrosa, C., Jacquet, C., Lafitte, C., Fournier, J., Mercier, L., Salamagne, S., Briand, X., Esquerre-Tugaye, M.-T. and Dumas, B. 2004. Gene expressing profiling and protection of *Medicago truncatula* against a fungal infection in response to an elicitor from green alga *Ulva* spp. *Plant Cell Environ.* 27(7), 917–28. doi:10.1111/j.1365-3040.2004.01197.x.

Colavita, G. M., Spera, N., Blackhall, V. and Sepulveda, G. M. 2011. Effect of seaweed extract on pear fruit quality and yield. *Acta Hortic.* 909(909), 601–7. doi:10.17660/ActaHortic.2011.909.72.

Colla, G. and Rouphael, Y. 2015. Biostimulants in horticulture. *Sci. Hortic.* 196, 1–2. doi:10.1016/j.scienta.2015.10.044.

Colla, G., Hoagland, L., Ruzzi, M., Cardarelli, M., Bonini, P., Canaguier, R. and Rouphael, Y. 2017a. Biostimulant action of protein hydrolysates: unraveling their effects on plant physiology and microbiome. *Front. Plant Sci.* 8, 2202. doi:10.3389/fpls.2017.02202.

Colla, G., Cardarelli, M., Bonini, P. and Rouphael, Y. 2017b. Foliar applications of protein hydrolysate, plant and seaweed extracts increase yield but differentially modulate fruit quality of greenhouse tomato. *HortScience* 52(9), 1214–20. doi:10.21273/HORTSCI12200-17.

Crouch, I. J. and Van Staden, J. 1991. Evidence for rooting factors in a seaweed concentrate prepared from *Ecklonia maxima*. *J. Plant Physiol.* 137(3), 319–22. doi:10.1016/S0176-1617(11)80138-0.

Crouch, I. J. and van Staden, J. 1993. Evidence for the presence of plant growth regulators in commercial seaweed products. *Plant Growth Regul.* 13(1), 21–9. doi:10.1007/BF00207588.

de Corato, U., Salimbeni, R., de Pretis, A., Avella, N. and Patruno, G. 2017. Antifungal activity of crude extracts from brown and red seaweeds by a supercritical carbon dioxide technique against fruit postharvest fungal diseases. *Postharvest Biol. Technol.* 131, 16–30. doi:10.1016/j.postharvbio.2017.04.011.

de Felício, R., Dealbuquerque, S., Young, M. C. M., Yokoya, N. S. and Debonsi, H. M. 2010. Trypanocidal, leishmanicidal and antifungal potential from marine red alga *Bostrychia tenella* J. Agardh (*Rhodomelaceae*, *Ceramiales*). *J. Pharm. Biomed. Anal.* 52(5), 763–9. doi:10.1016/j.jpba.2010.02.018.

Digruber, T., Sass, L., Cseri, A., Paul, K., Nagy, A. V., Remenyik, J., Molnar, I., Vass, I., Toldi, O., Gyuricza, C. and Dudits, D. 2018. Stimulation of energy willow biomass with triacontanol and seaweed extract. *Ind. Crops Prod.* 120, 104–12. doi:10.1016/j.indcrop.2018.04.047.

Dobrzański, A., Anyszka, Z. and Elkner, K. 2008. Reakcja marchwi na ekstrakty pochodzenia naturalnego z alg z rodzaju *Sargassum* - Algaminoplant i z Leonardytu - HumiPlant. *J. Res. App. Agric. Engin.* 53, 53–8.

du Jardin, P. 2015. Plant biostimulants: definition, concept, main categories and regulation. *Sci. Hortic.* 196, 3–14. doi:10.1016/j.scienta.2015.09.021.

Ehteshamul-Haque, S., Abid, M., Sultana, V., Ara, J. and Ghaffar, A. 1996. Use of organic amendments on the efficacy of biocontrol agents in the control of root rot and root knot disease complex of okra. *Nematol. Medit.* 24, 13–6.

El-Migeed, A. A. A., El-Sayed, A. B. and Hassan, H. S. A. 2004. Growth enhancement of olive transplants by broken cells of fresh green algae as soil application. *Minufia J. Agric. Res.* 29, 723–37.

El-Moniem, E. A. A. and Abd-Allah, A. S. E. 2008. Effect of green algae cells extract as foliar spray on vegetative growth, yield and berries quality of superior grapevines. *Am. Euras. J. Agric. Environ. Sci.* 4, 427–33.

El-Sharony, T. F., El-Gioushy, S. F. and Amin, O. A. 2015. Effect of foliar application with algae and plant extracts on growth, yield and fruit quality of fruitful mango trees cv. Fagri Kalan. *J. Hortic.* 2, 4.

Ertani, A., Francioso, O., Tinti, A., Schiavon, M., Pizzeghello, D. and Nardi, S. 2018. Evaluation of seaweed extract from *Laminaria* and *Ascophyllum nodosum* spp. as biostimulants in *Zea mays* L. using a combination of chemical, biological and morphological approaches. *Front. Plant Sci.* 9, 428. doi:10.3389/fpls.2018.00428.

Esserti, S., Smaili, A., Rifai, L. A., Koussa, T., Makroum, K., Belfaiza, M., Kabil, E. M., Faize, L., Burgos, L., Alburquerque, N. and Faize, M. 2017. Protective effect of three brown seaweed extracts against fungal and bacterial diseases of tomato. *J. Appl. Phycol.* 29(2), 1081–93. doi:10.1007/s10811-016-0996-z.

Eyras, M. C., Rostagno, C. M. and Defosse, G. E. 1998. Biological evaluation of seaweed composting. *Compost Sci. Util.* 6(4), 74–81. doi:10.1080/1065657X.1998.10701943.

Fan, D., Hodges, D. M., Zhang, J., Kirby, C. W., Ji, X., Locke, S. J., Critchley, A. T. and Prithiviraj, B. 2011. Commercial extract of the brown seaweed *Ascophyllum nodosum* enhances phenolic antioxidant content of spinach (*Spinaciaoleracea* L.) which protects *Caenorhabditis elegans* against oxidative and thermal stress. *Food Chem.* 124(1), 195–202. doi:10.1016/j.foodchem.2010.06.008.

Fei, H., Crouse, M., Papadopoulos, Y. and Vessey, J. K. 2017. Enhancing the productivity of hybrid poplar (Populus × hybrid) and switchgrass (*Panicum virgatum* L.) by the application of beneficial soil microbes and a seaweed extract. *Biomass Bioenergy* 107, 122–34. doi:10.1016/j.biombioe.2017.09.022.

Fornes, F., Sánchez-Perales, M. and Guardiola, J. L. 2002. Effect of a seaweed extract on the productivity of 'de Nules' clementine mandarin and *Navelina orange*. *Bot. Mar.* 45, 486–9.

Galal, H. R. M., Salem, W. M. and El-Deen, F. N. 2011. Biological control of some pathogenic fungi using marine algae extracts. *Res. J. Microbiol.* 6, 645–57.

Goñi, O., Quille, P. and O'Connell, S. 2018. *Ascophyllum nodosum* extract biostimulants and their role in enhancing tolerance to drought stress in tomato plants. *Plant Physiol. Biochem.* 126, 63–73. doi:10.1016/j.plaphy.2018.02.024.

Guardiola, J. L. and Garcia-Luis, A. 2000. Increasing fruit size in citrus. Thinning and stimulation of fruit growth. *Plant Growth Regul.* 31(1), 121–32. doi:10.1023/A:1006339721880.

Gutierrez-Gamboa, G., Garde-Cerdan, T., Souza-Da Costa, B. and Moreno-Simunovic, Y. 2018. Strategies for the improvement of fruit set in *Vitis vinifera* L. CV. 'carmenere' through different foliar biostimulants in two different locations. *Cienc. Téc. Vitiv* 33(2), 177–83.

Hamed, S. M., Abd Elrhman, A. A., Abdel-Raouf, N. and Ibraheem, I. B. M. 2018. Role of marine macroalgae in plant protection and improvement for sustainable agriculture technology. *Beni Suef Univ. J. Basic Appl. Sci.* 7(1), 104–10. doi:10.1016/j.bjbas.2017.08.002.

Hernández-Herrera, R. M., Virgen-Calleros, G., Ruiz-López, M., Zañudo-Hernández, J., Délano-Frier, J. P. and Sánchez-Hernández, C. 2014a. Extracts from green and brown seaweeds protect tomato (*Solanum lycopersicum*) against the necrotrophic fungus *Alternaria solani*. *J. Appl. Phycol.* 26(3), 1607–14. doi:10.1007/s10811-013-0193-2.

Hernandez-Herrera, R. M., Santacruz-Ruvalcaba, F., Ruiz-Lopez, M. A., Norrie, J. and Hernandez-Carmona, G. 2014b. Effect of liquid seaweed extracts on growth of tomato seedlings (*Solanum lycopersicum* L.). *J. Appl. Phycol.* 26(1), 619–28. doi:10.1007/s10811-013-0078-4.

Hong, D. D., Hien, H. M. and Son, P. N. 2007. Seaweeds from Vietnam used for functional food, medicine and fertilizer. *J. App. Phycol.* 19(6), 817–26. doi:10.1007/s10811-007-9228-x.

Horoszkiewicz-Janka, J. and Jajor, E. 2006. Wpływ zaprawiania nasion na zdrowotność roślin na jęczmienia, pszenicy i rzepaku w początkowych fazach rozwoju. *J. Res. App. Agric. Engg.* 51, 47–53.

Horoszkiewicz-Janka, J. and Michalski, T. 2006. Wpływ zabiegów ochrony roślin na wykształcenie ziarna jęczmienia i owsa. *Progr. Plant Prot.* 46, 417–23.

Hu, X., Jiang, X., Hwang, H., Liu, S. and Guan, H. 2004. Promotive effects of alginate-derived oligosaccharide on maize seed germination. *J. Appl. Phycol.* 16(1), 73–6. doi:10.1023/B:JAPH.0000019139.35046.0c.

Ibraheem, B. M. I., Hamed, S. M., Abd Elrhman, A. A., Farag, F. M. and Abdel-Raouf, N. 2017. Antimicrobial activities of some brown macroalgae against some soil borne plant pathogens and *in vivo* management of *Solanum melongena* root diseases. *Aust. J. Basic Appl. Sci.* 11, 157–68.

Ioannou, E. and Roussis, V. 2009. Natural products from seaweeds. In: *Plant-derived Natural Products: Synthesis, Function, and Application*. Springer, Berlin, Germany, pp. 55–81.

Jaulneau, V., Lafitte, C., Corio-Costet, M. F., Stadnik, M. J., Salamagne, S., Briand, X., Esquerré- Tugayé, M. T. and Dumas, B. 2011. An Ulva *armoricana* extract protects plants against three powdery mildew pathogens. *Eur. J. Plant Pathol.* 131(3), 393–401. doi:10.1007/s10658-011-9816-0.

Jayaraman, J., Norrie, J. and Punja, Z. K. 2011. Commercial extract from the brown seaweed *Ascophyllum nodosum* reduces fungal diseases in greenhouse cucumber. *J. Appl. Phycol.* 23(3), 353–61. doi:10.1007/s10811-010-9547-1.

Jeannin, I., Lescure, J. C. and Morot-Gaudry, J. F. 1991. The effects of aqueous seaweed sprays on the growth of maize. *Bot. Mar.* 34, 469–73.

Jeyaseelan, E. C., Kothai, S., Kavitha, R., Tharmila, S. and Thavaranjit, A. C. 2012. Antibacterial activity of some selected algae present in the Costal Lines of Jaffna Peninsula. *Int. J. Pharm. Biol. Arch.* 3, 352–6.

Ji, R. T., Dong, G. Q., Shi, W. M. and Min, J. 2017. Effects of liquid organic fertilizers on plant growth and rhizosphere soil characteristics of chrysanthemum. *Sustainability* 9(5), 841. doi:10.3390/su9050841.

Khan, W., Rayirath, U. P., Subramanian, S., Jitesh, M. N., Rayorath, P., Hodges, D. M., Critchley, A. T., Craigie, J. S., Norrie, J. and Prithiviraj, B. 2009. Seaweed extracts as biostimulants of plant growth and development. *J. Plant Growth Regul.* 28(4), 386–99. doi:10.1007/s00344-009-9103-x.

Khan, A. S., Ahmad, B., Jaskani, M. J., Ahmad, R. and Malik, A. U. 2012. Foliar application of mixture of amino acids and seaweed (*Ascophylum nodosum*) extract improve growth and physicochemical properties of grapes. *Int. J. Agric. Biol.* 14, 383–8.

Kumar, G. and Sahoo, D. 2011. Effect of seaweed liquid extract on growth and yield of *Triticum aestivum* var. Pusa Gold. *J. App. Phycol.* 23(2), 251–5. doi:10.1007/s10811-011-9660-9.

Kumar, C. S., Raju, D., Sarada, V. L. and Rengasamy, R. 2008. Seaweed extracts control the leaf spot disease of the medicinal plant *Gymnema sylvestre*. *Indian J. Sci. Technol.* 1, 93–4.

Kuwada, K., Wamocho, L. S., Utamur, M., Matsushita, I. and Ishii, T. 2006. Effect of red and green algal extracts on hyphal growth of Arbuscular Mycorrhizal Fungi, and on mycorrhizal development and growth of papaya and passionfruit. *Agron. J.* 98(5), 1340–4. doi:10.2134/agronj2005.0354.

Lalopua, V. M. N. and Latumahina, M. 2015. Secondary metabolites chemical compounds of edible macroalgae in Maulu waters. *Int. J. Latest Res. Sci. Technol.* 4, 59–62.

Liu, M., Liu, Y., Cao, M. J., Liu, G. M., Chen, Q., Sun, L. and Chen, H. 2017. Antibacterial activity and mechanisms of depolymerized fucoidans isolated from Laminaria japonica. *Carbohydr. Polym.* 172, 294–305. doi:10.1016/j.carbpol.2017.05.060.

Lizzi, Y., Coulomb, C., Polian, C., Coulomb, P. J. and Coulomb, P. O. 1998. Seaweed and mildew: what does the future hold ? Encouraging laboratory results. *Phytoma* 508, 29–30.

Lordan, S., Ross, R. P. and Stanton, C. 2011. Marine bioactives as functional food ingredients: potential to reduce the incidence of chronic diseases. *Mar. Drugs* 9(6), 1056–100. doi:10.3390/md9061056.

Masny, A., Basak, A. and Zurawicz, Z. 2004. Effect of foliar application of Kelpak and Goemar BM 86 preparations on yield and fruit quality in two strawberry cultivars. *J. Fruit Ornamental Plant Res.* 12, 23–7.

Matysiak, K., Kaczmarek, S., Kierzek, R. and Kardasz, P. 2010. Ocena działania ekstraktów z alg morskich oraz mieszaniny kwasów huminowych i fulwowych na kiełkowanie i początkowy wzrost rzepaku ozimego (*Brassica napus* L.). *J. Res. App. Agric. Engin.* 55, 28–32.

McGloughlin, M. N. 2010. Modifying agricultural crops for improved nutrition. *Nat. Biotechnol.* 27(5), 494–504. doi:10.1016/j.nbt.2010.07.013.

Michalak, I., Chojnacka, K., Dmytryk, A., Wilk, R., Gramza, M. and Rój, E. 2016a. Evaluation of supercritical extract of algae as biostimulants of plant growth in field trials. *Front. Plant Sci.* 7, 1591. doi:10.3389/fpls.2016.01591.

Michalak, I., Górka, B., Wieczorek, P. P., Rój, E., Lipok, J., Łęska, B., Messyasz, B., Wilk, R., Schroeder, G., Dobrzyńska-Inger, A. and Chojnacka, K. 2016b. Supercritical fluid extraction of algae enhances levels of biologically active compounds promoting plant growth. *Eur. J. Phycol.* 51(3), 243–52. doi:10.1080/09670262.2015.1134813.

Mzibra, A., Aasfar, A., El Arroussi, H., Khouloud, M., Dhiba, D., Meftah Kadmiri, I. M. and Bamouh, A. 2018. Polysaccharides extracted from Moroccan seaweed: a promising source of tomato plant growth promoters. *J. Appl. Phycol.* 30(5), 2953–62. doi:10.1007/s10811-018-1421-6.

Nelson, W. R. and van Standen, J. 1985. 1-aminocyclopropane-1-carboxylic acid in seaweed concentrate. *Bot. Mar.* 28, 415–7.

Nicola, S., Tibaldi, G. and Fontana, E. 2009. Fresh-cut produce quality: implications for a systems approach. In: Shewfelt, R. L., Prussia, S. E., Florkowski, W. J. and Brueckner, B. (Eds), *Postharvest Handling (Second Edition). A Systems Approach*. Academic Press, pp. 247–82.

Patier, P., Yvin, J. C., Kloareg, B., Liénart, Y. and Rochas, C. 1993. Seaweeds liquid fertilizer from *Ascophyllum nodosum* contains elicitors of plant B-glycanases. *J. Appl. Phycol.* 5(3), 343–9. doi:10.1007/BF02186237.

Paulert, R., Ebbinghaus, D., Urlass, C. and Moerschbacher, B. M. 2010. Priming of the oxidative burst in rice and wheat cell cultures by ulvan, a polysaccharide from green macroalgae, and enhanced resistance against powdery mildew in wheat and barley plants. *Plant Pathol.* 59(4), 634–42. doi:10.1111/j.1365-3059.2010.02300.x.

Pérez, M. J., Falqué, E. and Domínguez, H. 2016. Antimicrobial action of compounds from marine seaweed. *Mar. Drugs* 14(3), 52. doi:10.3390/md14030052.

Pesando, D. and Caram, B. 1984. Screening of marine algae from the French Mediterranean coast for antimicrobial and antifungal activity. *Bot. Mar.* 27, 381–6.

Polo, J. and Mata, P. 2017. Evaluation of a biostimulant (pepton) based in enzymatic hydrolyzed animal protein in comparison to seaweed extracts on root development, vegetative growth, flowering, and yield of gold cherry tomatoes grown under low stress ambient field conditions. *Front. Plant Sci.* 8, 2261. doi:10.3389/fpls.2017.02261.

Prabhavathi, V. and Rajam, M. V. 2007. Mannitol-accumulating transgenic eggplants exhibit enhanced resistance to fungal wilts. *Plant Sci.* 173(1), 50–4. doi:10.1016/j.plantsci.2007.04.004.

Raj, T. S., Graff, K. H. and Suji, H. A. 2016. Bio chemical characterization of a brown seaweed algae and its efficacy on control of rice sheath blight caused by *Rhizoctonia solani* Kuhn. *Int. J. Trop. Agric.* 34, 429–39.

Raj, T. S., Nishanthi, P., Graff, K. H. and Suji, H. A. 2018. Seaweed extract as a biostimulant and a pathogen controlling agent in plants. *Int. J. Trop. Agric.* 36, 563–80.

Ramkissoon, A., Ramsubhag, A. and Jayaraman, J. 2017. Phytoelicitor activity of three Caribbean seaweed species on suppression of pathogenic infections in tomato plants. *J. Appl. Phycol.* 29(6), 3235–44. doi:10.1007/s10811-017-1160-0.

Rao, P. S. and Parekh, K. S. 1981. Antibacterial activity of Indian seaweed extracts. *Bot. Mar.* 24, 577–82.

Rathore, S. S., Chaudhary, D. R., Boricha, G. N., Ghosh, A., Bhatt, B. P., Zodape, S. T. and Patolia, J. S. 2009. Effect of seaweed extract on the growth, yield and nutrient uptake of soybean (*Glycine max*) under rainfed conditions. *S. Afr. J. Bot.* 75(2), 351–5. doi:10.1016/j.sajb.2008.10.009.

Rayorath, P., Khan, W., Palanisamy, R., MacKinnon, S. L., Stefanova, R., Hankins, S. D., Critchley, A. T. and Prithiviraj, B. 2008. Extracts of the brown seaweed *Ascophyllum nodosum* induce gibberellic acid (GA 3)-independent amylase activity in barley. *J. Plant Growth Regul.* 27(4), 370–9. doi:10.1007/s00344-008-9063-6.

Rengasamy, K. R. R., Kulkarni, M. G., Stirk, W. A. and Van Staden, J. 2015. Eckol – a new plant growth stimulant from the brown seaweed *Ecklonia maxima*. *J. App. Phycol.* 27(1), 581–7. doi:10.1007/s10811-014-0337-z.

Ronga, D., Biazzi, E., Parati, K., Carminati, D., Carminati, E. and Tava, A. 2019. Microalgal biostimulants and biofertilisers in crop productions. *Agronomy* 9(4), 192, doi:10.3390/agronomy9040192.

Roussos, P. A., Denaxa, N. K. and Damvakaris, T. 2009. Strawberry fruit quality attributes after application of plant growth stimulating compounds. *Sci. Hortic.* 119(2), 138–46. doi:10.1016/j.scienta.2008.07.021.

Sangha, J. S., Kelloway, S., Critchley, A. T. and Prithivira, B. 2014. Seaweeds (macroalgae) and their extracts as contributors of plant productivity and quality: the current status of our understanding. *Adv. Bot. Res.* 71, 192–3.

Sas Paszt, L., Malusa, E., Sumorok, B., Canfora, L., Derkowska, E. and Gluszek, S. 2015. The influence of bioproducts on mycorrhizal occurrence and diversity in the rhizosphere of strawberry plants under controlled conditions. *Adv. Microbiol.* 5, 40–53.

Scaglia, B., Pognani, M. and Adani, F. 2017. The anaerobic digestion process capability to produce biostimulant: the case study of the dissolved organic matter (DOM) vs. auxin-like property. *Sci. Total Environ.* 589, 36–45. doi:10.1016/j.scitotenv.2017.02.223.

Shahnaz, L. and Shameel, M. 2009. Chemical composition and bioactivity of some benthenic algae from Karachi coast of Pakistan. *Int. J. Algae* 11, 377–93.

Sharma, H. S. S., Fleming, C., Selby, C., Rao, J. R. and Martin, T. 2014. Plant biostimulants: a review on the processing of macroalgae and use of extracts for crop management to reduce abiotic and biotic stresses. *J. Appl. Phycol.* 26(1), 465–90. doi:10.1007/s10811-013-0101-9.

Sharma, L., Banerjee, M., Malik, G. C., Gopalakrishnan, V. A. K., Zodape, S. T. and Ghosh, A. 2017. Sustainable agro-technology for enhancement of rice production in the red and lateritic soils using seaweed based biostimulants. *J. Cleaner Prod.* 149, 968–75. doi:10.1016/j.jclepro.2017.02.153.

Sharma, S., Chen, C., Khatri, K., Rathore, M. S. and Pandey, S. P. 2019. *Gracilaria dura* extract confers drought tolerance in wheat by modulating abscisic acid homeostasis. *Plant Physiol. Biochem.* 136, 143–54. doi:10.1016/j.plaphy.2019.01.015.

Silkina, A., Bazes, A., Mouget, J. L. and Bourgougnon, N. 2012. Comparative efficiency of macroalgal extracts and booster biocides as antifouling agents to control growth of three diatom species. *Mar. Pollut. Bull.* 64(10), 2039–46. doi:10.1016/j.marpolbul.2012.06.028.

Singh, A. and Chaudhary, B. 2010. Preliminary phyco-chemical analysis and in vitro antibacterial screening of *Pithophora oedogonia* (Mont.) Wittrock: A freshwater green alga forming mats in the water bodies. *J. Algal Biomass Util.* 1, 33–41.

Sivasankari, S., Venkatesalu, V., Anatharaj, M. and Chandrasekaran, M. 2006. Effect of seaweed extract on the growth and biochemical constitutes of *Vigna sinensis*. *Bioresour. Technol.* 97(14), 1745–51. doi:10.1016/j.biortech.2005.06.016.

Spinelli, F., Giovanni, F., Massimo, N., Mattia, S. and Guglielmo, C. 2009. Perspectives on the use of a sea weed extract to moderate the negative effects of alternate bearing in apple trees. *J. Hort. Sci. Biotechn.* 17, 131–7.

Spinelli, F., Fiori, G., Noferini, M., Sprocatti, M. and Costa, G. 2010. A novel type of seaweed extract as a natural alternative to the use of iron chelates in strawberry production. *Sci. Hortic.* 125(3), 263–9. doi:10.1016/j.scienta.2010.03.011.

Stoop, J. M. H., Williamson, J. D. and Pharr, D. M. 1996. Mannitol metabolism in plants: a method for coping with stress. *Trends Plant. Sci.* 1(5), 139–44. doi:10.1016/S1360-1385(96)80048-3.

Sultana, V., Ehteshamul-Haque, S., Ara, J. and Athar, M. 2005. Comparative efficacy of brown, green and red seaweeds in the control of root infecting fungi and okra. *Int. J. Environ. Sci. Technol.* 2(2), 129–32. doi:10.1007/BF03325866.

Sultana, V., Ara, J. and Ehteshamul-Haque, S. 2008. Suppression of root rotting fungi and root knot nematode of chili by seaweed and *Pseudomonas aeruginosa*. *J. Phytopathol.* 156(7–8), 390–5. doi:10.1111/j.1439-0434.2007.01369.x.

Sultana, V., Baloch, G. N., Ara, J., Ehteshamul-Haque, S., Tariq, R. M. and Athar, M. 2011. Seaweeds as an alternative to chemical pesticides for the management of root diseases of sunflower and tomato. *J. Appl. Bot. Food Qual.* 84, 162–8.

Tuney, I., Cadirci, B. H., Unal, D. and Sukatar, A. 2006. Antimicrobial activities of the extracts of marine algae from the coast of Urla (Izmir, Turkey). *Turk. J. Biol.* 30, 171–5.

Vasantharaja, R., Abraham, L. S., Inbakandan, D., Thirugnanasambandam, R., Senthilvelan, T., Ayesha Jabeen, S. K. A. and Prakash, P. 2019. Influence of seaweed extracts on growth, phytochemical contents and antioxidant capacity of cowpea (*Vigna unguiculata* L. Walp). *Biocat. Agricult. Biotechnol.* 17, 589–94. doi:10.1016/j.bcab.2019.01.021.

Vera, J., Castro, J., Gonzalez, A. and Moenne, A. 2011. Seaweed polysaccharides and derived oligosaccharides stimulate defense responses and protection against pathogens in plants. *Mar. Drugs* 9(12), 2514–25. doi:10.3390/md9122514.

White, P. J. and Broadley, M. R. 2009. Biofortification of crops with seven mineral elements often lacking in human diets–iron, zinc, copper, calcium, magnesium, selenium and iodine. *New Phytol.* 182(1), 49–84. doi:10.1111/j.1469-8137.2008.02738.x.

Zodape, S., Mukherjee, S., Reddy, M. and Chaudhary, D. 2009. Effect of *Kappaphycus alvare-zii* (Doty) Doty ex Silva. extract on grain quality, yield and some yield components of wheat (*Triticum aestivum* L.). *Int. J. Plant Prod.* 3, 7–101.

Chapter 5

Biostimulant action of protein hydrolysates on crops

Giuseppe Colla, University of Tuscia, Italy; Youssef Rouphael, University of Naples Federico II, Italy; Mariateresa Cardarelli, Consiglio per la ricerca in agricoltura e l'analisi dell'economia agraria, Centro di ricerca Orticoltura e Florovivaismo, Italy; Luigi Lucini, Università Cattolica del Sacro Cuore, Italy; and Andrea Ertani, University of Turin, Italy

1 Introduction

Protein hydrolysates (PHs) are a category of plant biostimulants containing a mixture of polypeptides, oligopeptides, and amino acids that are manufactured from protein sources using partial hydrolysis (Colla et al., 2015). PHs are mainly manufactured by chemical (acid or alkali treatment), thermal and/or enzymatic hydrolysis of proteins contained in agro-industrial by-products from animal (i.e. collagen, feathers, blood) or plant origin (i.e. vegetable by-products, legume seeds, alfalfa hay). Nowadays, PHs obtained by agro-industrial by-products represent a sustainable solution to the problem of waste disposal, making their production interesting from environmental and economic points of view (Colla et al., 2015). PHs are usually available on the market as liquid products, soluble powder, or in granules. Actually, more than 90% of the PH market in agriculture is based on products manufactured through chemical/thermal hydrolysis of proteins from animal origin (e.g. collagen from leather industry in Europe,

http://dx.doi.org/10.19103/AS.2020.0068.06

India, and China; fish by-products in the United States) as a result of the high availability of raw material and the low production cost. Plant-derived PHs are less common because they have been recently introduced to the biostimulant market and the number of companies producing this type of biostimulants is limited due to the complexity of the production technology based on enzymatic hydrolysis. Recently, plant-derived PHs are gaining a lot of interest in comparison to animal-derived PHs due to the growing restrictions on the use of animal-derived PHs in food crops (e.g. Regulation (EU) n. 354/2014 on organic production forbids the application of animal-derived PHs on edible parts of the crops) and the high efficacy of these biostimulant products (Colla et al., 2015). Several researches demonstrated that foliar or root applications of PHs can enhance growth and yield of several crops (Kunicki et al., 2010; Lisiecka et al., 2011; Paradikovic et al., 2011; Colla et al., 2014; Ertani et al., 2014). The above findings have been related to the PHs-induced biostimulant actions, such as the increase of nutrient uptake and assimilation by plants (Cerdán et al., 2009; Ertani et al., 2009; Halpern et al., 2015) and/or the enhancement of crop tolerance to abiotic stress. PHs can increase nutrient uptake of treated plants by (1) increasing soil microbial activity and soil enzymatic activities, (2) improving micronutrient availability, in particular, Fe, Zn, Mn, and Cu, (3) enhancing root surface area, especially fine-root system, and (4) enhancing nitrate reductase, glutamine synthetase, and Fe(III)-chelate reductase activities (Cerdán et al., 2009; Ertani et al., 2009; García-Martínez et al., 2010; Colla et al., 2014; Lucini et al., 2015). PHs can also alter the hormonal balance of the plant, thereby affecting plant development due to the presence of specific peptides, amino acids (e.g. phenylalanine, glutamate), and precursors of phytohormone biosynthesis, such as tryptophan (Colla et al., 2014). Moreover, several scientific papers indicated that the application of plant-derived PHs can elicit auxin- and gibberellin-like responses, and thus promote crop growth and productivity (Schiavon et al., 2008; Ertani et al., 2009; Matsumiya and Kubo, 2011; Colla et al., 2014). PHs can also enhance the quality of fruits and vegetables increasing their health-related phytochemicals (i.e. carotenoids, flavonoids, polyphenols) (Parrado et al., 2007; Paradikovic et al., 2011; Gurav and Jadhav, 2013; Ertani et al., 2014), and limit potentially the toxic compounds, such as nitrates (Liu and Lee, 2012). PHs have also successfully been used for improving crop tolerance to environmental stresses, such as thermal stress, salinity, drought, alkalinity, and nutrient deficiency (Botta, 2013; Cerdán et al., 2013; Ertani et al., 2013; Colla et al., 2014; Petrozza et al., 2014; Lucini et al., 2015; Visconti et al., 2015). Repeated applications of PHs at low dosages have proven to be useful to minimize the negative impact of environmental stressors on crops (Ertani et al., 2014) with variable results depending on genotype, phenological stages, environmental conditions, cultural practices, time and mode of biostimulant applications (Kauffman et al., 2007; Kunicki et al., 2010; Ertani et al., 2014).

Phytotoxic effects and growth depression of vegetable crops have been described after multiple applications of some animal-derived PH products (Cerdán et al., 2009; Lisiecka et al., 2011), whereas no phytotoxicity and growth reduction was observed in tomato and basil plants after foliar applications of plant-derived PHs (Cerdán et al., 2009; Colla et al., 2015). The above findings are related to an unbalanced amino acid composition, higher concentration of free amino acids, degree of racemization, and salinity (Colla et al., 2015). In this book chapter, we describe the production process of PHs and their resulting bioactive compounds and the biostimulant effects of PHs in improving soil/ rhizosphere nutrient availability, nutrient-use efficiency, crop tolerance to abiotic stress, and product quality of crops.

2 Bioactive compounds

To date, there is no general agreement on the bioactive compounds that are responsible for biostimulant effects in PHs. It is likely that several compounds might act synergistically to exert the observed biological properties. Nonetheless, some major categories have been identified and proposed such as free amino acids, peptides, and phytohormones. PH composition and quality are strongly affected by the source of protein (animal or plant origin) and the hydrolysis process (chemical, thermal, and/or enzymatic hydrolysis). Chemical hydrolysis of proteins is done under strong acid or alkaline conditions, hence via, a very aggressive process carried out at high temperature (>100°C) and pressure (>200 kPa). Thermal hydrolysis (160–180°C) under high pressure (>600 kPa) has been also proposed for producing PHs from animal-derived proteins (Nurdiawati et al., 2019). Such strong conditions (high temperature and pressure, acid or alkaline conditions) lead to a high degree of hydrolysis and the corresponding products show a high content of free amino acids rather than peptides. Moreover, chemical hydrolysis leads to the destruction of several amino acids (total degradation of tryptophan under acid hydrolysis; partial degradation of cysteine, serine and threonine; conversion of asparagine and glutamine into their acidic forms). Hydrolysis using proteolytic enzymes is milder than chemical hydrolysis and usually targets specific peptide bonds and does not require high temperatures (<60°C) in order to ensure the enzyme/s to be active. A specific enzyme to substrate concentration ratio is managed in order to control the degree of hydrolysis. Therefore, PHs from enzymatic hydrolysis are typically a mixture of free amino acids and peptides of different length, as a result of the conditions used. Combined chemical and enzymatic hydrolytic processes have been also proposed to achieve the desired degree of hydrolysis while preserving the structure of amino acids (Colla et al., 2015). Nonetheless, a critical aspect of chemical hydrolysis is the racemization, that is, the conversion of free amino acids from the L-form to D-form. Considering that amino acids in proteins of living organisms are in the L-form, D-amino acids need to be included in catabolism of plant cell otherwise they can have toxic effects, thus leading to the need for the involvement of detoxification mechanisms; this may induce growth depression and phytotoxic effects. Moreover, another drawback of chemical hydrolysis is the large use of acids/alkalis that leads to an increase in salinity of the hydrolysates. Regardless of the type of hydrolysis used, the amino acid composition in PHs also differs according to the protein source. Collagen-based PHs contain relatively higher amounts of glycine and proline, as compared to products from vegetal origin. Furthermore, when collagen is used as protein source, hydroxyproline and hydroxylysine (i.e. two non-standard amino acids) are present. However, plant-derived PHs such as legume-based products and fish-based PHs contain high amount of aspartic acid and glutamic acid. Furthermore, casein-derived products have a high amount of glutamic acid and proline. Liquid chromatography coupled to tandem mass spectrometry (LC–MS/MS, usually with high-resolution mass selectors), rather than MALDI-TOF mass spectrometry, is the preferred method for the separation and identification of peptides in complex mixtures such as PHs. The actual composition in terms of free amino acids and peptides profile is thought to impose the actual biostimulant properties as well as the phytotoxicity reported in some cases. In fact, the detrimental effect on plant growth of some animal-derived PHs produced through chemical hydrolysis has been related to their higher concentration of free amino acids (especially in D-form), less-balanced amino acid composition, and high salinity. In recent years, several bioactive peptides have been identified as endogenous signals acting as phytohormones in plants (Ghorbani et al., 2014). Similarly, exogenous peptides are expected to play a phytohormone-like activity in plants. For instance, a root hair-promoting peptide (RHPP induces root hair promotion and adventitious roots formation) has been isolated from a legume-derived PH (Matsumiya and Kubo, 2011). Such signaling peptide has a 12-amino acid sequence (Gly-Gly-Ile-Arg-Ala-Ala-Pro-T hr-Gly-Asn-Glu-Arg) and its activity was confirmed since root hair-promoting activity was conserved when it was chemically synthesized. Interestingly, RHPP and endogenous peptides seem to have no relationship within their secondary or tertiary structure. Indeed, RHPP contains four residues of α-helix breaking amino acids. Furthermore, the bioactivity was retained after a denaturing heat treatment. On the other hand, the biological activity of RHPP decreased when one residue of the C terminus was deleted. This suggests that the 12-amino acid residues chain length is fundamental for expressing root hair-promoting activity. Nonetheless, the mechanisms of RHPP biological activity seem to be different from that of endogenous peptides. The hormone-like activity of plant-derived peptides contained in PHs has been proposed in many scientific papers, where the products elicited auxin- and gibberellin-like activities and thus promoted crop performances (Colla et al., 2014; Ertani et al., 2013). Because many other signaling peptides have been identified in plant cells controlling growth, development, and stress responses of plants (Tavormina et al., 2015), it is expected that more signaling-peptide-based PHs will be developed in the near future. Besides free amino acids and peptides, other nitrogen-containing molecules, including betaines and polyamines have been reported in PHs. Although their degree of biological activity is somehow still under debate, they might contribute to the biostimulant activity. Among betaines, glycine betaine (trimethyl glycine) is an amino acid derivative being widely recognized as a stress-modulating compound. Interestingly, glycine betaine has been identified in PHs where it might take part in synergistic activity, together with free amino acids and peptides. Growth regulators such as auxins, gibberellins, cytokinins, and polyamines have been often identified in PHs especially from those coming from vegetal sources of proteins. However, the content of phytohormones in the final product is often not high enough to justify a direct effect on plant. In this respect, some indirect effects of amino acids can be postulated. The amino acid L-tryptophan is a precursor of indole compounds (thus including auxins), whereas L-methionine is known as a precursor of ethylene. The actual profile of other such compounds depends on the starting plant material as well as on the hydrolyzation process used to produce PH. Thermolabile compounds (e.g. vitamins) are also mainly destroyed during chemical /thermal hydrolysis rather than during the protein concentration step prior to hydrolysis.

3 Effects of protein hydrolysates on germination, growth, and yield of crops

The beneficial effects of PHs on crop performances in terms of seed germination, plant growth, and productivity have been known for a long time (Sharma and Kothari, 1993). Seed treatments with PHs can represent an interesting technology to improve the seed quality, especially for organic producers where the use of synthetic substances is forbidden. It has been reported that PHs such as vegetal-PH 'Trainer' can exert a significant auxin-like activity and gibberellin-like activity (Colla et al., 2014) promoting seed germination and seedling vigor. Moreover, wheat seed treatment with a collagen hydrolysate stimulated seed metabolism by increasing endogenous gibberellic acid, thus enhancing emergence and seedling biomass, and reducing abnormal seedlings (Gaidau et al., 2013). Similarly, seed treatments with a PH obtained by enzymatic hydrolysis of keratin enhanced germination percentage by 10% and germination energy by 18% in corn compared to water-treated control seeds (Bryndina et al., 2019). Seed treatments with the vegetal-PH 'BioST VPH' containing a root-stimulating peptide and amino acids has been also successfully used to stimulate germination, early root growth, crop stress tolerance and promote adhesion of fungicides/micronutrients on seed surface in corn and soybean (Bonini et al.,

2017). Moreover, the use of PHs can provide additional advantages, such as the limitation of dust formation and reduction of microbial inoculant detachment from the seed surface during handling. A recent patent (n. 201531523/3 presented on October 22, 2015) proposed by Agrotecnologías Naturales SL (Tarragona, Spain) reported that a soybean-derived PH was able to more than double the number of polyethylene microspheres (having 75–90 μm of diameter, and used as substitute for arbuscular mycorrhizal fungi spores) that stuck to the seed surface of wheat, corn, and soybean in comparison with water. Moreover, mechanical vibration of treated seeds showed that adding the soybean-derived PH to the microsphere/water suspension enhanced the adhesion strength of the microspheres by about 96%, 36%, and 21% in wheat, corn, and soybean seeds, respectively (Colla et al., 2017a). Several research studies demonstrated that application of PHs under open field and protected cultivations can enhance plant growth and productivity in many crops such as corn, kiwifruit, lettuce, lily, papaya, passionfruit, pepper, and tomato (Colla et al., 2015). Foliar applications of a legume-derived PH 'Trainer' showed to enhance the fruit yield of cherry tomato grown in open field as a result of the higher fruit number (Caruso et al., 2019a). Greenhouse tomato yield was also enhanced in plants of both cultivar Akyra and Sir Elyan when a legume-derived PH 'Trainer' was foliarly sprayed, especially using high rate (5.0 ml L^{-1}) of biostimulant product (Rouphael et al., 2017). Interestingly, the higher marketable production observed on plants treated with the legume-derived PH at 5.0 ml L^{-1} and to a lesser extent at 2.5 ml L^{-1}, in comparison to the untreated plants, was due to an increase in the number of fruits per plant in the small fruit size cultivar Akyra and to a higher fruit weight in the medium fruit size cultivar Sir Elyan. Similarly, Colla et al. (2017b) reported a fruit yield increase when tomato cultivar Sir Elyan was foliar sprayed with a legume-derived PH 'Trainer' at 3 ml L^{-1} in comparison with untreated control. Growth and yield of leafy vegetables were also significantly enhanced by PH applications. For instance, foliar applications of a legume-derived PH 'Trainer' increased the leaf surface expansion, leaf number, and leaf yield of perennial wall rocket grown in three greenhouse crop cycles (autumn–winter; winter; and winter–spring) compared to the non-treated control (Caruso et al., 2019b). In a greenhouse trial, Rouphael et al. (2018a) reported a significant increase in fresh yield of PH-treated spinach plants over untreated ones as a result of an increase in the leaf area and not in the number of leaves per plant. Soilless-grown lettuce yield was enhanced by foliar applications of a legume-derived PH 'Trainer' in comparison with untreated control plants. Animal-derived PHs were also used successfully to increase crop yield even though the effects were less consistent over the growing cycles. Foliar applications of animal-derived PH 'Terra Sorb Complex' increased the yield of soybean compared to the control, especially when the number of foliar applications during the growing cycle were two instead of one (Kocira, 2019). Morales-Pajan and Stall (2003) reported that foliar sprays of an animal-derived PH 'Siapton' enhanced fruit yield in papaya especially when the biostimulant was sprayed more frequently (monthly intervals) starting from the beginning of flowering. Foliar sprays with animal-derived PH 'Siapton' were also effective in enhancing carrot yield in the first growing cycle. However, the carrot yield increase induced by animal-derived PH was not evident in the following two growing cycles (Grabowska et al., 2012). No significant variation of yield was recorded in spinach and strawberry plants treated with animal-derived PH 'Siapton' in comparison with untreated control (Marfà et al., 2009; Kunicki et al., 2010). Greenhouse applications of an animal-derived PH 'Siapton', and carob germ hydrolysate increased both plant height as well as number of flowers per plant in tomato compared with untreated plants, though only those sprayed with carob germ hydrolysate enhanced the fruit number per plant after eighteen weeks (Parrado et al., 2008). PHs have also been demonstrated to enhance crop performances in ornamental species. Application of two PHs derived from animal epithelia and alfalfa enhanced the diameter of flower buds, leaf area, stem quality, and root biomass of lily in comparison to untreated plants (De Lucia and Vecchietti, 2012). Moreover, in a bedding trial assessing the effects of four foliar application rates of legume-derived PH 'Trainer' (0, 1, 3, or 5 mL L^{-1}) on growth, ornamental quality, leaf gas exchanges, and mineral composition of three bedding plants (*Begonia tuber* hybrida, *Pelargonium peltatum,* and *Viola cornuta*). Cirillo et al. (2018) observed a species-dependent enhancement in growth and ornamental quality in PH-treated plants. Protein hydrolysate at 1 mL L^{-1} concentration enhanced several growth and quality parameters (plant height, leaf area and number, and shoot dry weight) in *Pelargonium peltatum*, whereas PH applications positively affected only plant height of *Viola cornuta*, PH applications at the rates of 1 and 3 ml L^{-1} enhanced the number of flowers per plant of *Begonia tuber* hybrida. The better crop performances observed in *Pelargonium peltatum* and *Begonia tuber* hybrida treated with PH were attributed to an increase of net photosynthetic rate and stomatal conductance and, only for *Pelargonium peltatum,* to an enhancement of leaf N content. In another trial on greenhouse-potted snapdragon (Cristiano et al., 2018), three animal-derived PH rates (0, 0.1, and 0.2 g L^{-1}) and two application methods (substrate drench and foliar spray) were compared in terms of growth, root morphology, nitrogen content, and leaf gas exchange. Plant height, number of shoots, total shoot length, number of leaves, total leaf area, number of flowers, and total aboveground dry weight were significantly increased by the biostimulant application compared to the control, regardless of the dose. The above findings were associated with an increase of N content of roots and leaves, photosynthetic rate, transpiration rate, and stomatal conductance in PH-treated plants. The foliar spray compared to substrate drench had a significant effect only on flower dry weight, while substrate drench had a positive effect on dry weight, root morphology, leaf-N and root-N content, transpiration rate, stomatal conductance, concentration of CO_2 in intracellular spaces, as well as on the efficiency of Photosystem II. The effects of PHs on productivity of the main crops are summarized in Table 1.

4 Soil nutrient availability and nutrient-use efficiency

A growing interest in crop production is the development of cropping systems and management practices that improve the ability of crops to absorb nutrients, which could minimize the fertilizer inputs and potential for nutrient losses. Besides the ability of crops to take up nutrients, it is also important to consider the increase in performance of cropping systems and the efficiency of crops to utilize nutrients for maximum yield. Nutrient-use efficiency (NUE) has been proposed as a critical concept for evaluating crop production systems. NUE is generally defined as the yield of harvestable product per unit of nutrient available from the soil and fertilizer (Moll et al., 1982). It is the product of two physiological factors: (i) nutrient uptake efficiency, defined as the amount of nutrient uptake by the crop per unit of nutrient available to the crop and (ii) nutrient utilization efficiency, defined as the harvestable product per unit of nutrient uptake by the crop (Colla et al., 2011). PHs can improve NUE by enhancing both the uptake and utilization efficiency of nutrients (Colla et al., 2015). In soil treated with PHs, nutrient availability for plant uptake is increased by (1) the formation of metal-amino acids/peptides which prevents nutrient insolubilization (Fe, Zn, Mn, Cu); (2) reduction of micronutrients in forms more available for plant uptake (e.g. Cu^{2+} to Cu^{+} or Fe^{3+} to Fe^{2+}), thanks to the reductant activity of some amino acids; and (3) stimulation of nutrient-solubilizing microbes such as siderophore-producing Trichoderma species, and beneficial microbial communities involved in nutrient mineralization in the soil (Halpern et al., 2015). PHs can enhance nutrient uptake by (1) promoting root growth and especially fine roots which are the most physiologically active and absorptive roots (Colla et al., 2014); (2) stimulating root enzymes involved in nutrient uptake (e.g. ferric-chelate reductase activity; (Cerdán et al., 2009); and (3) upregulating the genes encoding for nutrient transporters (e.g. nitrate transporters). Moreover, PHs can increase translocation of micronutrients within the plant as a result of the metal-amino acids/peptides complex formation. Many researches demonstrated that nicotianamine, a non-proteinogenic amino acid, is involved in the complexation of metal ions and plays a fundamental role in metal uptake, transport, and homeostasis in all plants (Curie et al., 2009; Schmidke and Stephan, 1995). PHs can also improve nutrient assimilation processes through the upregulation of genes encoding for enzymes involved in the assimilation of inorganic nutrients (e.g. nitrates) (Sestili et al., 2018). Schiavon et al. (2008) reported that an alfalfa-derived PH promoted nitrogen assimilation in plants via a coordinated regulation

Table 1 Effects of protein hydrolysates (PHs) on relative crop yield increase over the control treatment

Crop	Growing conditions	Type of PH	Application method and timing	Yield increase	Reference
Bean	Open field (3 crop cycles: spring-summer)	Animal-derived PH	Foliar sprays (1 time per crop cycle at BBCH 12-13 stage or 2 times per crop cycle at BBCH 12-13, and BBCH 61 stage using a 5.0 ml L^{-1} for each application)	=	Kocira et al. (2020)
Carrot	Open field (3 crop cycles: spring-summer)	Animal-derived PH	Foliar sprays (2 times per crop cycle in the plant stage of 5-7 leaves and when root had a diameter of 5-6 mm using 1.5 or 3.0 L ha^{-1} for each application)	+11% (first year using low PH rate) = (second and third year)	Grabowska et al. (2012)
Endive	Open field (6 crop cycles in 2 years: 4 crop cycles in spring-summer, 2 crop cycles in summer-autumn)	Animal-derived PH	Foliar sprays (2 times per crop cycle) the second and third week after transplanting	=	Gajc-Wolska et al. 2012)
Grapevine	Open field (five years)	Soy-derived PH	Foliar sprays (3 times per year from fruit-set to bunch closure using 1.6 or 6.4 g L^{-1} for each application)	+5 (low PH rate) +31 (high PH rate)	Boselli et al. (2019)
		Casein-derived PH	Foliar sprays (3 times per year from fruit-set to bunch closure using 1.6 or 6.4 g L^{-1} for each application)	+14 (low PH rate) +27 (high PH rate)	Boselli et al. (2019)
		Lupine-derived PH	Foliar sprays (3 times per year from fruit-set to bunch closure using 1.6 or 6.4 g L^{-1} for each application)	+9 (low PH rate) +21 (high PH rate)	Boselli et al. (2019)
Lettuce	Greenhouse (1 crop cycle: spring)	Legume-derived PH	Foliar sprays (4 times at weekly intervals using 2.5 ml L^{-1} for each application)	+50%	Colla et al. (2013)
Papaya	Open field (3 crop cycles: spring-summer)	Animal-derived PH	Foliar sprays (7 times per crop cycle at monthly intervals starting at the beginning of flowering using 3.0 g L^{-1} for each application)	+22%	Morales-Pajan and Stall (2003)

(Continued)

Table 1 (*Continued*)

Crop	Growing conditions	Type of PH	Application method and timing	Yield increase	Reference
Rocket	Greenhouse (3 crop cycles: autumn-winter; winter; and winter-spring)	Legume-derived PH	Foliar sprays (3 times per crop cycle at seven-day intervals starting when the leaves were 6 cm long using 3.0 ml L^{-1} for each application)	+13%	Caruso et al. (2019b)
Soybean	Open field (3 crop cycles: spring-summer)	Animal-derived PH	Foliar sprays (2 times per crop cycle at BBCH 13–15 stage and BBCH 61 stage using 5 ml L^{-1} each application)	+32%	Kocira (2019)
Spinach	Greenhouse (1 crop cycle: spring)	Legume-derived PH	Foliar sprays (4 times at weekly intervals starting 17 days after sowing using 2.5 ml L^{-1} for each application)	+58%	Rouphael et al. (2018a)
	Open field (4 crop cycles: spring and autumn)	Animal-derived PH	Foliar sprays (2 times per crop cycle at 2-leaf and 4-leaf stage using 1.5 or 3.0 L ha^{-1} for each application)	=	Kunicki et al. (2010)
Strawberry	Open field (1 crop cycle: winter-spring)	Animal-derived PH	Fertigation (5 times starting immediately after transplanting and after 16 days, 29 days, 44 days, and 78 days using 2.5 ml L^{-1} for each application)	=	Marfà et al. (2009)
Tomato	Open field (2 crop cycles: spring-summer)	Legume-derived PH	Foliar sprays (4 times per crop cycle at 7-day interval from early growth of first fruit truss using 3.0 ml L^{-1} for each application)	+19%	Caruso et al. (2019a)
	Greenhouse (1 crop cycle: summer-autumn)	Legume-derived PH	Foliar sprays (9 times at 10-day intervals starting 15 days after sowing using 2.5 or 5.0 ml L^{-1} for each application)	+8 (low PH rate) +21 (high PH rate)	Rouphael et al. (2017)
	Greenhouse (1 crop cycle: summer-autumn)	Legume-derived PH	Foliar sprays (4 times at 10-day intervals starting from fruit set of the first truss using 3.0 ml L^{-1} for each application)	+7	Colla et al. (2017b)

of C and N metabolism. Specifically, this PH increased the activity of malate dehydrogenase, isocitrate dehydrogenase, and citrate synthase enzymes functioning in the tricarboxylic acid cycle and enzymes (nitrate reductase, nitrite reductase, glutamine synthetase, glutamate synthase, and aspartate aminotransferase) involved in N metabolic assimilation processes. Comparable results were obtained in the work conducted by Colla and coauthors (2014), who tested the effect of a legume-derived protein hydrolysate 'Trainer' on nitrogen uptake and growth of maize and tomato. The authors found that the product enhanced nitrogen uptake by stimulating the root growth and nitrogen assimilation and consequently crop performance. The above findings explain the increase in nutrient content such as nitrogen, phosphorus, potassium, calcium, magnesium, and iron in plant tissues generally observed after PH applications (Carillo et al., 2019; Caruso et al., 2019a,b; Cerdán et al., 2009; Rouphael et al., 2017, 2018a). However, it must be considered that nitrogen content increase in PH-treated plants can result not only from biostimulant action but also from fertilizer action (direct plant uptake of amino acids/ peptides or PH-derived inorganic nitrogen compounds), especially when high rates of PHs are applied in low-input cropping systems. In a greenhouse spinach trial (Carillo et al., 2019), eight treatments were compared derived from the factorial combination of two biostimulant applications (control or foliar application of a legume derived PH 'Trainer') and four mineral-N fertilization levels (0, 15, 30, or 45 kg N ha^{-1}). Interestingly, PH application increased fresh yield by 33.3% and 24.9% under limited N availability (at 0 and 15 kg N ha^{-1}, respectively) compared to untreated plants. Moreover, foliar applications of the legume-derived PH enhanced NUE by 25%, 16%, and 8% at 15, 30, and 45 kg N ha^{-1}, respectively (Fig. 1a). Similar results were obtained from Di Mola et al. (2019a) who tested eight treatments derived from the factorial combination of two biostimulant applications (control or foliar application of a legume-derived PH 'Trainer') and four mineral-N fertilization levels (0, 60, 80, or 100 kg N ha^{-1}). Cumulative rocket fresh yield of three harvests increased linearly as nitrogen fertilizer rate increased with a greater slope in PH treatment compared to control treatment (0.130 vs. 0.094). Moreover, foliar applications of the legume-derived PH enhanced NUE by 29%, 27%, and 36% at 60, 80, and 100 kg N ha^{-1}, respectively (Fig. 1b). Similarly, Colla et al. (2013) confirmed that PHs increased NUE especially under low-nutrient availability, in a greenhouse trial, foliar sprays at weekly intervals with 2.5 ml L^{-1} of 'Trainer' increased yield, greenness readings (i.e., SPAD index), and N uptake by 50%, 11%, and 11%, respectively, in comparison to control plants.

Moreover, substrate application of a microgranular biostimulant containing PH of vegetal origin enhanced NUE of tomato transplants in nursery (Rouphael et al., data not published). The NUE enhancement was due to an increase in both N-uptake efficiency and N-utilization efficiency (Table 2).

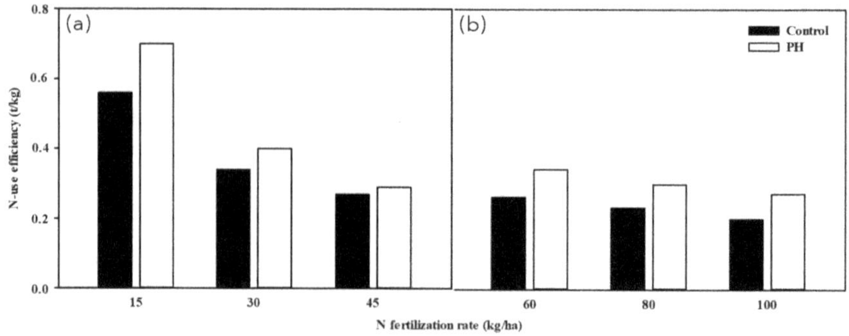

Figure 1 Effect of foliar applications of legume-derived protein hydrolysate (PH) on nitrogen use efficiency (expressed as ton of fresh shoot biomass per kg of applied N fertilizer) of spinach (a) and rocket (b) grown in greenhouse under three nitrogen levels (modified from Carillo et al., 2019 and Di Mola et al., 2019a).

Table 2 Effect of a microgranular biostimulant on nitrogen-use efficiency, nitrogen uptake efficiency, and nitrogen utilization efficiency of tomato transplants (the values in parentheses indicate percentage change as compared with control).

PH rate (g L^{-1})	N-use efficiency[1]	N-uptake efficiency[2]	N-utilization efficiency[3]
0.0	10.9	0.17	62.9
0.5	12.3 (+13.2%)	0.18 (+5.3%)	67.8 (+7.4%)

[1]N-use efficiency calculated as the mg of plant dry weight per mg of N supplied in the substrate.
[2]N-uptake efficiency calculated as the mg of N uptaked by the plants per mg of N supplied in the substrate.
[3]N-utilization efficiency calculated as the mg of plant dry weight per mg of N uptaked by the plants.

5 Crop tolerance to abiotic stress

Abiotic stresses, such as sub/supra-optimal temperatures, drought, salinity, and heavy metals, are major factors limiting crop productivity and sustainability worldwide. A general effect of stress in plant tissues is the generation of reactive oxygen species (ROS) that trigger oxidative stress (Blokhina et al., 2003). Plants control the concentrations of ROS in cells by enzymatic and non-enzymatic (proline) antioxidants (antioxidant defense system); enzymatic antioxidants include superoxide dismutase (SOD), catalase (CAT), ascorbate peroxidase, guaiacol peroxidase, and glutathione reductase. SODs catalyze the dismutation of the superoxide anion (O_2^-) to yield hydrogen peroxide (H_2O_2). The level of H_2O_2 is further tightly regulated by the concerted activity of CAT and peroxidases (Blokhina et al., 2003). PHs can mitigate the negative effect of environmental stress on crops by enhancing antioxidant defense system especially when PHs are applied before the stress event occurs. Moreover, PHs can be a source of antioxidants for plant cells since some peptides have substantially higher antioxidant activity (Colla et al., 2015). This is the reason why the same PH generally exerts a protective action against many environmental stresses (Colla et al., 2015). Additional PH-mediated mechanisms involved in stress tolerance of crops are: (1) the increase of root growth and root to shoot ratio; (2) the improvement of plant nutrition; (3) the enhancement of cell membrane stability; (4) the accumulation of osmoprotectants; and (5) the change of hormonal status. In the last few years, increases in soil salinity have become a serious issue for agriculture, especially in coastal areas. High concentrations of salts can cause impairment of several morphological, physiological, and biochemical traits in plants (Rouphael et al., 2017) mainly via osmotic (i.e. water deficit stress) and ionic (i.e. Na^+ and Cl^-) effects on plant nutrient uptake and root to shoot translocation, nitrogen metabolism, photosynthesis, and protein synthesis (Tester and Davenport, 2003). Visconti et al. (2015) addressed the question of whether the application of an animal-derived PH containing Ca 'Stressal' was able to counteract salt stress in fruit trees such as persimmon. The experimental results demonstrated that PH application alleviated the salt stress in persimmon by lowering chloride uptake and translocation to aerial parts, thus reducing leaf necrosis symptoms (Visconti et al., 2015). Greater tolerance to salt stress was associated with the composition of the PH, particularly compatible solutes such as proline and glycine betaine which stimulated the plant's metabolism to build up compatible solutes and enhanced the ability to exclude Cl^- ions. Leaf sodium content was roughly two orders of magnitude lower than the leaf chloride content with no differences between calcium-PH-treated and -untreated plants. Application of an alfalfa-derived PH in hydroponically grown maize was able to mitigate the negative impact of salinity on crop by increasing the potassium and proline concentrations in plant tissues (Ertani et al., 2012). Proline synthesis has been reported to improve the adaptation of several plant species to excess NaCl by inducing the expression of salt-stress-responsive proteins (Khedr et al., 2003). Mutants of *Arabidopsis thaliana* deficient in production of proline have stress-sensitive phenotypes (Szabados and Savouré, 2010). These plants can have their phenotype rescued with exogenous application of L-proline, an amino acid normally well-represented in PHs. Similar results on crop tolerance improvement to salinity were obtained in another trial carried out by Ertani et al. (2013) where application of an alfalfa-derived PH in hydroponically grown maize under saline conditions also enhanced plant biomass and reduced the activity of antioxidant enzymes as a result of decreased stress, but prompted the enhancement of proline and flavonoids content, as well as phenylalanine ammonia-lyase activity (PAL) and gene expression, compared to salt-stressed controls plants. The increase in flavonoid content was in line with the variation in activity of the enzyme PAL. The gene encoding this protein is responsive to a number of abiotic and biotic stresses in many plant species and can be upgraded by biostimulant application (Schiavon et al., 2010; Ertani et al.,

2011). The results of this study demonstrated that PAL is salt-stress responsive, and its activity and gene expression were both upregulated under salinity stress likely because of the indoleacetic acid content in the PH. The stimulation of the phenlylpropanoid pathway by the PH was also partially due to the improvement of N assimilation. In a lettuce trial, Lucini et al. (2015) reported that substrate drench, and especially foliar spray plus substrate drench applications of a legume-derived PH 'Trainer', aided plants to preserve higher photochemical activity of the photosystem II, and attain better nutritional status of shoots under saline conditions, leading to superior crop performance. Moreover, metabolomic analysis of leaf tissues was carried out to reveal the metabolic differences in the PH-treated and control leaves of salt-stressed lettuce plants. The metabolome analysis revealed that PH-mediated improvement in lettuce salinity tolerance was associated to oxidative stress mitigation, change of hormonal balance, as well as production of secondary metabolites including glucosinolate, sterols, and terpenes. Drought stress is one of the most severe conditions for plants, especially in the face of the emerging problem of global warming. Paul et al. (2019) addressed the question of whether the applications of legume-derived PH 'Trainer' (foliar spray or substrate drench) were able to sustain tomato growth under limited water availability. A high-throughput image-based phenotyping approach and metabolomic analysis were used to elucidate the biostimulant action of PH application method on morpho-physiological traits and metabolic profile of tomato grown under limited water availability in a controlled environment. The experimental results demonstrated that PH application especially as substrate drench enhanced crop growth (digital biomass and relative growth rate) under limited water availability in comparison with control plants. Moreover, PH-treated tomato plants exhibited an increased transpiration-use efficiency and an improved tolerance to ROS-mediated oxidative imbalance. Such tolerance involved a coordinated action of salicylic acid, hydroxycinnamic amide signaling, carotenoids, and prenyl quinone radical scavenging, as well as reduced tetrapyrrole biosynthesis (Paul et al., 2019). In a grapevine experiment, Boselli et al. (2019) reported that foliar applications of soybean- and casein-derived PH, three times on 'Corvina' vines every 10 days, from fruit set (BBCH 71) until bunch closure (BBCH 79), significantly decreased the stress index in treated vines, showing the ability of the PHs to reduce stomatal conductance and hence the transpiration demand, thus ameliorating the tolerance to water stress. Under water stress conditions, foliar spray with legume-derived PH alone 'Trainer' or enriched with urea 'Stimtide' both at 3 ml L^{-1} during pre-veraison on 'Montepulciano' vines up-regulated photosynthesis-related proteins, and metabolites involved in plant growth regulations, and promoted the delay of leaf senescence (Bavaresco et al., 2020). The potential for PH application to reduce the detrimental impact of thermal stress in several crops have also been reported (Botta, 2013). Enhancement of plant tolerance to sub- and supra-optimal temperature conditions was noticed in lettuce and perennial ryegrass when the animal-derived PH 'Terra-Sorb foliar' was applied (Botta, 2013). In these trials, PH-treated lettuce plants subjected to three cold stress treatments showed higher fresh weight compared to untreated plants, along with higher stomatal conductance. Moreover, PH could maintain higher membrane permeability at lower temperatures and increase wall storage space causing a decrease of cold-derived damages. In another experiment on strawberry plants exposed to frost during the night (up to −6°C), five applications into the soil near each plant of an animal-derived PH 'Siapton' at 4 L ha^{-1} did not increase the plant survival and fruit yield in comparison with control plants (Marfà et al., 2009). The above findings may be explained by the different type of PHs, application mode, timing, and magnitude of cold stress used in the lettuce and strawberry trials. Moreover, PH-treated ryegrass plants exposed to high temperatures (36°C) showed a superior photosynthetic efficiency and maintained higher levels of chlorophylls and carotenoids in comparison with control plants. Application of PH-biostimulants could be also considered as an effective tool for achieving high productivity under adverse soil conditions or such as alkalinity or high level of heavy metals. In a tomato trial, Cerdán et al. (2013) reported that the application of PHs, particularly those derived from plant origin, enhanced tomato seedling growth under alkaline conditions due to an increase in leaf and root Fe reductase activities when applied to roots. Other studies (Sharma and Dietz, 2006; Sytar et al., 2013) reported that some amino acids (e.g. asparagine, cysteine, and glutamine) and peptides (e.g. glutathione and phytochelatins) could enhance crop tolerance to a range of toxic heavy metals (Cu, Zn, As, Cd, and Ni) through metal chelation and binding.

6 Product quality

One of the main challenges for fruit and vegetable producers is to meet the growing demand for foods of high and predictable quality (Rouphael and Kyriacou, 2018b). Protein hydrolysates can contribute toward improving quality traits of edible products such as external attributes, such as color and internal properties such as nutritive molecules (sugars, vitamins, minerals), as well as non-nutritive phytochemicals (phenolic compounds, flavonoids, terpenoids) (Colla et al., 2015). Moreover, PHs can reduce the content of undesired compounds such as nitrates and heavy metals which may represent a risk for human health. The positive effects of PHs on nutritional quality are due to (1) the enhancement of carbohydrate supply through photosynthesis, (2) the increase of nutrient uptake and assimilation (e.g. nitrite and nitrate reductase as well as glutamate synthase and glutamine synthetase activities), (3) the activation of secondary metabolism (e.g. PAL), and (4) the reduction of plant uptake of heavy metals and other pollutants (Colla et al., 2015). In a greenhouse tomato trial, foliar applications of a legume-derived PH 'Trainer' increased the nutritional quality of fruits by enhancing antioxidant activity, and the amount of soluble solids, mineral elements (K and Mg), lycopene, and ascorbic acid (Rouphael et al., 2017). Similar results were also reported by Colla et al. (2017a) in another greenhouse tomato trial using the same PH. In a field-grown tomato trial, foliar applications of a legume-derived PH 'Trainer' resulted in cherry tomatoes having higher dry matter, total soluble solids, K, Mg, total phenol and ascorbic acid as well as lycopene content, and lipophilic antioxidant activity than the non-treated control (Caruso et al., 2019a). Moreover, foliar applications of a legume-derived PH 'Trainer' on greenhouse baby lettuce enhanced the antioxidant capacity and total ascorbic acid of leaves and did not increase nitrate content as recorded in seaweed extracts and tropical plant extract treatments (Di Mola et al., 2019b). Similarly, Rouphael et al. (2018a) reported that foliar applications of legume-derived PH 'Trainer' on greenhouse spinach improved chlorophyll biosynthesis and elicited an increase in bioactive compounds (total phenols and ascorbic acid). Moreover, PH enhanced the K and Mg concentrations and did not result in increased nitrate accumulation as observed with the other two seaweed-based biostimulant treatments. Caruso et al. (2019b) also reported that foliar treatments with a legume-derived PH 'Trainer' of rocket plants enhanced leaf dry matter, oxalic and citric acids, Ca and P concentrations, phenols and ascorbic acid content as well as antioxidant activity, but did not increase nitrate content. Ertani et al. (2014) carried out a greenhouse pot experiment with the aim of assessing the impacts of two rates of biostimulants, one coming from alfalfa plants (25 and 50 ml L^{-1}) and another from red grapes (50 and 100 ml L^{-1}), on nutraceutical properties of habanero type pepper. PH-treated plants provided fruits with high concentrations of chlorogenic acid and antioxidant activity, whereas both alfalfa and red grape PH-treated fruits were highly enriched with capsaicin. A significant improvement in total soluble solids, proteins, total phenolics, flavonoids, as well as antioxidant activity was observed even in fruit tree crops. Parrado et al. (2007) reported that foliar applications of enzymatically treated vegetable extract coming from agricultural wastes on red grapevine enhanced the total phenolic and anthocyanin concentration by 22% and 70%, respectively, over control plants. Similarly, Boselli et al. (2019) observed that three foliar applications of PHs derived from lupine, soybean, and casein in grapevine significantly increased the total anthocyanin content of berries in comparison with control treatment with highest values using casein-derived PH at 1.6 g L^{-1}. Moreover, total soluble solids' content was also significantly increased by PH applications, with total soluble solids' values ranging from 19.05 (control treatment) up to 22.80 °Brix (lupine-derived PHs at 1.6 g L^{-1}). The beneficial effects of PH were also detected on the phytochemical profile of the aromatic plant lemon balm (Mehrafarin et al., 2015). The authors reported that a foliar application of 'Aminolforte' and 'Fosnutren' at 2 L ha^{-1}, increased contents of citronellal, neral, deltacadinene, germacrene, and geranial compared to untreated plants. Some authors (Andarwulan and Shetty, 1999a) evaluated the effects of fish-derived PH in preventing hyperhydricity and stimulating the production of phenolics compounds in oregano tissue cultures via stimulation of pentose phosphate route. The results showed that the fish-derived PH increased the production of free phenolics and rosmarinic acid in oregano cultures. Other studies also exhibited that fish-derived PHs can be used to stimulate secondary metabolism inducing a greater production of epoxy-pseudoisoeugenol-(2-methylbutyrate) in anise root culture (Andarwulan and Shetty, 1999b).

7 Conclusion and future trends

In conclusion, the use of PHs as plant biostimulants can represent a promising tool for the amelioration of nutrient availability, NUE, and quali-quantitative traits of crops, especially under abiotic stress conditions. The positive effects of PHs on plant growth and metabolism could be ascribed to small peptides and amino acids, and to a lesser extent to other compounds such as phytohormones, carbohydrates, phenols, etc. However, the cause-effect relationships between specific bioactive compounds, alone or in combination, and crop response are still not well known and only recently, thanks to the use of 'omics' science, is becoming possible to clarify the mode of action of PHs and their fractions. Among the omics technologies, the most promising are the high-throughput phenotyping technologies and metabolomics which allow to accurately evaluate the biostimulant activity of PHs in term of morpho-physiological traits and understand the metabolic pathways triggered by the PH application and their fractions. Moreover, elucidating the biostimulant × plant species × environment interaction is also crucial, in order to select the best combinations in terms of yield and quality, and to optimize application parameters (e.g., timing, mode, and rate). Finally, it is also important to understand the type of interactions that occur between biostimulant products and chemical inputs, such as plant-protection products and fertilizers, in order to integrate the use of PHs into a more sustainable farming system.

8 Where to look for further information

Guidelines for future research on PHs can be summarized as follows:

1) identification of bioactive compounds in PHs through molecular fractionation and bioassay screening of each fraction;

2) definition of PH composition which improves desired crop traits (e.g. rooting, flowering, fruit set, produce quality) and crop response to single, multiple individual, and combined environmental stresses;
3) deepening the knowledge on the PH-mediated effects on plant microbiome including beneficial and pathogenic microorganisms;
4) application of omics technologies to better understand the mode of action of PHs and their fractions, and to study the PH-mediated effects on plant microbiome;
5) conducting agronomic trials under realistic growing conditions to optimize rate, timing, and application method of PHs for different crops and environmental conditions.

The chapter's authors are affiliated with the following institutions:

- Department of Agriculture and Forest Sciences (DAFNE), University of Tuscia, Viterbo (Italy) with teaching and research competences in the fields of agriculture, forestry and nature conservation sciences, and agro-biotechnology (http://www.unitus.it/en/dipartimento/dafne).
- Department of Agriculture (DIA), University of Naples Federico II, Portici (Italy) covering a wealth of teaching courses and research subjects related to agriculture, food science and technology, and environment (http://www.agraria.unina.it).
- Consiglio per la ricerca in agricoltura e l'analisi dell'economia agraria Federico II, Centro di ricerca Orticoltura e Florovivaismo, Pontecagnano Faiano (Italy) conducting research activities with integrated and multidisciplinary approaches on breeding and biodiversity, agronomic techniques, and sustainable plant protection of vegetables, flowers, aromatic and ornamental plants (https://www.crea.gov.it).
- Department for Sustainable Food Process (DiSTAS), Università Cattolica del Sacro Cuore, Piacenza (Italy) comprises teaching and research activities related to chemical, environmental, microbiological and technological aspects of production and transformation of agricultural commodities including sustainable agricultural practices, improvement of functional characteristics, quality, traceability and authenticity of food commodities (https://dipartimenti.unicatt.it/distas-home).
- Department of Agricultural, Forest and Food Sciences, University of Turin (Italy) with teaching and research in the fields of primary production and transformation of plants and animals, as well as the interactions of these activities in the territory and the environment (https://www.disafa.unito.it/do/home.pl).

9 Acknowledgement

This work was supported by MIUR (Minister for education, University and Research), Law 232/2016, 'Department of Excellence'.

10 References

Andarwulan, N. and Shetty, K. 1999a. Influence of fish protein hydrolysates in combination with acetyl salicylic acid on hyperhydricity reduction and phenolic synthesis in oregano (*Origanum vulgare*) tissue cultures. *Journal of Food Biochemistry* 23(6), 619–35. doi:10.1111/j.1745-4514.1999.tb00590.x.

Andarwulan, N. and Shetty, K. 1999b. Stimulation of novel phenolic metabolise, epoxypseudoisoeugenol-(2-methylbutyrate) (EPB), in transformed anise (*Pimpinella anisum* L.) root cultures by fish protein hydrolysates. *Food Biotechnology* 14, 1-20.

Bavaresco, L., Lucini, L., Squeri, C., Zamboni, M. and Frioni, T. 2020. Protein hydrolysates modulate leaf proteome and metabolome in water stressed grapevines. Scientia Horticulturae, in press. doi.org/10.1016/j.scienta.2020.109413.

Blokhina, O., Virolainen-Arne, E. and Fagerstedt, K. V. 2003. Antioxidants, oxidative damage and oxygen deprivation stress: a review. *Annals of Botany* 91(Spec No), 179–94. doi:10.1093/aob/mcf118.

Bonini, P., Long, D. H., Canaguier, R., Colla, G. and Leman, J. 2017. Seed treatments with endophytic fungi and biostimulant compounds enhance yield of corn and soybean. *The 3rd World Congress on the Use of Biostimulants in Agriculture*, 27-30 November 2017, Miami (US), NewAG International. Book of Abstracts, p. 147.

Boselli, M., Bahouaoui, M. A., Lachhab, N., Sanzani, S. M., Ferrara, G. and Ippolito, A. 2019. Protein hydrolysates effects on grapevine (*Vitis vinifera* L., cv. Corvina) performance and water stress tolerance. *Scientia Horticulturae* 258. doi:10.1016/j.scienta.2019.108784.

Botta, A. 2013. Enhancing plant tolerance to temperature stress with amino acids: an approach to their mode of action. *Acta Horticulturae* 1009(1009), 29–35. doi:10.17660/ActaHortic.2013.1009.1.

Bryndina, L., Ilyina, N., Baklanova, O. and Moiseyeva, E. 2019. Comparative evaluation of biostimulator efficiency on corn seeds germination: keratin protein and preparation Ribav Extra. *IOP Conference Series: Earth and Environmental Science* 392, 012068. doi:10.1088/1755-1315/392/1/012068.

Carillo, P., Colla, G., Fusco, G. M., Dell'Aversana, E., El-Nakhel, C., Giordano, M., Pannico, A., Cozzolino, E., Mori, M., Reynaud, H., Kyriacou, M. C., Cardarelli, M. and Rouphael, Y. 2019. Morphological and physiological responses induced by protein hydrolysate-based biostimulant and nitrogen rates in greenhouse spinach. *Agronomy* 9(8), 450. doi:10.3390/agronomy9080450.

Caruso, G., De Pascale, S., Cozzolino, E., Cuciniello, A., Cenvinzo, V., Bonini, P., Colla, G. and Rouphael, Y. 2019a. Yield and nutritional quality of vesuvian Piennolo tomato PDO as affected by farming system and biostimulant application. *Agronomy* 9(9), 505. doi:10.3390/agronomy9090505.

Caruso, G., De Pascale, S., Cozzolino, E., Giordano, M., El-Nakhel, C., Cuciniello, A., Cenvinzo, V., Colla, G. and Rouphael, Y. 2019b. Protein hydrolysate or plant extract-based biostimulants enhanced yield and quality performances of greenhouse perennial wall rocket grown in different seasons. *Plants* 8(7), 208. doi:10.3390/plants8070208.

Cerdán, M., Sánchez-Sánchez, A., Oliver, M., Juárez, M. and Sánchez-Andreu, J. J. 2009. Effect of foliar and root applications of amino acids on iron uptake by tomato plants. *Acta Horticulturae* 830(830), 481–8. doi:10.17660/ActaHortic.2009.830.68.

Cerdán, M., Sánchez-Sánchez, A., Jordá, D. J., Juárez, M. and Andreu, J. S. 2013. Effect of commercial amino acids on iron nutrition of tomato plants grown under lime-induced iron deficiency. *Journal of Plant Nutrition and Soil Science* 176(1), 1–8. doi:10.1002/jpln.201390005.

Cirillo, C., Rouphael, Y., Pannico, A., El-Nakhel, C., Colla, G. and De Pascale, S. 2018. Application of protein hydrolysate-based biostimulant as new approach to improve performance of bedding plants. *Acta Horticulturae* 1215(1215), 443–8. doi:10.17660/ActaHortic.2018.1215.80.

Colla, G., Rouphael, Y., Mirabelli, C. and Cardarelli, M. 2011. Nitrogen-use efficiency traits of mini-watermelon in response to grafting and nitrogen-fertilization doses. *Journal of Plant Nutrition and Soil Science* 174(6), 933–41. doi:10.1002/jpln.201000325.

Colla, G., Svecová, E., Cardarelli, M., Rouphael, Y., Reynaud, H., Canaguier, R. and Planques, B. 2013. Effectiveness of a plant-derived protein hydrolysate to improve crop performances under different growing conditions. *Acta Horticulturae* 1009(1009), 175–9. doi:10.17660/ActaHortic.2013.1009.21.

Colla, G., Rouphael, Y., Canaguier, R., Svecova, E. and Cardarelli, M. 2014. Biostimulant action of a plant-derived protein hydrolysate produced through enzymatic hydrolysis. *Frontiers in Plant Science* 5, 448. doi:10.3389/fpls.2014.00448.

Colla, G., Nardi, S., Cardarelli, M., Ertani, A., Lucini, L., Canaguier, R. and Rouphael, Y. 2015. Protein hydrolysates as biostimulants in horticulture. *Scientia Horticulturae* 196, 28–38. doi:10.1016/j.scienta.2015.08.037.

Colla, G., Cardarelli, M., Bonini, P. and Rouphael, Y. 2017a. Foliar applications of protein hydrolysate, plant and seaweed extracts increase yield but differentially modulate fruit quality of greenhouse tomato. *HortScience* 52(9), 1214–20. doi:10.21273/HORTSCI12200-17.

Colla, G., Hoagland, L., Ruzzi, M., Cardarelli, M., Bonini, P., Canaguier, R. and Rouphael, Y. 2017b. Biostimulant action of protein hydrolysates: unraveling their effects on plant physiology and microbiome. *Frontiers in Plant Science* 8, 2202. doi:10.3389/fpls.2017.02202.

Cristiano, G., Pallozzi, E., Conversa, G., Tufarelli, V. and De Lucia, B. 2018. Effects of an animal-derived biostimulant on the growth and physiological parameters of potted snapdragon (*Antirrhinum majus* L.). *Frontiers in Plant Science* 9, 861. doi:10.3389/fpls.2018.00861.

Curie, C., Cassin, G., Couch, D., Divol, F., Higuchi, K., Le Jean, M., Misson, J., Schikora, A., Czernic, P. and Mari, S. 2009. Metal movement within the plant: contribution of nicotianamine and yellow stripe 1-like transporters. *Annals of Botany* 103(1), 1–11. doi:10.1093/aob/mcn207.

De Lucia, B. and Vecchietti, L. 2012. Type of biostimulant and application method effects on stem quality and root system growth in LA Lily. *European Journal of Horticultural Science* 77, 1–10.

Di Mola, I., Ottaiano, L., Cozzolino, E., Senatore, M., Giordano, M., El-Nakhel, C., Sacco, A., Rouphael, Y., Colla, G. and Mori, M. 2019a. Plant-based biostimulants influence the agronomical, physiological, and qualitative responses of baby rocket leaves under diverse nitrogen conditions. *Plants* 8(11), 522. doi:10.3390/plants8110522.

Di Mola, I., Cozzolino, E., Ottaiano, L., Giordano, M., Rouphael, Y., Colla, G. and Mori, M. 2019b. Effect of vegetal- and seaweed extract-based biostimulants on agronomical and leaf quality traits of plastic tunnel-grown baby lettuce under four regimes of nitrogen fertilization. *Agronomy* 9(10), 571. doi:10.3390/agronomy9100571.

Ertani, A., Cavani., L., Pizzeghello, D., Brandellero, E., Altissimo, A., Ciavatta, C. and Nardi, S. 2009 Biostimulant activity of two protein hydrolyzates in the growth and nitrogen metabolism of maize seedlings. *Journal of Plant Nutrition and Soil Science* 172(2), 237–44. doi:10.1002/jpln.200800174.

Ertani, A., Francioso, O., Tugnoli, V., Righi, V. and Nardi, S. 2011. Effect of commercial Lignosulfonate-Humate on *Zea mays* L. metabolism. *Journal of Agricultural and Food Chemistry* 59(22), 11940–8. doi:10.1021/jf202473e.

Ertani, A., Pizzeghello, D., Baglieri, A., Cadili, V., Tambone, F., Gennari, M. and Nardi, S. 2012. Agro-industrial residues and their biological activity on maize (*Zea mays* L.) metabolism. *Journal of Geochemical Exploration* 129, 103–11.

Ertani, A., Schiavon, M., Muscolo, A. and Nardi, S. 2013. Alfalfa plant-derived biostimulant stimulate short-term growth of salt stressed *Zea mays* L. plants. *Plant and Soil* 364(1–2), 145–58. doi:10.1007/s11104-012-1335-z.

Ertani, A., Pizzeghello, D., Francioso, O., Sambo, P., Sanchez-Cortes, S. and Nardi, S. 2014. *Capsicum chinensis* L. growth and nutraceutical properties are enhanced by biostimulants in a long-term period: chemical and metabolomic approaches. *Frontiers in Plant Science* 5, 375. doi:10.3389/fpls.2014.00375.

Gaidau, C., Niculescu, M., Stepan, E., Epure, D. G. and Gidea, M. 2013. New mixes based on collagen extracts with bioactive properties, for treatment of seeds in sustainable agriculture. *Current Pharmaceutical Biotechnology* 14(9), 792–801. doi:10.2174/1389201014666131227112020.

Gajc-Wolska, J., Kowalczyk, K., Nowecka, M., Mazur, K. and Metera, A. 2012. Effect of organic-mineral fertilizers on the yield and quality of Endive (*Cichorium endivia* L.). *Acta Scientiarum Polonorum: Hortorum Cultus* 11, 189–200.

García-Martínez, A. M., Díaz, A., Tejada, M., Bautista, J., Rodríguez, B., Santa María, C., Revilla, E. and Parrado, J. 2010. Enzymatic production of an organic soil biostimulant from wheat condensed distiller solubles: effects on soil biochemistry and biodiversity. *Process Biochemistry* 45(7), 1127–33. doi:10.1016/j.procbio.2010.04.005.

Ghorbani, S., Fernandez, A., Hilson, P. and Beeckman, T. 2014. Signaling peptides in plants. *Cell and Developmental Biology* 3(2), 141. doi:10.4172/2168-9296.1000141.

Grabowska, A., Kunicki, E., Sękara, A., Kalisz, A. and Wojciechowska, R. 2012. The effect of cultivar and biostimulant treatment on the carrot yield and its quality. *Vegetable Crops Research Bulletin* 77(1), 37–48. doi:10.2478/v10032-012-0014-1.

Gurav, R. G. and Jadhav, J. P. 2013. A novel source of biofertilizer from feather biomass for banana cultivation. *Environmental Science and Pollution Research International* 20(7), 4532–9. doi:10.1007/s11356-012-1405-z.

Halpern, M., Bar-Tal, A., Ofek, M., Minz, D., Muller, T. and Yermiyahu, U. 2015. The use of biostimulants for enhancing nutrient uptake. *Advances in Agronomy* 130, 141–74. doi:10.1016/bs.agron.2014.10.001.

Kauffman, G. L., Kneival, D. P. and Watschke, T. L. 2007. Effects of biostimulant on the heat tolerance associated with photosynthetic capacity, membrane thermostability, and polyphenol production of perennial ryegrass. *Crop Science* 47(1), 261–7. doi:10.2135/cropsci2006.03.0171.

Khedr, A. H. A., Abbas, M. A., Wahid, A. A. A., Quick, W. P. and Abogadallah, G. M. 2003. Proline induces the expression of salt-stress-responsive proteins and may improve the adaptation of *Pancratium maritimum* L. to salt-stress. *Journal of Experimental Botany* 54(392), 2553–62. doi:10.1093/jxb/erg277.

Kocira, S. 2019. Effect of amino acid biostimulant on the yield and nutraceutical potential of soybean. *Chilean Journal of Agricultural Research* 79(1), 17–25. doi:10.4067/S0718-58392019000100017.

Kocira, A., Lamorska, J., Kornas, R., Nowosad, N., Tomaszewska, M., Leszczyńska, D., Kozłowicz, K. and Tabor, S. 2020. Changes in biochemistry and yield in response to biostimulants applied in bean (*Phaseolus vulgaris* L.). *Agronomy* 10(2), 189. doi:10.3390/agronomy10020189.

Kunicki, E., Grabowska, A., Sękara, A. and Wojciechowska, R. 2010. The effect of cultivar type, time of cultivation, and biostimulant treatment on the yield of spinach (*Spinacia oleracea* L). *Folia Horticulturae* 22(2), 9–13. doi:10.2478/fhort-2013-0153.

Lisiecka, J., Knaflewski, M., Spizewski, T., Fraszczak, B., Kaluzewicz, A. and Krzesinski, W. 2011. The effect of animal protein hydrolysate on quantity and quality of strawberry daughter plants cv. 'Elsanta'. *Acta Scientiarum Polonorum Horticulture* 10, 31–40.

Liu, X. Q. and Lee, K. S. 2012. Effect of mixed amino acids on crop growth. In: Aflakpui, G. (Ed.), *Agricultural Science*. InTech Europe Publisher, Rijeka, Croatia, pp. 119–58.

Lucini, L., Rouphael, Y., Cardarelli, M., Canguier, R., Kumar, P. and Colla, G. 2015. The effect of a plant-derived biostimulant on metabolic profiling and crop performance of lettuce grown under saline conditions. *Scientia Horticulturae* 182, 124–33. doi:10.1016/j.scienta.2014.11.022.

Marfà, O., Cáceres, R., Polo, J. and Ródenas, J. 2009. Animal protein hydrolysate as a biostimulant for transplanted strawberry plants subjected to cold stress. *Acta Horticulturae* 842(842), 315–8. doi:10.17660/ActaHortic.2009.842.57.

Matsumiya, Y. and Kubo, M. 2011. Soybean peptide: novel plant growth promoting peptide from soybean. In: El-Shemy, H. (Ed.), *Soybean and Nutrition*. InTech Europe Publisher, Rijeka, Croatia, pp. 215–30.

Mehrafarin, A., Qavami, N., Tahmasebi, Z., Naghdi Badi, H., Abdossi, V. and Seifsahandi, M. 2015. Phytochemical and morpho-physiological responses of lemon balm (*Melissa officinalis* l.) to biostimulants application. *Journal of Medicinal Plants* 14, 29–42.

Moll, R. H., Kamprath, E. J. and Jackson, W. A. 1982. Analysis and interpretation of factors which contribute to efficiency of nitrogen utilizatilon. *Agronomy Journal* 74(3), 562–4. doi:10.2134/agronj1982.00021962007400030037x.

Morales-Pajan, J. P. and Stall, W. M. 2003. Papaya (*Carica papaya*) response to foliar treatments with organic complexes of peptides and amino acids. *Proceedings of the Florida State Horticultural Society* 116, 30–2.

Nurdiawati, A., Suherman, C., Maxiselly, Y., Akbar, M. A., Purwoko, B. A., Prawisudha, P. and Yoshikawa, K. 2019. Liquid feather protein hydrolysate as a potential fertilizer to increase growth and yield of patchouli (*Pogostemon cablin* Benth) and mung bean (*Vigna radiata*). *International Journal of Recycling of Organic Waste in Agriculture* 8(3), 221–32. doi:10.1007/s40093-019-0245-y.

Paradikovic, N., Vinkovic, T., Vrcek, I. V., Zuntar, I., Bojic, M. and Medic-Saric, M. 2011. Effect of natural biostimulants on yield and nutritional quality: an example of sweet yellow pepper (*Capsicum annuum* L.) plants. *Journal of the Science of Food and Agriculture* 91(12), 2146–52. doi:10.1002/jsfa.4431.

Parrado, J., Escudero-Gilete, M. L., Friaza, V., Garcia-Martinez, A., González-Miret, M. L., Bautista, J. D. and Heredia, F. J. 2007. Enzymatic vegetable extract with bioactive components: influence of fertilizer on the colour and anthocyanins of red grapes. *Journal of the Science of Food and Agriculture* 87(12), 2310–8. doi:10.1002/jsfa.2989.

Parrado, J., Bautista, J., Romero, E. J., Garcia-Martinez, A. M., Friaza, V. and Tejada, M. 2008. Production of a carob enzymatic extract: potential use as a biofertilizer. *Bioresource Technology* 99(7), 2312–8. doi:10.1016/j.biortech.2007.05.029.

Paul, K., Sorrentino, M., Lucini, L., Rouphael, Y., Cardarelli, M., Bonini, P., Miras Moreno, M. B., Reynaud, H., Canaguier, R., Trtílek, M., Panzarová, K. and Colla, G. 2019. A combined phenotypic and metabolomic approach for elucidating the biostimulant action of a plant-derived protein hydrolysate on tomato grown under limited water availability. *Frontiers in Plant Science* 10, 493. doi:10.3389/fpls.2019.00493.

Petrozza, A., Santaniello, A., Summerer, S., Di Tommaso, G., Di Tommaso, D., Paparelli, E., Piaggesi, A., Perata, P. and Cellini, F. 2014. Physiological responses to Megafol (R) treatments in tomato plants under drought stress: a phenomic and molecular approach. *Scientia Horticulturae* 174, 185–92. doi:10.1016/j.scienta.2014.05.023.

Rouphael, Y., Colla, G., Giordano, M., El-Nakhel, C., Kyriacou, M. C. and De Pascale, S. 2017. Foliar applications of a legume-derived protein hydrolysate elicit dose dependent increases of growth, leaf mineral composition, yield and fruit quality in two greenhouse tomato cultivars. *Scientia Horticulturae* 226, 353–60. doi:10.1016/j.scienta.2017.09.007.

Rouphael, Y., Giordano, M., Cardarelli, M., Cozzolino, E., Mori, M., Kyriacou, M. C., Bonini, P. and Colla, G. 2018a. Plant- and seaweed-based extracts increase yield but differentially modulate nutritional quality of greenhouse spinach through biostimulant action. *Agronomy* 8(7), 126. doi:10.3390/agronomy8070126.

Rouphael, Y. and Kyriacou, M. C. 2018b. Quality and safety of fresh fruits and vegetables at harvest. *Scientia Horticulturae* 239, 78–9. doi:10.1016/j.scienta.2018.05.006.

Schiavon, M., Ertani, A. and Nardi, S. 2008. Effects of an alfaalfa protein hydrolysate on the gene expression and activity of enzymes of TCA cycle and N metabolism in *Zea mays* L. *Journal of Agricultural and Food Chemistry* 56(24), 11800–8. doi:10.1021/jf802362g.

Schiavon, M., Pizzeghello, D., Muscolo, A., Vaccaro, S., Francioso, O. and Nardi, S. 2010. High molecular size humic substances enhance phenylpropanoid metabolism in maize (*Zea mays* L.). *Journal of Chemical Ecology* 36(6), 662–9. doi:10.1007/s10886-010-9790-6.

Schmidke, I. and Stephan, U. W. 1995. Transport of metal micronutrients in the phloem of castor bean (*Ricinus communis*) seedlings. *Physiologia Plantarum* 95, 147–53.

Sestili, F., Rouphael, Y., Cardarelli, M., Pucci, A., Bonini, P., Canaguier, R. and Colla, G. 2018. Protein hydrolysate stimulates growth in tomato coupled with N-dependent gene expression involved in N assimilation. *Frontiers in Plant Science* 9, 1233. doi:10.3389/fpls.2018.01233.

Sharma, R. K. and Kothari, R. M. 1993. Recycled cereal proteins as foliar spray enhances quality and production of food crops. *Resources, Conservation and Recycling* 9(3), 213–21. doi:10.1016/0921-3449(93)90005-Z.

Sharma, S. S. and Dietz, K. J. 2006. The significance of amino acids and amino acid-derived molecules in plant responses and adaptation to heavy metal stress. *Journal of Experimental Botany* 57(4), 711–26. doi:10.1093/jxb/erj073.

Sytar, O., Kumar, A., Latowski, D., Kuczyńska, P., Strzałka, K. and Prasad, M. N. V. 2013. Heavy metal-induced oxidative damage, defense reactions, and detoxification mechanisms in plants. *Acta Physiologiae Plantarum* 35(4), 985–99. doi:10.1007/s11738-012-1169-6.

Szabados, L. and Savouré, A. 2010. Proline: a multifunctional amino acid. *Trends in Plant Science* 15(2), 89–97. doi:10.1016/j.tplants.2009.11.009.

Tavormina, P., De Coninck, B., Nikonorova, N., De Smet, I. and Cammue, B. P. 2015. The plant peptidome: an expanding repertoire of structural features and biological functions. *The Plant Cell* 27(8), 2095–118. doi:10.1105/tpc.15.00440.

Tester, M. and Davenport, R. 2003. Na^+ tolerance and Na^+ transport in higher plants. *Annals of Botany* 91(5), 503–27. doi:10.1093/aob/mcg058.

Visconti, F., de Paz, J. M., Bonet, L., Jordà, M., Quinones, A. and Intrigliolo, D. S. 2015. Effects of a commercial calcium protein hydrolysate on the salt tolerance of *Diospyros kaki* L. cv. "Rojo Brillante" grafted on *Diospyros lotus* L. *Scientia Horticulturae* 185, 129–38. doi:10.1016/j.scienta.2015.01.028.

Chapter 6

Silicon as a biostimulant in agriculture

Wendy Zellner, The University of Toledo, USA; and Lawrence Datnoff, Louisiana State University Agricultural Center, USA

1 Introduction

The fine line between health and disease for any cellular organism depends on their ability to maintain an internal balance as the external environment continually changes. This constant struggle to maintain balance is a well-established process termed homeostasis and is a key factor in how biostimulants work. When certain factors disequilibrate due to drastic changes in the environment, diseases occur. In plants, diseases can be classified as either abiotic or biotic. Abiotic diseases are not caused by biological agents and include certain conditions such as drought, temperature extremes, salt, nutrient imbalances and heavy metal toxicity. Biotic stress includes diseases brought on by other organisms, such as plant pathogenic fungi, bacteria and viruses. As changes to the internal environment in plants occur, a vast number of pathways and processes are altered in order to return to homeostasis.

http://dx.doi.org/10.19103/AS.2020.0068.07

Silicon (Si) is a mineral element that is well known to protect plants against several environmental changes. While the exact mechanism is not known, beneficial responses across a broad range of crops highlight Si's role in maintaining the internal homeostasis, or balance, of plants. All plants accumulate Si to either macronutrient or micronutrient concentrations. This allows Si to work as a biostimulant due to its ability to reduce abiotic stresses, including osmotic and nutritional imbalances, while enhancing root growth and biomass production, especially in an ever-changing environment where this nutrient is limiting, such as soil-less substrates and hydroponic production. Further understanding of Si use in plants is important as phytotoxicity has been reported when concentrations or application of certain Si products are supplied. However, using these fertilizers appropriately may expand Si's benefits outside the context of plants as recent studies are providing new insight into enhanced probiotic growth and responses as well as biofortification of foods for human and livestock consumption. While current efforts are underway to elevate Si to a nutrient in the United States, its role as a biostimulant has been well-documented in the literature for decades. Even though Si also plays a role in enhancing host defense mechanisms against infection by plant pathogens, and is marketed as a plant protectant (biopesticide), this chapter will focus on the function of Si in the context of abiotic stress reduction and growth-promoting factors in plants.

Biostimulants are a diverse group of products that enhance plant growth and vigor via complex or unknown modes of action. For instance, plant extracts and beneficial microbes provide a wide range of enzymes, metabolites, osmolytes, proteins and other material that are absorbed by plants resulting in improved health and growth. Many times, it is the combination of these factors, not just a single peptide or enzyme, that allows for protection, making it difficult to tease apart the mechanism(s) leading to improved growth. Silicon fertilizers and products are unique in that they are not yet considered as a plant nutrient by Association of American Plant Food Control Officials (AAPFCO) standards, yet they provide benefits to plant health and growth by maintaining many of the homeostatic pathways. These responses are most obvious in a challenging environment and in soils or soil-less substrates that are low in soluble Si.

Prior to and following Si's isolation in 1824 by the Swedish chemist, Jöns Jacob Berzelius, the element was studied by many biologist and chemist throughout Europe to determine its effect on agriculture crops. These include studies by de Saussure (1804; Hill, 2012), Davy (1814), Struve (1835), Cruger (1857), Wiesner (1867) and Liebig (1840) (Prychid et al., 2003; Liang et al., 2015a). Elemental Si is extremely abundant and consists of 28% of the earth's crust. Even though Si concentrations in soils are high, the majority is complexed as aluminum- or crystalline silicates (Tubana et al., 2016). These forms are not readily available for biological absorption and use. Silicic acid (H_4SiO_4) is considered as the biologically absorbed form of Si for plants and animals, and is present in soil solutions, water and foods at varying concentrations (Tubana et al., 2016; Frantz et al., 2010; Powell et al., 2005).

2 Silicon availability in fertilizers and growing media

The major form of plant-available Si in the growing medium is H_4SiO_4. The concentration is heavily dependent on different factors within the growing media that include solution pH, minerology of both the liquid and solid portions of the medium, as well as the variety of plant grown. Silicic acid begins to polymerize into polysilicates as the concentration in solution climbs to 2.7 mM Si (75.8 mg Si·L^{-1}, ppm).

Soil solutions typically contain Si between 0.1 and 0.6 mM Si (2.8–17.13 ppm Si) (Knight and Kinrade, 2001). Peat-based substrates tend to contain extremely low concentrations of Si (0.5 mg Si·kg^{-1}) compared to the range of amorphous silica in soils of 1–30 $mg-^{1}$ (Frantz et al., 2010; Tubana et al., 2016). Irrigation water tends to contain Si concentrations ranging from 1 to 5 ppm Si, depending on the source (Frantz et al., 2010). These concentrations are quite dynamic and change throughout a growing season.

How Si is supplied to the growing plant has a significant impact on whether beneficial responses are observed. Si needs to be provided in a form that is either readily available or can be released at a rate needed by the plant. For instance, the release rate of H_4SiO_4 from quartz is extremely slow and this is why sand is not a good source of plant-available Si. Total Si in this instance does not reflect the amount of H_4SiO_4 released for plant uptake.

Currently, Si fertilizers sold in the United States either report total Si or percent soluble Si (formally 'plant-available' Si). The soluble Si is based on the concentration of Si released from the fertilizer following a 5-day extraction in a sodium carbonate-ammonium nitrate solution (Sebastian et al., 2013). The 5-day method was developed as an indicator test to determine if a fertilizer provided soluble Si within a range of 0.2–8.4% Si. While the percentage of Si is reported with this method, the value is not intended to provide a quantitative value in order to compare one Si fertilizer with another. As with many extractions, the 5-day method works well with many fertilizer compositions containing Si but can also have material bias (Buck et al., 2010; Pereira et al., 2007; Sebastian et al., 2013; Zellner et al., 2015). Most notably, particle size has a large influence on Si release. When comparing grow-out studies using *Poa trivialis* 'Darkhorse', there was a marginal correlation of (R^2 = 0.6) between the percentage of Si calculated from plant uptake and the 5-day method (Fig. 1a). The 5-day method requires the processing of the sample to pass through a sieve smaller than 300 μm. However, in the grow-out trials, the original starting size of the sample was used. The particle size bias is clearly shown in slag A with particle sizes

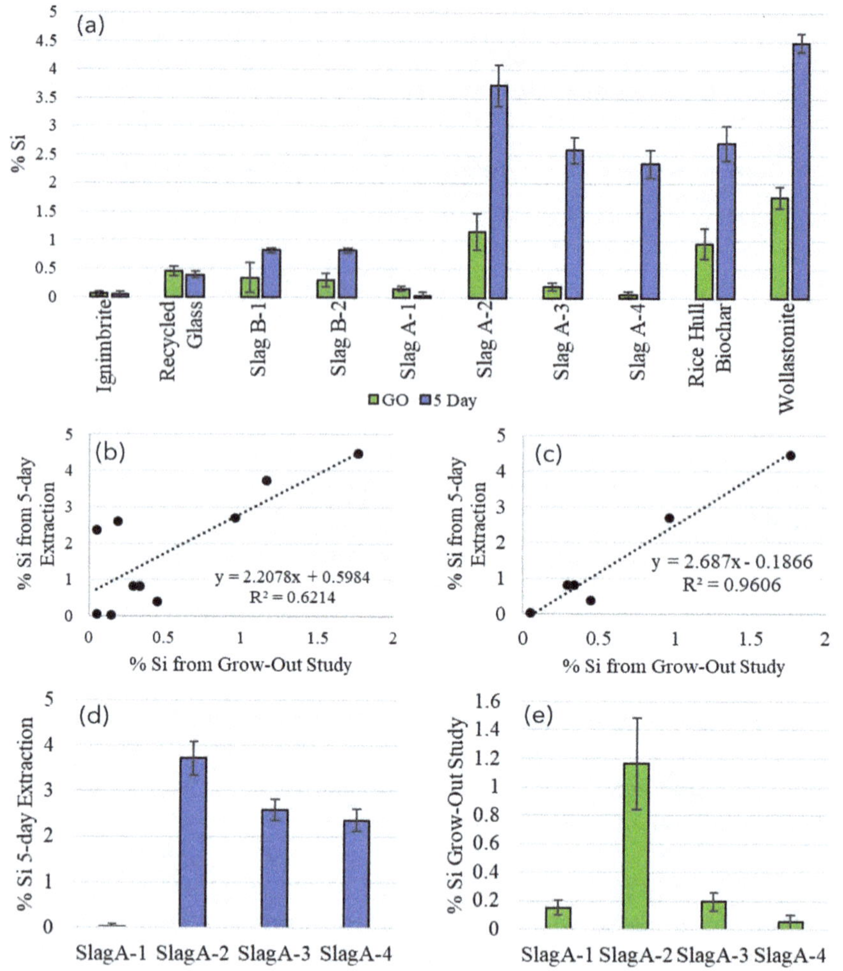

Figure 1 Comparing calculated % Si from grow-out trials in *Poa trivialis* 'Darkhorse' to 5-day extraction method. Different fertilizers were tested for their ability to release H_4SiO_4 (A). Slag B-1 and -2 are the same material used in the grow-out trial at half and equal rates as other fertilizers, respectively. Slag A-2:6 are the same material, processed to different mesh sizes of <300 μm, 200/300 SGN and 280 SGN particle sizes, respectively. Slag A-1 is a non-Si containing slag. Correlation of the % Si from the grow-out and 5-day method increased with the exclusion of slagB-3:6 (compare B to C) due to Si release as measured by plant-uptake significantly decreased with increasing mesh size (E), while the 5-day method reported similar % Si in both (D).

ranging from a powder (slag A-1 and A-2) to a 280 SGN particle size (slag A-4). The 5-day extraction shows similar percentage of Si for slag A-3 and A-4, which are half the concentration lower than slag A-2 (Fig. 1d). More interestingly, plant uptake was drastically lower with the larger particle size slag showing the same Si uptake as the non-Si containing slag A-1 (Fig. 1e). When these data were removed from the analysis, the correlation increased drastically ($R^2 = 0.96$) (Fig. 1c). It is important to note that the concentration had no effect on the Si accumulation as a different calcium-silicate slag B applied at half and equal rates as the other fertilizers had similar calculated percentage of Si from the grow-out trials. The 5-day method provides a good starting point for identifying Si fertilizers, but further work is needed to provide more reliable, quantitative predictions.

There are a number of commercially available liquid and solid forms of Si materials on the market. Of the liquids, there are potassium, sodium and calcium silicates, formulations containing a combination of mono- and poly-silicic acid as well as humic and fulvic acids. The concentration of total Si in liquids tends to represent the amount of soluble Si to the plants, and these products tend to be highly concentrated, have high pHs and require dilution prior to application and are not always compatible for mixing with other liquid fertilizers. Other liquid products contain Si at lower doses in combination with other nutrients and/or plant extracts such as seaweed. Liquid products can be applied as a drench, included in hydroponic growth solutions, or applied as foliar sprays depending on the cropping system and intended outcome of the product.

Solid Si fertilizers are diverse in their mineralogy and are derived from a wide range of sources, including mined rocks and sediment, recycled materials and plant by-products. Of the mined rocks, wollastonite is the most used for its Si content with 6% plant-soluble Si based on the 5-day extraction method. This mineral is used in agriculture applications as well as in the textile, food additive, ceramic and other industries. Wollastonite is a metamorphic rock that has a calcium silicate (Ca_2SiO_4) composition. Another mined material is diatomaceous earth mined from sedimentary rocks. These deposits contain fossilized diatoms and can vary significantly in Si composition depending on the species of the diatoms present due to differences in physical and morphological characteristics, independent of geological or geographical origin (Arnaud et al., 2005; Korunic, 1998). There are other forms of mined silica containing rocks that do not directly provide Si to growing plants. Ignimbrite is an igneous rock that contains amorphous silica deposits, however, in a soil-less system, such as peat, the material does not provide soluble Si to the plant, nor does the material release a significant amount of H_4SiO_4 during the 5-day extraction (Fig. 1a).

Recycled materials are also dynamic in their composition and Si-availability, based largely on their starting material. During the steel-making process, Si is sometimes added as a fluxing agent to remove impurities and is discarded as slag. These by-products vary greatly depending on the type of steel processing and differences in starting material, additives and temperature treatments

(Klimczyk et al., 2012). Silicon in these slags has a Ca_2SiO_4 mineralogy in addition to other trace nutrients (Scott et al., 1986). Another product used in the hydroponics industry is recycled glass, an amorphous silicate material that has the ability to release Si in a recycling nutrient solution, like those used in nutrient film technology (NFT) or deep-water culture cultivations. Plant material may also provide soluble Si as the silica deposits within the tissue can release H_4SiO_4. Rice hulls contain high concentrations of Si and provide a significant amount of Si to the growing media (Frantz et al., 2010). However, like other fertilizers, Si composition and concentration are dependent on the plant type and tissue as well as the environment in which the plant was grown. Furthermore, not all phytoliths release Si at the same rate. For example, pine needles have a high concentration of Si, but the combination of phenolic compounds inhibits its release (Bartoli and Wilding, 1980; Fraysse et al., 2006). Other materials like miscanthus and wheat straw also provide soluble Si but in varying degrees (Frantz et al., 2010). Seaweed extract may also release soluble Si to the growing media as the plants have high concentrations of the element, however, little work has been tested on this material's release rate.

3 Plant accumulation, transport and deposition of silicon

Classically, plants have been placed into categories as high- intermediate- or low- and non-silicon accumulators based on shoot concentrations of Si and the Si:Ca ratio. Strigel (1912) was one of the first scientists to observe that Si and Ca were the most variable in foliar concentrations between monocotyeldons and dicotyledons. Taking this further, Japanese researchers developed criteria based on a 0.5% cut off, as plants that passively take up 10 ppm Si from a solution with a transpiration coefficient of 500 would equal this value (Ma and Takashashi, 2002). With these two criteria, a non-accumulator would have a value below 0.5, a low accumulator in the range of 0.5-2, an intermediate accumulator in the range of 2-4 and a high-accumulator above 4 for both their %Si concentration and Si:Ca ratio, respectively (Takahashi et al., 1990). It is important to note that a high-accumulator in this context is considered a plant that accumulates an extraordinary large amount of Si from its environment, often referred to as a hyperaccumulator when referencing other nutrients (Takahashi et al., 1990). No other nutrient is classified using ratios or based on the accumulation in hyper accumulator species, and this nomenclature should be adjusted to reflect our current understanding of Si absorption and accumulation in plants, independent of other nutrients.

Plants absorb nutrients at varying rates and the tissue concentration determines whether the nutrient is considered a macronutrient or micronutrient. Concentrations of macronutrients are present in foliar tissue at or above 1000 ppm (or 0.1% dry weight). While micronutrient concentrations are usually below 500 ppm (0.05%), with the exception of iron and manganese in some plant species. Epstein and Bloom (2005) have Si listed as a macronutrient along with sulfur at 0.1% concentration in dry weight. Plants that accumulate Si at 0.1% or greater should be considered as macro-Si accumulators, while others should be viewed as micro-accumulators. The majority of plants tested accumulate Si in the macronutrient range, including over 70% of dicots (Table 1). When determining whether a plant accumulates Si, past studies solely focused on measuring Si concentration foliarly and not in other plant organs such as roots (Hodson et al., 2005; Takahashi et al., 1990; Lewin and Reimann, 1969). However, clover, coffee, pepper and tomatoes may concentrate as much, if not more, Si in the roots compared to leaves (Carre-Missio et al., 2009; French-Monar et al., 2010; Huang et al., 2011). Nevertheless, all plants absorb measurable concentrations of Si from the environment and bioaccumulate the nutrient in their tissue.

In general, angiosperms accumulate Si in the range of 0.005–11% Si (Hodson et al., 2005). Except for Poaceae, the majority of angiosperm accumulate Si between 0.5–1%, with a higher frequency of macro-accumulators in monocots and lower angiosperm (Table 2). A very narrow range of monocots within the Poaceae family accumulate extremely high concentrations of Si in their leaves with a range of 0.1–11% Si. These include rice, wheat and to a lesser extent, corn. Outside of this family, other monocots accumulate Si in the range of 0.01–7% Si. Higher ranges of Si accumulation (greater than 30 000 ppm) seem to be the exception, not the rule for monocots. Si accumulation in dicots ranges from 0 to 11% with over 40 species reported as macro-Si accumulators (Fig. 2). This includes European beech that were reported to accumulate greater than 7% Si. Trees within Rosales, including American elm, Fiddle leaf fig, Red mulberry and slippery elm as well as sugar maples belonging to Sapindales, were reported to contain greater than 1% Si in their leaves. Other dicot, macro-Si accumulators are cucumber, garden mums, certain sunflower species, vinca, verbena and zinnia. Many of the remaining monocots and dicots have Si values exceeding micronutrient concentrations, and plants that have foliar concentrations at or above 0.1% Si should be considered macro-Si accumulators. Even plants that maintain foliar concentrations of Si below 500 ppm still are within the range of zinc concentrations. This shows that plants absorb variable levels of Si within ranges of macronutrients and micronutrients.

Variability in Si accumulation is seen between genotypes as well. There are multiple studies in rice that show differences between Japonica and Indica varieties and they have used these differences to determine QTLs responsible for these higher Si concentrations (Deren et al., 1992; Ma and Takashashi, 2002; Winslow et al., 1997; Winslow, 1992; Wu et al., 2006). More recently, this variability among varieties within the same species was observed in Petunia and tomato (Boldt and Altland, 2017; Zellner, unpublished data). Tomato varieties

Table 1 Macro Silicon Accumulators. Species in bold had more than one observation and the average % Si is reported

Order	Family	Genus	Species	%Si
Monocots				
Alismatales	Juncaginaceae	*Triglochin*	*maritima*	1.12
Arecales	Arecaceae	*Sabal*	*etonia*	1.47
		Serenoa	*repens*	1.72
		Caryota	*mitis*	1.29
		Washingtonia	*filifera*	1.43
		Sabal	*minor*	0.99
		Serenoa	*repens*	2.45
		Trachycarpus	*fortunei*	1.41
Poales	Cyperaceae	*Carex*	*atherodes*	1.27
		Carex	*biwensis*	1.22
		Carex	*cinica*	2.54
		Carex	*dispalata*	2.41
		Carex	***filifolia***	**1.43**
		Carex	*flavicans*	1.03
		Carex	*parciflora*	1.73
		Carex	*sempeirnens*	1.08
		Carex	*subspathacea*	1.26
		Carex	*thunbergii*	1.67
		Cladium	*jamaicense*	1.10
		Cyperus	*alternifolius*	3.52
		Cyperus	*papyrus*	1.75
		Cyperus	*surinamensis*	1.10
		Eleocheris	*cełlulosa*	1.25
		Eleocharis	*parvula*	2.87
		Eleocharis	*uniglumis*	7.00
		Fimbristylis	*spadicea*	1.53
		Rynchospora	*płumosa*	1.80
		Scirpus	*americanus*	1.33
		Scirpus	*cyperinus*	1.55
		Scirpus	*olneyi*	1.19
		Scirpus	*robustus*	2.44
		Scirpus	*validus*	1.11
	Juncaceae	*Juncus*	*gerardii*	1.40
Zingiberales	Musaceae	*Musa*	*basjoo*	1.08
Dicots				
Apiales	Apiaceae	*Ferula*	*varia*	1.33
		Lilaeopsis	*chinensis*	1.93
Asterales	Asteraceae	*Achillea*	*millefolium*	1.96

Order	Family	Genus	Species	%Si
Asterales	Asteraceae	*Aster*	*macrophyllus*	1.22
		Chrysanthemum	*morifolium*	1.04[a]
		Helianthus	*maximilianii*	2.34
		Helianthus	*tuberosus*	2.24
		Inula	*helenium*	1.03
		Zinnia	*elegans*	1.27[a]
Brassicales	Brassicaceae	*Nasturtium*	*officinale*	1.15
Caryophyllales	Amaranthaceae	*Amaranthus*	*spp.*	2.34
	Polygonaceae	*Polygonum*	*aviculare*	1.97
		Polygonum	*punctatum*	1.00
Cucurbitales	Cucurbitaceae	*Cucumis*	*sativus*	1.02[a]
Ericales	Ericaceae	*Calluna*	*vulgaris*	2.50
		Vaccinium	*myrtillus*	1.00
	Polemoniaceae	*Phlox*	*subulata*	1.11
	Primulaceae	*Glaux*	*maritima*	1.40
Fabales	Fabaceae	*Cajanus*	*cajan*	1.31
		Glycine	*max*	1.38
		Trifolium	*subterraneum*	1.45
Fagales	**Fagaceae**	***Fagus***	***sylvatica***	**9.25**
Lamiales	Lamiaceae	*Betonica*	*foliosa*	1.14
		Mentha	*piperita*	1.54
		Melissa	*officinalis*	1.61
		Thymus	*marschallianus*	2.21
	Verbenaceae	*Lantana*	L.	1.91[b]
		Verbena	L.	2.04[b]
Ranunculales	Ranunculaceae	*Eranthis*	*giganteus*	1.66
Rosales	**Celtidaceae**	***Celtis***	***occidentalis***	2.86
	Elaeagnaceae	*Hippophae*	*rhamnoides*	2.34
	Moraceae	*Ficus*	*lyrata*	1.42
		Morus	***rubra***	**1.61**
	Rosaceae	*Aronia*	*melanocarpa*	1.44
		Potentilla	*erecta*	1.17
	Ulmaceae	*Ulmus*	*rubra*	1.00
		Ulmus	***americana***	**1.96**
	Urticaceae	***Urtica***	***dioica***	**1.76**
Sapindales	Sapindaceae	*Acer*	*saccharum*	1.23
Saxifragales	Crassulaceae	*Rhodiola*	*linearifolia*	2.94
		Sedum	*hybridum*	3.59

Data are adapted from Hodson et al. (2005) as well as Frantz et al (2008)[a] and unpublished data[b].

Table 2 NPA spacing in different plant species

Order	Family	Species	NPA/NPV	Spacing
Amborellales	Amborellaceae	*Amborella trichopoda*		108
Arecales	Arecaceae	*Elaeis guineensis*		108
		Phoenix dactylifera		108
Asparagales	Asparagaceae Orchidaceae	*Asparagus officinalis*	NPS/NPA	108
		Apostasia Shenzhenica		110/114
		Dendrobium catenatum		110
		Plwlaenopsis equestris		110
Poales	Bromeliaceae	*Ananas comosus*		108/110
	Poaceae	*Aegilops tauschiisubsp. Tauschii*		108
		Brachypodium distachyon		108
		Dichanthelium oligosanthes		108
		Hordeum vulgare		108
		Oryza brachyantha		108
		Oṇza sativa Japonica Croup		108
		Panicum hallii		108
		Panicum miliaceum		108
		Setaria italica		108
		Sorghum bicolor		108
		Triticum urartu		108
		Zea mays		108
	Pooideae	*Lolium perenne*		108
Zinġiberales	Musaceae	*Musa acuminata subsp. Malaccensis*		108
Dicots				
Apiales	Apiaceae	*Daucus carota subsp. Sativus*		108
Asterales	Asteraceae	Artemisia annua		108
		Cynara cardunculus var. scoJymus		108
		Helianthus annus		108
		Lactuca sativa		115
Brassicales	Caricaceae	*Brachypodium distachyon*		108
	Clemaceae	*Tarenaya hassleriana*		108
Caryophyllales	Amaranthaceae	*Chenopodmm quinoa*	NPA/NPV	108
		Spinacia oleracea	NPA/NPT	108
Cucurbitales	Cucurbitaceae	*Cucumis melo*	NPA/NPV	108
		Cucumis melo		108
		Cucumis sativus	NPA/NPV	108
		Cucumis sativus		108

Order	Family	Species	NPA/NPV	Spacing
		Cucurbita flcifolia		108
		Cucurbita maxima	NPA/NPV	108
		Cucurbita maxima		108
		Cucurbita moschata	NPA/NPV	108
		Cucurbita moschata		108
		Cucurbita pepo subsp. Pepo	NPA/NPV	108
		Cucurbita pepo subsp. pepo		108
		Momordica charantia	NPA/NPV	108
		Momordica charantia		108
Ericales	Actiridiaceae	*Actinidia chinensis var chinensis*		108
Fabales	Fabaceae	*Abrus precatorius*		108
		Arachis duranensis		108
		Arachis hypogaea		108
		Archis ipaensis		108
		Cajanus cajan		108
		Cicer arietinum		108
		Glycine max		108
		Glycine soja		108/**109**
		Lupinus angustifolius		108
		Medicago truncatula		108
		Mucuna pruriens		108
		Phaseolus vulgaris		108
		Trifolium repens		108
Fabales	Fabaceae	*Vigna angularis*		108
		Vigna radiata var. radiata		108
Fagales	Fagaceae	*Quercus suber*		108
Gentianales	Rubiaceae	*Coffea Arabica*		108
		Coffea canephora		108
		Coffea eugenioides		108
Lamiales	Bignoniaceae	*Handroanthus impetiginosus*		108
	Gesneriaceae	*Dorcoceras hygrometricum*		108
	Pedaliaceae	*Sesamum indicum*		108/**109**
	Phrymaceae	*Erythranthe guttata*		108
Malpighiales	Euphorbuaceae	*Hevea brasiliensis*		108
		Jatropha curcas		108
		Manihot esculenta		108
		Ricinus communis		108

(*Continued*)

Table 2 (*Continued*)

Order	Family	Species	NPA/NPV	Spacing
	Salicaceae	*Popuhis euphratica*		108
		Populus trichocarpa		108
		Salix purpurea		108
Malvales	Malvaceae	*Durio zibethinus*		108
		Durio zibethinus	NPG/NPA	108
		Gossypium arboreum		108
		Gossypium hirsutum		108
		Gossypium hirsutum	NPA/NIR	106
		Gossypium raimondii		108
		Gossypium raimondii	NPA/NIR	106/107
		Herrania umbratica		108
		Theobroma cacao		108
		Theobroma cacao		108
Myrtales	Myrtaceae	*Eucalyptus grandis*	NPA/NRT	**109**
Proteales	Nelumbonaceae	*Nelumbo nucifera*		108
Ranunculaes	Papaveraceae	*Papaver somniferum*		108
	Papaveraceae	*Papaver somniferum*	NPA/NPV	108
	Ranunculaceae	*Aquilegia coerulea*		108
	Papaveraceae	*Macleaya cordata*	NPA/NPV	108
Rosales	Cannabaceae	*Parasponia andersonii*		108
		Parasponia andersonii	NPA/NPV	108
		Parasponia andersonii		108
		Trema orientate		108
		Trema orientate	NPA/NPV	108
	Moraceae	*Morus notabilis*		108
		Morus notabilis	NPA/NPV	108
	Rosaceae	*Fragaria vesca subsp. Vesca*		108
		Malus domestica		108
		Prunus avium		108
		Prunus cerasifera		108
		Prunus persica		108
		Punica granatum		108
	Rhamnaceae	*Pyrus X bretschneideri*		108
		Rosa chinensis		108/110
		Ziziphus jujiiba		108
Solanales	Convolvulaceae	*Ipomoea nil*		**109**
	Solanaceae	*Nicotiana attenuata*		108

Order	Family	Species	NPA/NPV	Spacing
		Nicotiana tabacum		108
		Nicotiana sylvestris		108
		Solanum lycopersicum		**109**
		Solanum pennelii		**109**
		Solanum tuberosum		**109**
Sapindales	Rutaceae	*Citrus Clementina*		108
		Citrus sinensis		108
		Citrus unshiu		108
Vitales	Vitaceae	*Vitis vinifera*		108

Differences in NPA residues are listed. Amino acid lengths of 109 are reported to interfere with Si transport (Deshmukh et al., 2015).

were identified that contained extraordinarily high concentrations of Si in their root tissue under 2 mM K_2SiO_3 treatment (up to 4%). These varieties also had higher foliar Si (Fig. 3). Genotypic differences were observed in root tissue under both high and Si-deprived conditions but not in leaves under low Si. This is likely due to many of the plants having non-detectable levels of Si in their leaf tissue. The genotypic differences observed in rice, petunia and tomato suggests that there are likely varying degrees of Si accumulation in different varieties and cultivars within other species.

3.1 Si transport via passive and active transport systems

Plants are able to vary their Si absorption and movement similar to other nutrients in that they all contain transporters that are differentially regulated and modulated based on the needs and developmental growth stages. Nutrient absorption from the media occurs due to the presence of transporters within the plasma membrane. The concentration and subcellular location of these proteins determine how much and where various solutes will be shipped within the cells and throughout various plant organs.

Two types of Si transporters were identified from rice mutants that had the ability to grow in the presence of germanium (Ma et al., 2006, 2007). They were named Lsi1 and Lsi2 for 'Low Si' in rice. Lsi1 was identified as a passive transporter, belonging to the Nodulin-26 like intrinsic protein (NIP), which is a subgroup of the larger aquaporin family of transporters in plants. Functional Lsi1 (NIP2) Si transporters have been characterized in both monocots and dicots. These plants include corn, crookneck squash, cucumber, rice, soybean, tobacco and wheat (Mitani et al., 2009a,b, 2011; Sun et al., 2017; Ma et al., 2006; Deshmukh et al., 2013; Zellner et al., 2019; Montpeptit et al., 2012).

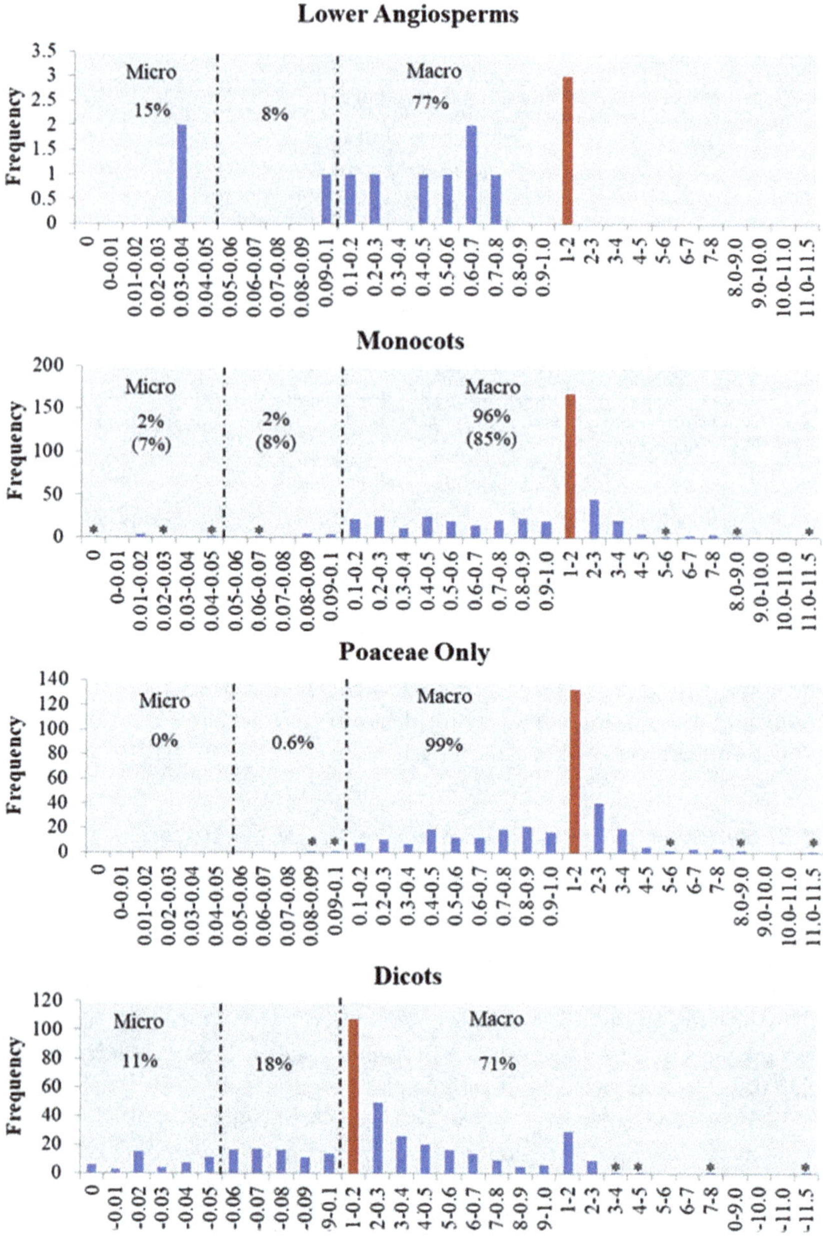

Figure 2 % Si frequency distribution in angiosperm (adapted from Hodson et al., 2005). Lower angiosperm, monocots and dicots show Si distribution between 0.01and 11% Si. Micronutrient and macronutrient concentrations are indicated with the percentage of plants observed in that range. Percentages in parenthesis are from monocots other than Poaceae. Note that the highest % Si in plants across all three categories falls into the macronutrient range. The red bar indicates the highest frequency for % Si within the group. Asterisks denote two or fewer plants within the specified range.

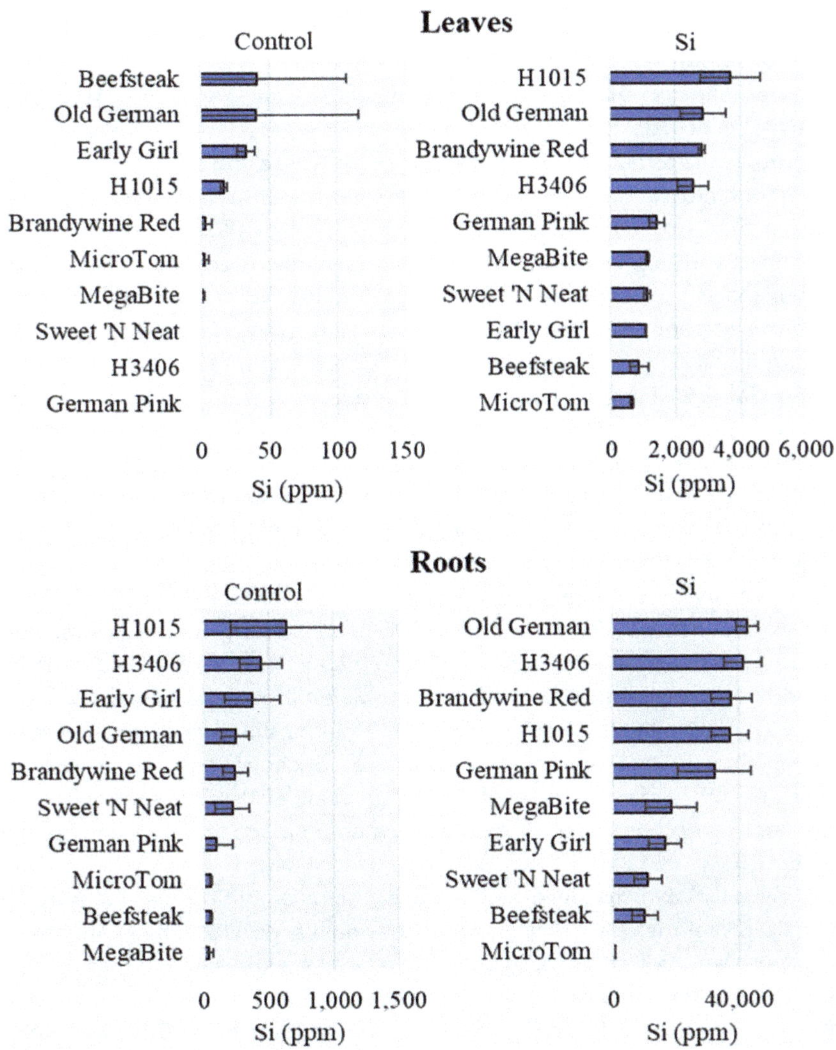

Figure 3 Si concentration (mg Si·kg^{-1} dry tissue) in leaves and roots of hydroponically grown tomatoes supplemented with 0 (control) or 2 mM K_2SiO_3 (Si). Standard deviation is represented by the error bars.

Putative NIP2 proteins have been identified from genomic data in strawberry, however, the protein has not been tested for its transport capabilities (Ouellette et al., 2017). NIP proteins have unique selectivity filter residues of G-S-G-R that vary from other characterized NIPs. As Lsi1 homologs were discovered in other plant species, they were denoted as NIP2 or Lsi1-like proteins. For the remainder of this article, we will refer to Lsi1 and Lsi1-like proteins as NIP2. These, like other aquaporins, are permeable to a range of solutes, such as

OsNIP2;1 permeability to arsenite (Ma et al., 2008), methylated arsenic (Li et al., 2009) and selenite (Zhao et al., 2010).

In addition to the selectivity filter residues, spacing of the asparagine-proline-alanine (NPA) motif at 108 aa residues is needed for proper function of the pore, as the tomato NIP2;1 has a spacing of 109 aa and does not allow for Si movement into *Xenopus* oocytes (Deshmukh et al., 2015). Interestingly, potato also has this additional aa, however the distant relative tobacco NtNIP2;1 has the correct 108 aa spacing and is a functional Si transporter (Zellner et al., 2019). Further analysis of the deposited transcripts within NCBI showed only five other species that contained the 109 aa NPA spacing, while six species contained the predicted NPA spacing higher than 109 aa and one species lower than 108 aa (Table 2). Further studies are needed to determine the functionality of these annotated genes. Another aa change from proline to leucine in the last transmembrane domain causes the crookneck squash CmNIP2;1 transporter to be non-functional, even with the correct 108 aa NPA spacing (Mitani et al., 2011). These data lay caution when determining the functionality of a transporter solely on sequencing data.

Interestingly, petunia and members of the brassica family, including *Arabidopsis thaliana*, do not contain a NIP gene with the GSGR filter residues in their genomic profiles, yet they still have the ability to absorb and transport Si and have shown beneficial responses to Si fertilization as will be discussed below. Additionally, studies using Os*NIP2;1* mutant still have detectable levels of Si in their foliar tissue, even though the concentrations are significantly lower than the wild type (Ma et al., 2006; Lin et al., 2019). This strengthens the argument that other transport mechanisms are in place for Si movement into plants.

Unlike Lsi1, which is involved in passive diffusion of Si across membranes, the Lsi2 protein is one component of the multi-subunit Arsenite-Antimonite Efflux (Ars) family of active transporters. In prokaryotes, this operon has multiple genes that contain varying combinations of arsA, B, C, D and R (Cervantes et al., 1994; Fekih et al., 2018). As (III) efflux is carried out by the arsB/arsA complex, ArsB is a membrane-spanning active transporter that uses energy from ATP hydrolysis via arsA-ATPase activity. Studies have shown that arsB can also actively efflux substrate in the absence of arsA, using a proton-motive force created by pH and/or electrochemical membrane gradients (Dey and Rosen, 1995; Yang et al., 2012). Homologs to arsB and arsA are present within a variety of plant species.

To date, arsB has been characterized from barley, crookneck squash, corn and rice (Chiba et al., 2009; Mitani et al., 2009a, 2011; Yamaji and Ma, 2007). In plants, only the pore portion of this complex independent of arsA-ATPase, has been studied. The presence of an arsA homolog in plants suggests that the arsB/arsA complex is likely a conserved transport system and that arsA is the energy component responsible for efficient efflux of substrates through the arsB transporter. To date, no literature is present identifying or describing how these proteins interact in order to allow for efflux of substrate from the cell. As observed for other arsB-containing cells, substrate efflux of As (III) is possibly in the absence of arsA; however, efficiency of efflux was significantly enhanced in cells containing both arsB and arsA (Dey and Rosen, 1995; Saltikov and Olson, 2002). Data are needed to determine whether or not this arsA/B combination enhances Si efflux in plants.

Cellular location of NIP2 and arsB suggests preferential Si transport via symplastic movement in roots. Root hairs possess NIP2 proteins, suggesting the passive transporter is involved in Si absorption from the growing medium. These proteins are found on the distal side of seminal root cells in barley, corn and rice (Chiba et al., 2009; Mitani et al., 2009a; Yamaji and Ma, 2007). Lateral roots of corn show NIP2;1 located throughout the cell layers, while still more heavily located near the distal side. CmLsi1 in roots seems to have less organized distal location of NIP2;1 and the protein is present throughout the root cells (Mitani et al., 2011). ArsB tends to be localized around barrier cells, such as the exodermis and endodermis, restricting apoplastic Si transport and forcing the element to move intracellularly (Ma et al., 2007; Mitani et al., 2009b). Casparian bands are hydrophobic cell-wall layers that act as diffusion barriers preventing the apoplastic flow, reducing the loss of solutes and water back out of the root into the rhizosphere. Si induces Casparian band formation in the exodermis of both macro-Si accumulators, rice and corn, as well as a micro-Si accumulator, onion (Fleck et al., 2015). Cellular root architecture is dependent on plant species, root age and external stimuli present. Specifically, in rice, the passive and active transporters are arranged on opposite sides of root cells, while in corn and barley, this arrangement was not observed (Mitani et al., 2009b). Once Si passes the endodermis, it is loaded into the xylem where it can be transported to growing roots and shoots.

Other transporters are likely involved in the movement of Si into roots from the growing media, as studies using the rice lsi1 mutant show improvement in growth and increased Si deposition following Si treatment (Isa et al., 2010). Recent use of nanosilica particles is discussed below and plants may absorb these larger oligomers via endocytoses (Chang et al., 2013; Tripathi et al., 2017).

3.2 Si location and deposition in plant tissue: beyond biosilicification

Plant species vary in their use of Si in foliar tissue. Many plants that accumulate foliar Si concentrations above 0.5% do so because they possess biosilicification mechanisms that require Si. For instance, grasses like rice and wheat, have specialized silica cells that deposit Si into phytoliths, which are important for structural support (Currie and Perry, 2007; Kumar et al., 2017). Grasses also accumulate amorphous Si in their hulls, protecting the developing seeds. Other plants such as zinnia, verbena and New Guinea impatiens deposit Si at the base of trichomes and onto scales above the epidermal tissue (Frantz et al., 2005, 2008). It has also been shown that Si concentrations on epidermal surfaces of *Arabidopsis thaliana* appear only in the presence of powdery mildew infection and Si treatment (Fauteux et al., 2006). This suggests plants have different strategies for Si deposition and/or usage and can help explain why foliar concentrations vary among different plant species. Certain plants, like tobacco and onion, contain micronutrient concentrations of Si in foliar tissue. The deposition of Si above the epidermal surface or within scales and trichomes has not been demonstrated in these species and it would be important to understand if these micro-accumulators undergo any type of biosilicification.

While Si does move through the xylem, deposition does not just occur at the ends of transpiration streams, as once thought. Si can also be absorbed by cells and measurable concentrations were observed in different plant species that vary in their foliar concentrations. Si was shown to be present in the hypodermal layer of palm leaves as well as their vacuoles (Lins et al., 2002). Tobacco also was shown to have Si within their vacuoles (Neumann and De Figueiredo, 2002). Si is deposited at the base of different types of trichomes (Dayanandan and Kaufman, 1976). The function of Si at this location has yet to be determined, however, Si supplementation has shown responses in composition of trichome metabolites. Incorporation of $MgCa_2SiO_4$ to a Brazilian 'clayey' Oxisol soil at 400 $kg{\cdot}ha^{-1}$ significantly increased *Artemisia annua* glandular trichome size and the concentration of extracted artemisinin from the leaves (Rostkowska et al., 2016). However, at higher and lower doses of Si, the artemisinin concentration was not affected. Cannabis accumulates Si to appreciable levels in leaves. As cannabis has been declassified by the recent Farm Bill, studying the role of Si in growth and cannabinoid production in these plants should increase.

Root uptake of H_4SiO_4 is not the only proposed uptake mechanism and a number of beneficial effects of Si via foliar application have been reported (Laane, 2018). The effects of foliar applications could be due to runoff of Si from leaves and subsequent root uptake as described above. Additional evidence from other nutritional studies suggests that small neutral compounds, such as H_4SiO_4, have the ability to migrate through the cuticle matrix (Franke, 1967; Fernandez and Brown, 2013; Fernandez et al., 2013). This migration might occur prevalently around cracks and crevasses adjacent to trichomes and other foliar structures that could give Si access to the epidermal layer. H_4SiO_4 may also gain entry through hydathodes by dissolving into xylem sap during guttation events. Once at the epidermal plasma membrane, foliar-expressed Si transporters, such as Lsi6 in rice (Yamaji et al., 2008), could provide a pore for

H_4SiO_4 movement into these cells. This mechanism likely would be less efficient than root-applied uptake and could explain why extremely high levels of H_4SiO_4 (up to 1500 ppm) are needed for beneficial effects to be observed. This form of treatment may be advantageous for certain cropping systems, such as orchards or established groves of trees, where immediate action is needed to protect plants from an invading biotic agent (such as fungi or insects) or adverse environmental condition (drought or short frost). It should also be noted that with foliar sprays, application is suggested for every seven days. This demonstrates that Si is not efficiently transported via the phloem and a continual supply is needed to provide useful benefits.

In addition to Si movement through the epidermal matrix, root uptake may be triggered when the leaves are sprayed. Hydroponically grown *N. tabacum* leaves sprayed with distilled water had increased Si accumulation in leaves compared to non-sprayed controls (unpublished data). This suggests that the application of liquid to leaves can be sensed (through either mechano/osmo/or moisture sensors) by the plant and produce a signal-stimulating movement of Si into leaves. Thus, foliar-applied Si likely provides benefits by several different mechanisms: by creating a superficial Si barrier on the exterior of the cuticle, Si migration through the cuticle matrix or hydathodes toward the epidermal plasma membrane with subsequent uptake, or the application itself inducing Si uptake into the shoots from the roots. Data is lacking to provide a clear picture of the mechanism(s) underlying this type of treatment.

While plants vary in their foliar Si concentrations and deposition mechanisms, the ability of Si to protect plants against a number of environmental stressors seems to be conserved across a wide range of species. One of the most documented is the ability of Si to protect against fungal infection. As this chapter focuses on the role of Si as a biostimulant and not a plant protectant, these defense mechanisms will not be discussed herein as they are well-reviewed in the literature (Pozza et al., 2015; Rodrigues and Datnoff, 2015; Wang et al., 2017). The focus will be on the role of Si in protecting plants against abiotic stress, including osmotic stress from drought and salinity, heavy metal toxicity and nutrient imbalances.

4 Silicon and plant abiotic defense: drought and salinity

Plants experience external disequilibrium daily due to fluctuations in humidity, temperature, nutrient availability, as well as oxygen and carbon dioxide concentrations. All these situations alter the internal balance within plants leading to induction of short- and long-term strategies to return to or maintain homeostasis. Changes occur rapidly within the plants to adjust for these variations and Si is involved in reducing the stress caused by extreme fluctuations in environmental conditions.

Different external stimuli trigger similar defense pathways or responses as these external variations expose the cellular environment to the same type of stress. For example, both drought and salinity result in cells experiencing higher solute concentrations, even though the two causes are very different. This is why many studies with Si and drought, or salinity stress show similar increases in cellular osmolytes, changes in transpiration and altered root growth in order to adjust to solute imbalances. In addition, malondialdehyde (MDA) is an end product of lipid peroxidation and is used to quantify the amount of cellular damage due to reactive oxygen species (ROS) induced under a wide range of stress (Ayala et al., 2014).

Enzymes are additional players in abiotic stress responses, as they reduce the reactive oxygen species generated when cellular distribution of solutes is disrupted. Enzyme activity following stress is dependent on the type, duration and intensity of the stress, as well as the plant species. Key enzymes involved in ROS scavenging are SOD, CAT, POD and APX. SOD detoxifies superoxide into hydrogen peroxide that can be detoxified by CAT into water and oxygen (Das and Roychoudhury, 2014). Other enzymes and non-enzymatic proteins also detoxify different free radicals.

An excellent review by Manivannan and Ahn (2017) provides an insightful interpretation of transcriptional changes induced by Si treatment across a wide range of abiotic and biotic stress. Their models depict modulation of photosynthetic pathways, transcription factors and their targeted genes, AQP changes and polyamine biosynthesis. Even though the majority of transcriptional changes were described from high-Si accumulators, the two macro-Si accumulators, tomato and *Arabidopsis* have a non-functional and non-classical NIP2 AQP, respectively, yet still show beneficial responses and genetic changes with Si treatment. This suggests that they too possess Si-sensing mechanisms, allowing for absorption and use of Si to modulate stress-tolerant pathways.

The main cellular response to osmotic stress caused by salinity or drought is adjustments to osmolyte concentrations used to regulate osmosis and maintain salt concentrations within a tolerable range (Burg and Ferraris, 2008). Generally, osmotic stress increases the concentrations of proline, free amino acids and polyamines. Changes in enzymatic activity often occur with long-term osmotic stress and are due to the production of free radicals as the cell adjusts metabolic processes to deal with the changes in solute concentrations. In addition to enzymatic changes, high concentrations of ROS can weaken and destroy cell membranes through lipid peroxidation. This is measured either by electrolyte leakage or the concentration of malondialdehyde (MDA), a reactive aldehyde that can be produced following lipid peroxidation. While most of these pathways are conserved, specific osmolytes and enzymes vary based on plant species and duration and intensity of the stress. In addition, researchers use both NaCl and polyethylene glycol (PEG) to induce osmotic stress in plants.

While both of these throw off water balance, they may also trigger additional pathways separate from one another, as Na+ is an important electrolyte used for various mechanisms within plants.

Short-term exposure to 100 mM NaCl (2 h) in *Sorghum bicolor* treated with 1.67 mM sodium silicate (Na_2SiO_3) resulted in increased leaf elongation, photosynthetic rate, relative water content, stomatal conductance, transpiration rate, water potential, hydraulic conductance and increased SOD and CAT activity, decreased POD activity and no change to APX (Liu et al., 2015).

In dill amended with Na_2SiO_3 and exposed to salt stress, Si enhanced enzymatic activity for CAT, SOD and POD, but not for APX (Shekari et al., 2017). Si-treated plants also had higher chlorophyll a and b with salt stress and lower MDA, with no difference in proline or soluble sugar content. There was also a decrease in shoot fresh and dry weight, and root dry weight with Si treatment in the absence of salt stress. Sodium-potassium levels were also altered with Si addition.

Silicon can alter the distribution of potassium (K) and sodium (Na) in salt-stressed plants. In purslane leaves treated with Na_2SiO_3 and 120 mM salt stress, K concentrations were increased to control levels while Na was slightly decreased (Kafi and Rahimi, 2011). Similarly, in leaves of aloe, potassium silicate (K_2SiO_3) application in the presence of 100 mM salt stress increased K while decreasing Na concentrations (Xu et al., 2015). The authors further investigated the role of ATPases and found that PM-H+-ATPase activity was recovered to control levels in salt-stressed plants treated with Si, however, tonoplast ATPase and PPase activity was increased with Si treatment following salt stress, but not as high as non-stressed plants.

X-ray microanalysis of aloe roots grown with and without 1.5 mM K_2SiO_3in the presence or absence of 100 mM NaCl had similar concentrations of Na, Cl and K in epidermal, cortical and stellar cells (Xu et al., 2015). This suggests that Si is not simply interacting with NaCl in the growing media, but having an effect internally within the plants, allowing for tolerance to salt. Si-treated aloe under salt stress had thicker leaves with higher fresh weight compared to salt treatment alone, suggesting that mitigation of stress was in part due to retention of water within the plants.

Cucumbers exposed to 65 mM NaCl and treated with 0.83 mM Na_2SiO_3 had increased fresh weights, root/shoot ratio, chlorophyll content and net photosynthetic rate (Wang et al., 2015). Additionally, transpiration rates, leaf water content, osmotic potential and root hydraulic conductance were increased, while a slightly higher K/Na ratio was reported due to decreased Na with slightly higher K concentrations in leaves. The cucumbers also had significant increases in free and conjugated putrescine and decreased concentrations of conjugated spermidine. Polyamines play a role as osmolytes during osmotic stress, such as ionic, salinity and drought, in plants.

A salt sensitive pepper showed increased dry weights, leaf area as well as enhanced photosynthetic and transpiration rates when challenged with 75 and 150 mM NaCl and treated with 2 mM K_2SiO_3 (Altuntas et al., 2018).

In another pepper cultivar, 'Bugwang', 1.8 K_2SiO_3 addition enhanced growth parameters, including shoot and root length, shoot diameter, number of roots and fresh weight above controls with and without 50 mM salt stress (Manivannan et al., 2016). Increased net photosynthesis, stomatal conductance and transpiration rate were observed in Si-treated plants as well as differences in protein concentrations using a 2D-gel electrophoresis. Enzymatic activity of SOD, GPX, CAT and APX was also reported with Si treatment in both NaCl-stressed and non-stressed plants compared to their respective controls.

In wild licorice, root length, plant height, stem diameter and plant dry weight were all increased with the addition of 2 mM K_2SiO_3 in the presence of 100 mM NaCl (Lang et al., 2019). In addition, endogenous GA_3, IAA and ABA concentrations were significantly increased with 2 mM K_2SiO_3 treatment in the presence of 100 mM NaCl. In gooseberry treated with 0.5% NaCl, H_4SiO_4 at 0.5 and 1.0 $g \cdot L^{-1}$ was able to enhance chlorophyll content, carotenoid concentrations and stomatal density (Rezende et al., 2018).

In honeysuckle stressed with 100 mM NaCl and treated with 0.5 $g \cdot L^{-1}$ K_2SiO_3, fresh weight was significantly increased, while there was no change in dry weight, suggesting a higher water content in these plants (Gengmaeo et al., 2015). The Na concentration was significantly decreased in leaves and roots of salt-stressed plants, while K concentrations were similar to non-amended plants. Silicon-protected thylakoid membranes under salt stress and a reduction in plastoglobules were observed with 0.5 $g \cdot L^{-1}$ K_2SiO_3 amendment. Interestingly, photosynthetic rate, transpiration rate and stomatal conductance were all significantly reduced in Si-amended plants with and without 100 mM NaCl treatments. The researchers also observed reduction in chlorogenic acid concentration under 200 mM NaCl treatment with the addition of K_2SiO_3. Enzymatic activity was altered with the addition of 0.5 $g \cdot L^{-1}$ K_2SiO_3, increasing SOD and CAT while decreasing POD. Not surprisingly, Si concentration in honeysuckle increased with increasing K_2SiO_3 in both roots and shoots.

Basil treated with silica nanoparticles had increased chlorophyll - a concentration under salt stress (Kalteh et al., 2014). Chlorophyll b was increased in the absence and presence of salt stress with nanosilica treatment, compared to control as well as proline concentration and fresh and dry weights.

Si supplied at 1 mM K_2SiO_3 enhanced germination rate of drought tolerant *Glycyrrhiza uralensis* that were stressed with 100 mM NaCl and 100 mM NaCl with 10% PEG (Zhang et al., 2017). Control plants amended with Si also had enhanced germination rates. The radicle length was also increased with these treatments, and radicle and hypocotyl diameter were increased in Si-fed NaCl-treated plants compared to non-amended plants. However, the combination of NaCl and PEG at 20% with the addition of Si had a negative effect on the germination rate. In this study, Si reduced H_2O_2, proline and soluble sugar content, hinting that the plants were experiencing less osmotic stress compared to controls.

Tomato amended with 2.5 mM K_2SiO_3 and exposed to 10% PEG-6000 showed reduction in stress with reduced free radical formation, a slight increase in photosynthetic and transpiration rate, increased water content, and reduction in lipid peroxidation and electrolytic leakage (Shi et al., 2016). There was no significant change in proline content between any of the four treatments. Following seven days of exposure to osmotic stress, Si-amended plants had higher SOD, CAT and ascorbic acid, compared to the non-amended. The authors also looked at expression of three PIPs and found that *SlPIP1;3* expression increased with Si treatment, compared to controls. After seven days of treatment, osmotic-stressed plants had significantly higher *PIP1;3* expression and Si amendment lowered the expression to Si-treated controls. The expression for both Si-treatments was significantly higher than that of the non-stressed control.

Tomato treated with 150 mM NaCl and treated with 2 mM Na_2SiO_3 had increased pigment, net photosynthetic and transpiration rate and stomatal conductance as well as increased leaf and root water and osmotic potential, respectively (Li et al., 2015). Tomatoes stressed with 10% PEG-6000 and treated with 2.5 mM K_2SiO_3 showed increase biomass, pigment formation and chlorophyll fluorescence parameters compared to control (Zhang et al., 2018).

In tobacco BY2 cells, ethylene perception was shown to be involved in the salt stress response as treatment with NaCl, 2 mM K_2SiO_3 and silver (Ag), which causes ethylene receptors to be unresponsive to ethylene, reduced cell viability while increasing MDA content (Liang et al., 2015b). The use of aminooxyacetic acid (AOA) as an inhibitor of ethylene biosynthesis also showed a similar, but less dramatic response. In the absence of salt stress, the treatment with Ag and K_2SiO_3 significantly reduces fresh weight and increases MDA, H_2O_2 and nitric oxide. This could suggest that ethylene pathways are used to modulate the Si response or keep the nutrient from inducing cell death in tobacco.

Rosa hybrid 'Rock Fire' grown in deep-water hydroponic solutions containing 50 mM NaCl and 1.8 mM K_2SiO_3 had similar weight and MDA content as control, while 50 mM NaCl alone caused a significant reduction in weight and increased MDA content (Soundararajan et al., 2018). Rose grown with NaCl and Si also had more root hairs compared to the NaCl treatment.

Regulation of drought stress in plants takes place through several different mechanisms at the whole-plant and cellular/organelle levels. Well-developed casparian bands protect the root system as a whole, while the release or production of osmolytes helps maintain constant turgor pressure during water-limiting events. Silicon has been shown to reduce drought stress and maintain turgor pressure under limited moisture in a variety of plants through changes in transpiration, osmolyte concentrations and enzymatic activity.

Addition of 2 mM Na_2SiO_3 to *Nicotiana rustica* significantly increased leaf and root fresh and dry weight in the absence of stress (Hajiboland et al., 2017). Chlorophyll a and b as well as carotenoid concentrations were higher in Si-amended plants compared to controls, with no change in anthocyanins. There was also a higher concentration of soluble sugar and a reduction of starch in Si-fed plants in both root and shoot tissue. Water stress maintained this trend. Proline, soluble proteins and free amino acids also were at higher concentration in Si-fed plants compared to control and water-stressed plants. With 60% field capacity (FC) watering, significant weight increases were still observed and Si-amended, water-stressed plants had similar weights to unstressed controls. Silicon treatment enhanced water-use efficiency under water reduction treatments and data were presented showing reduction in daily water consumption for these plants. In leaves, enzymatic activity of SOD and POD was increased with 30% FC treatment and Si-amendment. Catalase was slightly lower than control plants at 30% FC water regime. Root enzymatic activity was markedly different with CAT showing increased activity in non-stressed and 60% FC. Peroxidase activity was elevated for control and 60% FC Si-amended plants and APX was higher in Si-amended, non-stressed plants. In leaves, H_2O_2 concentrations were consistently lower between all Si-amended treatments compared to controls. In roots, drought stress and Si-amendment resulted in lower concentrations. MDA concentrations were elevated with Si-amendment in leaves with respect to drought treatments, while in the roots only 60% FC had a slight increase compared to non-Si-amended plants exposed to the same stress.

In general, Si amendment to the growing media resulted in osmotic stress tolerance by enhancing growth and enzymatic activity in addition to increasing concentrations of osmolytes such as proline, ultimately leading to reduced ROS and lipid peroxidation. This was observed in plants that accumulate Si at both macronutrient and micronutrient concentrations.

5 Silicon and plant abiotic defense: tolerance to heavy metals

There are several mechanisms that plants use to avoid heavy metal toxicity that include avoidance, chelation and compartmentalization. With the versatility of Si, many studies have shown that when added to the media, concentrations of the metal in plant tissue are significantly reduced. This could suggest that Si's first role is to chelate the metal in the growing media, reducing the metal concentration for plant absorption. While this mechanism aids in resistance, other internal processes are likely involved. This likely includes modulation of enzymatic and non-enzymatic factors that control the oxidative state of the cells.

Chromium uptake in roots of pea was reduced with nanosilica treatments, suggesting that part of the tolerance in pea is likely due to sequestration of the metal in the growing media, resulting in lower Cr (VI) movement into the shoots (Tripathi et al., 2015). Activities of SOD and CAT were also increased with nanosilica addition, while APX slightly increased in the roots of Si-treated peas. Glutathione reductase (GR) and dehydroascorbate reductase (DHAR) were also increased with the addition of silica nanoparticles. Nutrient concentration for all elements was significantly enhanced in both roots and shoots of nanosilica-treated peas, with the exception of sulfur. Interestingly, sulfur was significantly higher in Cr and nanosilica-treated plants in root tissue.

Root concentrations of aluminum drastically increased in Al-sensitive potatoes grown in 1.85 mM $AlCl_3$ and amended with 0.5 mM Na_2SiO_3, but the foliar levels were lower compared to the stressed controls (Dorneles et al., 2016). Silicon addition also increased the number of root branches in both varieties but did not result in any changes in biomass of roots or shoots under aluminum stress.

Indian mustard roots in the presence of 150 μM $NaAsO_2$ had increased ROS, including superoxide radical and nitric oxide (Pandey et al., 2016). These concentrations were reduced with 1.5 mM H_4SiO_4 treatment. Cysteine and proline concentrations in the roots of plants treated with both Si- and As- were higher than control and As treatments.

In flowering Chinese cabbage, the addition of 5 μM Na_2SiO_3 significantly increased the concentration of cadmium in the cell wall fraction while lowering apoplastic and symplastic concentrations that lead to lower Cd detected in the xylem sap (Wu et al., 2016). That in addition to higher root concentration of Cd in the presence of Si suggests that Cd reduction was due to immobilization of the metal in root tissue.

Three-day exposure of *A. thaliana* Sf-1 with 30 μM $CuSO_4$ reduced root and shoot growth as well as foliar nutrient concentrations, while addition of 1.0 mM K_2SiO_3 only had a slight decrease in growth compared to the controls (Li et al., 2008).

Longer exposure of 35 μM $CuSO_4$ to *Nicotiana tabacum* showed inducible Si accumulation in both roots and leaves when treated with 1.5 mM K_2SiO_3 (Flora et al., 2019). In addition, changes in expression of different metabolic and hormonal responsive genes were presented.

Si is able to alleviate heavy metal toxicity through sequestration both externally and internally within the plant, leading to lower intracellular metal concentrations. In addition, enhanced enzymatic activity associated with Si fertilization resulted in lower ROS, protecting the cells from damage in a wide range of macro- and micro-Si accumulators.

6 Silicon and plant abiotic defense: tolerance to other environmental stresses

In addition to the well-documented alleviation of osmotic and heavy metal stress, Si amendment also offers protection to other stresses, including nutrient deficiency and temperature extremes. How Si is part of a balanced nutrient plan has yet to be determined. Few studies have looked into the changes in nutrient content of plants with Si fertilization. In a study done by Aires et al. (2007) when looking at the effect of nitrogen and sulfur fertilization on elemental composition in broccoli, application of 91 ppm N significantly increased Si concentration in 11-day-old broccoli sprouts. Under iron-starvation, cucumber treated with 1.5 mM H_4SiO_4 and grown for 7 days in iron-free nutrient solution had higher chlorophyll content and root and shoot growth compared to the controls (Pavlovic et al., 2013).

Heat stress of *Salvia splendens* placed at 35°C for two days and treated with 100 ppm K_2SiO_3 exhibited a stress-induced Si accumulation in leaf tissue as concentrations increased from 130 ppm to 307 ppm Si and 107 ppm to 255 ppm Si, for Vista Red and Sizzler Red, respectively (Soundararajan et al., 2014).

Chilling stress to corn seedlings treated with H_4SiO_4 resulted in a significant increase in IAA, Zeatin, JA and SA in the shoot of non-chilled controls, while ABA and GA where even higher than non-stressed controls (Moradtalab et al., 2018). All root hormone levels with Si treatment under chilling stress were the same as non-stressed controls.

Silicon has been shown to enhance post-harvest quality for fruit. Lemon fruit dipped in 50 ppm K_2SiO_3 showed the least amount of chilling stress after 28 days at −0.5°C (Mditshwa et al., 2013). However, dips at 250 ppm K_2SiO_3 showed increased damage at day 21 and day 28 of chilling for both 2°C and −0.5°C. The authors attributed the protection of the lemon fruit to increased antioxidant activity.

Silicon enables plants to adapt faster to short-term fluctuations in their immediate environment, such as localized nutrient deficiencies or changing temperatures. The mechanisms controlling these responses in the context of Si have not yet been determined. The wide range of protection Si provides against abiotic stress suggests that the nutrient either has a role in a general stress response or that it plays multiple roles within plant stress and development.

7 Silicon and enhanced growth

While Si has the most visible response to stress tolerance, other growth parameters, such as plant weight, stem size and extended post-harvest parameters have been observed in different plant species. Use of stabilized H_4SiO_4 in Europe has been reported to increase the yield and growth in a number of different plant species with both foliar and drench applications (Laane, 2017). In these studies, however, very little if any Si tissue analysis was collected and correlated to the observed growth or yield responses.

7.1 Si and hormonal influences

Cytokinins are an important phytohormone involved in various cellular responses as well as promoting and inhibiting shoot and root growth, respectively, and reducing leaf senescence (Kieber and Schaller, 2018; Zwack and Rashotte, 2013). Silicic acid was shown to induce cytokinin in both wild type and transgenic *Arabidopsis* expressing the rice *NIP2;1* transcript (Markovich et al., 2017). The increase in cytokinin suppressed dark-induced leaf senesce in detached leaves treated with Si. The authors concluded that the soluble H_4SiO_4 and not a polymerized version is responsible for the induction of cytokinin and that this response is observed in both macro-and micro-Si accumulators. Studies in *Brassica napus* L. var. 'Citizen' showed an increase in active cytokinin and decrease in leaf senesce when treated with 1.7 mM Na_2SiO_3 following nitrogen starvation (Haddad et al., 2018). Si protection of the plants allowed for recovery after resupply of nitrogen compared to controls. Extended vase life of eight and two to four days was observed in carnations and lisianthus treated with 2.5 mM and 1.5–3.0 mM Si, respectively (Kazemi et al., 2012a,b). In addition, increased chlorophyll content and SOD activity with decreased proline, MDA and ACO (indication of ethylene production) were observed in flowers placed in Si solutions compared to controls. However, 1.5 mM Si treatment to cut *Argyranthemum* flowers only resulted in a day increase in shelf life but still exhibited decreased ACO and MDA content (Kazemi, 2012). In contrast, cowpea grown with metasilicic acid at greater than 0.08 $g{\cdot}L^{-1}$ had a reduction in the cytokinin zeatin ribose and an increase in abscisic acid (ABA) concentrations as well as stem and peduncle strength (Dakora and Nelwamondo, 2003).

ABA is another important hormone involved in regulating developmental processes and defense responses. In rice, application of 10 μM ABA to the hydroponics solution reduced Si uptake and *OsLsi1* expression in roots (Yamaji et al., 2008). Preliminary studies in tobacco show a similar reduction in leaf Si concentration with foliar application of 25 μM Si (Fig. 4). It is important to note that spraying leaves with water alone in this species increased Si concentration in leaves and proper controls are essential for determining whether a component of the spray (i.e. the hormone) or the physical treatment (i.e. spraying a solution on leaves) is responsible for the observed effect.

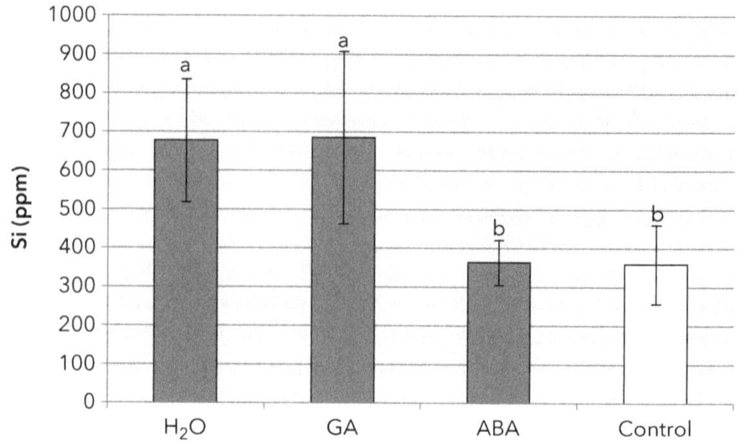

Figure 4 Silicon concentration of tobacco leaves grown hydroponically with 2.0 mM K_2SiO_3 and sprayed with water (H_2O), gibberellic acid (GA), abscisic acid (ABA) or not sprayed (control, white). Different letters correspond to significant differences with a p value <0.05 based on Tukey's HSD post hoc test.

7.2 Effects of Si on germination and enhanced growth

Throughout the literature there are studies that show that Si alone effects plant growth from germination to fruit ripening. As with abiotic stress, plant variety, Si source in addition to application timing and rate are main factors that influence the response. Enhanced germination rates, plant size and increased photosynthetic parameters are being documented in many different plant species with Si supplementation.

Siddiqui and Al-Whaibi (2014) found that using 8 g·L^{-1} (8 000 ppm) silica nanoparticles with an average size of 12 nm and surface area of 200 m^2·g^{-1} significantly increased the germination rate and seed vigor as well as seedling weight of tomato. Seedling growth of both native and hybrid orchids as well as increased epidermal and mesophyll thickness were observed with Ca_2SiO_4 application (Soares et al., 2012).

Indian mustard seedlings treated with 1.5 mM Si in the form of H_4SiO_4 had a significant increase in lateral root formation compared to control (Pandey et al., 2016). Addition of solid Si fertilizers or Si drenches between 10 and 20 ppm Si increased root weight for citrus seedlings (Matichenkov et al., 1999). In addition, the researchers found that water-extractable Si in the soil of healthy-appearing citrus trees was significantly higher than the soil directly under unhealthy trees with smaller leaves and thinner canopies.

Mustard amended with 0.7 mM Na_2SiO_3 showed increased leaf area, root and shoot length, and increased dry weights compared to controls (Ashfaque et al., 2017). In addition, in non-stressed plants, Si addition increased net photosynthesis as well as chlorophyll and carotenoid contents, while reducing Cr concentrations in root tissue compared to controls. Peas treated with 10 μM nanosilica particles made from Na_2SiO_3 were able to enhance growth parameters such as fresh and dry weights, and lengths of roots and shoots, independent of chromium stress (Tripathi et al., 2015). Total chlorophyll content, carotenoids and protein were also increased with treatment of nanosilica particles alone. The application of K_2SiO_3 affected overall growth of phalaenopsis orchids, where Si content of the plant ranged from 0.5% to 1.7% (Vendrame et al., 2010). Overall, Si applied between 0.5% and 1.0% increased fresh weight and dry weight and at 1.0% Si significantly increased dry weight of root, shoot and whole plant over the control. Addition of 100 ppm K_2SiO_3 to *Salvia splendens* increased chlorophyll content as well as growth characteristics such as shoot length, total leaf area and root length in non-stressed plants (Soundararajan et al., 2014).

Weights of deep-water cultured cotton, canola and wheat increased with 1.5 mM Na_2SiO_3 with a decrease in transpiration rates likely causing the observed decrease in Ca content following 12 days of treatment (Mehrabanjoubani et al., 2015). The study also shows increased potassium and iron uptake with Si treatment in all three species, with additional uptake of magnesium and zinc in canola and wheat and increase in phosphate by cotton and canola. Cotton was the only species to have decreased boron when treated with Si.

Additionally, sunflowers grown in media amended with an anhydrous form of K_2SiO_3 at 140 $g \cdot m^{-3}$ had an increase in height and diameter of both the stem and flower, while plants sprayed with 50, 100 and 150 ppm Na_2SiO_3 only had increased stem diameter (Kamenidou et al., 2008).

7.3 Si and micropropagation

Si is an important nutrient for tissue culture as it enhances growth and survival in this type of system (Sahebi et al., 2016; Sivanesan and Park, 2014). Similar to environmental stress, micropropagation places plant cells under stress. Si likely induces similar responses to reduce oxidative damage and maintain internal homeostasis to allow for enhanced growth and regeneration of the various plant organs.

Carnation micro-propagated with K_2SiO_3 or $CaSiO_3$ had increased stomatal density, PsbA protein and Si concentration compared to controls (Manivannan et al., 2018). Higher concentrations of Si and stomatal density were reported with K_2SiO_3 treatment compared to $CaSiO_3$. In addition, the number of shoots produced with K_2SiO_3 was significantly higher in K_2SiO_3 compared to $CaSiO_3$ and control.

The significantly fewer reports on Si and plant growth under non-stressed conditions are likely due to Si contamination from the growing environment. Plant use of Si for normal growth and development likely occurs at the micronutrient concentrations as is seen in animals. In contrast, the concentrations needed to elicit a defense response in plants are much higher. These Si concentrations are strongly plant dependent as developmental needs for rice and sugarcane have been demonstrated, while those in other monocots and dicots are still unclear.

8 Silicon and phytotoxicity

To date, there is an increasing number of studies raising concern to the possibility of Si-toxicity in plants. The main culprits are due to high-application rates of liquid fertilizers in a nutrient-limiting media or nanosilica particles of varying doses and sizes.

A certain class of liquid products available in Europe is stabilized silicic acid (sSiA) materials that utilize PEG or choline to maintain high concentrations of H_4SiO_4. The potency of these materials is likely much higher than solid materials that dissociate over time. *Verbena hybrida* treated with six applications of foliar-applied sSiA at 240 ppm Si had a decrease in florescent diameter and numbers, while *Salvia farinacea* treated with two applications at that rate had a decrease in plant and inflorescent diameter (Dębicz et al., 2016). Si concentrations in these plants were not increased with the application of the sSiA, so it is hard to determine if the decrease in sizes were due to Si or an osmotic stress from spraying the solution on the plants. It is important to note that in a follow-up study at lower doses of sSiA, there were increased plant sizes for *Gazania rigens*, *Salvia farinacea* and *Verbena* between 120 and 180 ppm Si applied weekly for five weeks (Debicz et al., 2017). This again stresses that sSiA and other liquid products may show phytotoxicity at higher application rates.

Other forms of liquid Si such as K_2SiO_3 also can have adverse effects at high concentrations. Sunflowers treated with weekly foliar drenches of 200 ppm (7.12 mM) K_2SiO_3 had decreased height, floral diameter and longer days to anthesis compared to non-Si treated plants (Kamenidou et al., 2008). Abnormal floral development was also observed at this treatment rate, but not at lower Si drenches of 50 and 100 ppm K_2SiO_3. The Si concentration in the 200 ppm treatment was significantly higher than all other treatments in the leaves, stems and flower and low concentrations of magnesium in leaves was reported. The imbalance of nutrients could have led to the observed abnormal floral development.

Using silica nanoparticles at certain rates can lead to phytotoxic responses in certain plants. Application of 43 nm nanosilica particles at concentrations above 400 $mg{\cdot}L^{-1}$ in agar media significantly reduced *A. thaliana* root elongation and leaf number but had no effect on germination rate (Lee et al., 2010). Photosynthetic rates significantly dropped in non-stress tomatoes treated with 1 mM and 2 mM nanosilica particles that ranged in size from 20–35 nm (Haghighi and Pessarakli, 2013). At 50 mM NaCl, 2 mM nanosilica particles significantly reduced the rate below controls and other Si treatments. Plants treated with 2 mM nano-Si had significantly reduced fresh and dry weight as well as root volume. An earlier study also showed that 2 mM nanosilica significantly reduced shoot dry weight and mean germination time of tomato compared to 1 mM nanosilica and control treatments (Haghighi et al., 2012). Silica nanoparticles of 50 nm size and surface range of 200–600 $m^2 \cdot g^{-1}$ at a concentration of 100 $mg \cdot L^{-1}$ made potato plants more susceptible to salt stress compared to control and treatments of 50 $mg \cdot L^{-1}$ (Gowayed et al., 2017). This suggests that higher concentrations of nanoparticles could make plants more susceptible to stress.

Taken together, Si does have the potential to induce phytotoxicity in plants. This is dependent on additives in the formulation, application rates and physiochemical properties of the fertilizer. Following labeled instructions of these products sold in the United States reduces the risk of these adverse side-effects, as the concentration of Si and other compounds are significantly below rates used in the aforementioned studies.

9 Regulatory issues

The agricultural industry and regulatory agencies have struggled in how to define Si and, therefore, they are unsure on whether Si should be classified as a fertilizer, a plant protectant or biostimulant. In fact, all three classifications are proposed, and in a few instances accepted. The problem with describing the element as a biostimulant as well as using the fertilizer label only allows these classifications to be used for a Si product that will be marketed for use in combating abiotic stresses such as heavy metal toxicities or temperature extremes, and not for plant disease control. If marketed for plant disease control, at least in the United States, the product must be registered as a plant protectant (pesticide).

Many countries still do not have a legislative way forward to classify Si. To the best of our current knowledge, Japan and Korea recognized the importance of Si in rice back in the 1950s, and they classified this element as being agronomically essential, and could be sold as a fertilizer. In 2004, Brazil was the third country that formally recognizes Si. The Brazilian Ministry of Agriculture, which regulates the commercial production of fertilizers, ruled that Si is a beneficial micronutrient.

In 2013, the American Association for Plant Food Control Officials (AAPFCO), the regulatory body that governs the labeling of fertilizers in the United States, recognized Si as a beneficial substance and is now sold as a fertilizer across the United States. In 2007, K_2SiO_3, (Sil-MATRIX^R, PQ Corporation), was registered by the EPA in the United States, and certified by OMRI as an organic pesticide for the control of powdery mildew, mites and aphids on high-value crops such as strawberries, wine grapes and others.

Clearly, Si is a cross-over element that may perform as a biostimulant or a plant protectant because the element affects both abiotic and biotic stresses. A large body of information also exists that demonstrates Si acts as a fertilizer because the element enhances plant growth, development and yield. Perhaps, a new category needs to be created so that Si may achieve universal acceptance by regulatory agencies worldwide.

10 Summary and future trends

Based on overwhelming evidence in a number of plant species under different external stimuli, Si likely plays a role in maintenance of homeostasis. As a consequence, Si-treated plants show lower levels of stress markers, such as MDA, while physiological parameters, such as transpiration, chlorophyll content and enzymatic activity are maintained similar to non-stressed plants. Clearly, Si works as a biostimulant to enhance the plant's internal responses to external stimuli and our understanding of how it functions as a biostimulant is greatly improving. Effective, practical means of application, affordable sources of Si and methods for identifying conditions under which Si will be beneficial are still needed for reducing abiotic stresses. However, research on the use of Si for abiotic stress reduction under field conditions is virtually nonexistent. Further, no soil tests for gauging the amounts of soluble Si have been calibrated for many agronomic or horticulture crops. Likewise, most analytical laboratories do not routinely assay plant tissue for Si. In fact, the current standard tissue digestion procedures used in most laboratories would render Si insoluble, making an analysis of the digested tissue meaningless. Thus, the two analytical tools most often used for determining the need for fertilization with plant nutrients are not widely available for Si. Nevertheless, as the need for environmentally friendly strategies for the management of abiotic stress increases, Si could provide a valuable tool. As researchers and growers become more aware of Si and its potential as a biostimulant, it is likely that this often-overlooked element will be recognized as a viable means for enhancing plant health and performance.

10.1 Effect of Si on plant-growth promoting rhizobacteria and other beneficial microbes

Si is beneficial to many multicellular and single cellular organisms. However, the role of Si on microbial processes, including symbiotic relationships with plants has not been fully explored. As the complex interconnections between plant and microbial networks are unraveled, Si will likely have an impact on the efficiency and mutual benefits provided by each partner.

In strawberry, addition of 3 mM Na_2SiO_3 increased *Rhizophagus clarus* hyphae and arbuscule percentages compared to controls during non-stressed and drought conditions (Moradtalab et al., 2019). Furthermore, when challenged with drought stress and treated with Si, the plants were able to increase root and shoot biomass as well as leaf-relative water content, while reducing leaf osmotic potential. The addition of *Rhizophagus clarus* and Si increased these parameters under normal-water conditions.

In field soils supplemented with nanosilica particles derived from rice hulls, an increase in microbial populations of both phosphate solubilizers and nitrogen fixers was observed with and without the presence of corn (Rangaraj et al., 2014). Additionally, Na_2SiO_3 and micro SiO_2 applications induced the microbial population of silicate solubilizers and micro SiO_2 and H_4SiO_4 addition enhanced the nitrogen fixers. The biomass of carbon and nitrogen were increased in nanosilica and H_4SiO_4 treatments but decreased with Na_2SiO_3 application prior to and after sowing with corn.

10.2 Seeing past silicic acid: availability and mobility of silica nanoparticles

Silica nanoparticles as described above induce both beneficial (4.1 and 4.2) and phytotoxic (5) responses in plants dependent on particle size and concentration. These particles are too large to move through the identified Si transporter pores and likely are transported via other pathways, such as endocytosis. It is possible that these particles at high concentrations induced defense responses similar to silicosis in mammals. Further studies in the field of nano-silica are needed to understand what concentrations are beneficial, how these particles move through plant tissue and how long they maintain their particle integrity.

10.3 Enhancing Si accumulation in lower accumulators

One strategy to increase Si responses is to increase Si concentrations in leaves of plants through transgenic expression of NIP2 transcripts. However, studies using transgenic *Arabidopsis thaliana* Col-0 expressing a wheat TaNIP2;1 exhibited necrotic lesions along the leaf edges (Montpeptit et al., 2012). This is likely due to the deposition of amorphous silica at the end of transpiration streams, disrupting homeostasis within the plants. The phenotype was alleviated when the transporter was only expressed in root tissue. This suggests that *Arabidopsis* does not have mechanisms to deal with high concentrations of Si in leaves and Si translocation from roots to shoots is likely tightly regulated to avoid Si-sensitivity. Interestingly, another study in which the rice NIP2;1 (OsLsi1) was constitutively expressed in *Arabidopsis* did not exhibit necrosis as described above but did have significant reduction in root growth and plant size compared to wild type (Markovich et al., 2017). This could be due to differences in growth conditions, including light duration and intensity as well as temperature. Additionally, constitutive expression of the active ArsB transcript (OsLsi2) in petunia showed more susceptibility to drought stress with a marked increase in MDA at 7 days after drought compared to controls (Tao et al., 2014). Transgenic petunia expressing the OsNIP2;1 passive transporter had a positive response to drought. Unfortunately, there was no data on plant size or weight, making it difficult to know what effect constitutive expression of these transporters had on the petunia phenotype. Taken together, the use of transgenics to increase Si concentrations in leaves of micro-Si accumulators above a certain threshold could have negative effects on plant growth and development. This suggests to date that Si transporters are not the best approach for producing transgenic plants with enhanced stress tolerance.

There are likely other factors than the Si transporters that regulate Si perception and movement in plants. These factors could be used as targets for improved stress-tolerance. Analyzing differences between varieties that accumulate varying concentrations of Si may identify important targets. For example, rice can accumulate up to 10% Si in their foliar tissue and Japonica varieties tend to accumulate higher concentrations than Indica (Wu et al., 2006; Deren et al., 1992; Winslow, 1992; Winslow et al., 1997; Ma and Takashashi, 2002). In addition, root H_4SiO_4 saturation occurred at a concentration of 0.8 mM (22.5 ppm Si) for rice. This could reflect the evolutionary behavior of the uptake mechanism in that soil concentrations of H_4SiO_4 are 0.1–0.6 mM (Epstein, 1994). Most notably in the studies by Dai et al. (2005) and Wu et al. (2006), QTLs involved in variety differences in Si accumulation were located on chromosomal areas separate from characterized passive and active Si transporters OsNIP2;1 (chromosome 2), OsLsi6 (chromosome 6), OsArsB (chromosome 3) and OsLsi3 (chromosome 10). This suggests other factors are involved in Si absorption and translocation in plants and that simply inserting the transporter may not be as advantageous as conventional breeding, since many of the QTLs indicated span multiple genes with varying functions. This greatly contrasts the theory of using genomic sequences for the presence or absence of NIP2;1 in the genome to determine whether a plant will or will not benefit from Si application and may give some insight into how tomato has the ability to respond positively with Si treatment even with the genomic context of a non-functional NIP2;1. Both studies by Wu et al. (2006) and Deren et al. (1992) show that the genetic background, not the Si concentration, was responsible for the variation in Si accumulation among the rice species tested even at varying Si concentrations.

Taken together, this suggests that classical breeding may provide a better source for enhancing Si concentration in micro-Si accumulators as opposed to manufacturing transgenic plants expressing Si transporters from macro-accumulating species.

10.4 Si biofortification of foods

Increasing the quality and nutritional value of a crop can add significant value, especially for specialty crops. Silicon is an essential trace nutrient for animals and is involved in collagen and bone formation and deposition among other things (Carlisle, 1986). An extensive list of Si content in food was put together by Powell and his colleagues (Powell et al., 2005). Not too surprisingly, grains had higher Si content in their edible parts compared to fruits and vegetables. Beer, specifically draft ale and draft lager compared to canned and bottled, were among the highest concentration of Si reported. High-Si concentration in foods does not necessarily mean higher available Si. Adding 100 ppm K_2SiO_3 to a floating hydroponics system significantly increased the bioavailability of Si in mizuna, tatsoi, basil, chicory and Swiss chard, but had no effect on purslane (D'Imperio et al., 2016). *In vivo* studies using human subjects found that cooked green beans provided more bioavailable Si compared to ripe bananas even though bananas have a higher Si content (Sripanyakorn et al., 2009).

In vitro studies testing the H_4SiO_4 release from edible plant parts has been performed using simulated gastrointestinal (SGI) conditions in order to determine Si bioavailability. Silicon dissolution was reported to be higher during the simulation of gastric digestion in the small intestine compared to that of the stomach, which is likely related to pH changes from acidic to near neutral between the two organs (Sripanyakorn et al., 2009). In other studies, using a modified in vitro dialysis model showed that higher Si content in solid foods resulted in lower percentages of bioavailable Si (Robberecht et al., 2008, 2009).

Silicon increases other valuable qualities, especially in tomato. Potassium silicate addition to tomatoes at 2.5 mM increased concentrations of β-carotene, lycopene and vitamin C, total solute solids, and fruit firmness in addition to decreasing the incidence of blossom end-rot (Stamatakis et al., 2003). More recently, Marodin et al. (2016) reported that the silicon source and dose had an effect on these parameters and noted that fruit from tomatoes treated with calcium and Na_2SiO_3 had higher lycopene concentration compared to K_2SiO_3. The authors also show that silicon content of fruits increases from 0 to 200 $kg{\cdot}ha^{-1}$ but remains similar for increased Ca_2SiO_4 and Na_2SiO_3 rates, while drastically decreasing with increasing K_2SiO_3.

Silicon efficacy has been demonstrated for decades and is well represented in the literature. As Si research moves forward, identification and understanding of cellular and molecular functions will provide further insight into the benefits of Si. Exploring different Si sources and uptake mechanisms may expand the list of fertilizers and functions of Si in plants, enhancing nutritional provisions for a wide variety of crops.

11 Abbreviations

AAPFCO:	Association of American Plant Food Control Officials
ABA:	abscisic acid
AQP:	aquaporin
APX:	ascorbate peroxidase
As:	arsenic
Ca_2SiO_4:	calcium silicate
CAT:	catalase
FC:	field capacity
GA:	gibberellic acid
K_2SiO_3:	potassium silicate
Lsi:	Low Silicon in Rice
MDA:	malondialdehyde
Na_2SiO_3:	sodium silicate
NFT:	nutrient film technology
NIP:	Nodulin 26-like Intrinsic Protein
NPA:	asparagine-proline-alanine
H_4SiO_4:	silicic acid
PEG:	polyethylene glycol
PIP:	plasma membrane intrinsic protein
POD:	peroxidase
QTL:	quality trait loci
ROS:	reactive oxygen species
SGN:	size guide number
Si:	silicon
sSiA:	stabilized silicic acid
SOD:	superoxide dismutase

12 Where to look for further information

Rodrigues, F. A. and Datnoff, L. E. 2015. *Silicon and Plant Diseases*. Springer International Publishing, Switzerland. doi:10.1007/978-3-319-22930-0.

Savvas, D. and Ntatsi, G. 2015. Biostimulant activity of silicon in horticulture. *Scientia Horticulturae* 196, 66–81.

Tubana, B. S., Babu, T. and Datnoff, L. E. 2016. A review of silicon in soils and plants and its role in US agriculture: history and future perspectives. *Soil Science* 181, 393–411.

13 References

Aires, A., Rosa, E., Carvalho, R., Haneklaus, S. and Schnug, E. 2007. Influence of nitrogen and sulfur fertilization on the mineral composition of broccoli sprouts. *Journal of Plant Nutrition* 30(7), 1035–46. doi:10.1080/01904160701394402.

Altuntas, O., Dasgan, H. Y. and Akhoundnejad, Y. 2018. Silicon-induced salinity tolerance improves photosynthesis leaf water status, membrane stability, and growth in pepper (*Capsicum annuum* L.). *HortScience* 53(12), 1820-6. doi:10.21273/HORTSCI13411-18.

Arnaud, L., Lan, H. T. T., Brostaux, Y. and Haubruge, E. 2005. Efficacy of diatomaceous earth formulations admixed with grain against populations of *Tribolium castaneum*. *Journal of Stored Products Research* 41(2), 121–30. doi:10.1016/j.jspr.2003.09.004.

Ashfaque, F., Inam, A., Inam, A., Iqbal, S. and Sahay, S. 2017. Response of silicon on metal accumulation, photosynthetic inhibition and oxidative stress in chromium-induced mustard (*Brassica juncea* L.). *South African Journal of Botany* 111, 153–60. doi:10.1016/j.sajb.2017.03.002.

Ayala, A., Munoz, M. F. and Arguelles, S. 2014. Lipid peroxidation: production, metabolism, and signaling mechanisms of malondialdehyde and 4-hydroxy-2-nonenal. *Oxidative Medicine and Cellular Longevity* 2014, 360438. doi:10.1155/2014/360438.

Bartoli, F. and Wilding, L. P. 1980. Dissolution of biogenic opal as a function of its physical and chemical properties. *Soil Science Society of America Journal* 44(4), 873–8. doi:10.2136/sssaj1980.03615995004400040043x.

Boldt, J. and Altland, J. 2017. Cultivar variation in silicon accumulation and distribution in *Petunia x hybrida*. *HortScience* 52(9), S198.

Buck, G. B., Korndorfer, G. H. and Datnoff, L. E. 2010. Extractors for estimating plant available silicon from potential silicon fertilizer sources. *Journal of Plant Nutrition* 34(2), 272–82. doi:10.1080/01904167.2011.533327.

Burg, M. B. and Ferraris, J. D. 2008. Intracellular organic osmolytes: function and regulation. *The Journal of Biological Chemistry* 283(12), 7309–13. doi:10.1074/jbc.R700042200.

Carlisle, E. M. 1986. Silicon as an essential trace element in animal nutrition. In *Silicon Biochemistry. Ciba Foundation Symposium* 121, 123–39. doi:10.1002/9780470513323.ch8.

Carre-Missio, V., Rodrigues, F. A., Schurt, D. A., Pereira, S. C., Oliveira, M. G. A. and Zambolim, L. 2009. Inefficiency of silicon in leaf rust control on coffee grown in nutrient solution. Tropical Plant Pathology 34(6), 416–21. doi:10.1590/S1982-56762009000600009.

Cervantes, C., Ji, G., Ramirez, J. L. and Silver, S. 1994. Resistance to arsenic compounds in microorganisms. *FEMS Microbiology Reviews* 15(4), 355–67. doi:10.1111/j.1574-6976.1994.tb00145.x.

Chang, F. P., Kuang, L. Y., Huang, C. A., Jane, W. N., Hung, Y., Hsing, Y. C. and Mou, C. Y. 2013. A simple plant gene delivery system using mesoporous silica nanoparticles as carriers. *Journal of Materials Chemistry B* 1(39), 5279–87. doi:10.1039/c3tb20529k.

Chiba, Y., Mitani, N., Yamaji, N. and Ma, J. F. 2009. HvLsi1 is a silicon influx transporter in barley. *The Plant Journal: for Cell and Molecular Biology* 57(5), 810–8. doi:10.1111/j.1365-313X.2008.03728.x.

Cruger, H. 1857. Westindische Fragmente, 9. El Cauto. *Botanische Zeitung* 15(1), 281–308.

Currie, H. A. and Perry, C. C. 2007. Silica in plants: biological, biochemical and chemical studies. *Annals of Botany* 100(7), 1383–9. doi:10.1093/aob/mcm247.

Dai, W. M., Zhang, K. Q., Duan, B. W., Zheng, K. L., Zhuang, J. Y. and Cai, R. 2005. Genetic dissection of silicon content in different organs of rice. *Crop Science* 45(4), 1345–52. doi:10.2135/cropsci2004.0505.

Dakora, F. D. and Nelwamondo, A. 2003. Silicon nutrition promotes root growth and tissue mechanical strength in symbiotic cowpea. *Functional Plant Biology* 30(9), 947–53. doi:10.1071/FP02161.

Das, K. and Roychoudhury, A. 2014. Reactive oxygen species (ROS) and response of antioxidants as ROS-scavengers during environmental stress in plants. *Frontiers in Environmental Science* 2, 53. doi:10.3389/fenvs.2014.00053.

Davy, H. 1814. *Elements of Agricultural Chemistry: In a Course of Lectures from the Board of Agriculture* (2nd edn.). J. G. Barnard, London, UK.

Dayanandan, P. and Kaufman, P. B. 1976. Trichomes of *Cannabis sativa* L. (Cannabaceae). *American Journal of Botany* 63(5), 578–91. doi:10.1002/j.1537-2197.1976.tb11846.x.

Dębicz, R., Pawlikowska, A., Wróblewska, K. and Bąbelewski, P. 2016. Influence of foliar treatment with silicon contained in the Actisil Hydro Plus preparation on the growth, flowering and chemical composition of *Gazania rigens* (L.) Gaertn., *Salvia farinacea* Benth and *Verbena hybrida* Voss. *Journal of Elementology* 21(3/2016), 681–92. doi:10.5601/jelem.2015.20.2.909.

Dębicz, R., Pawlikowska, A., Wróblewska, K. and Bąbelewski, P. 2017. The influence of silicon and multinutrient fertilizer on the quality and chemical composition of Gasania rigensi 'Kiss Yellow', Salvia farinacea 'Fairy Queen' and Verbena 'Obsession lilac' plants. Journal of Horticultural Rsearch 251(1), 35–45. doi:10.1515/johr-2017-0004.

Deren, C. W., Datnoff, L. E. and Snyder, G. H. 1992. Variable silicon content of rice cultivars grown on Everglades histosols. *Journal of Plant Nutrition* 15(11), 2363–8. doi:10.1080/01904169209364480.

Deshmukh, R. K., Vivancos, J., Guerin, V., Sonah, H., Labbe, C., Belzile, F. and Belanger, R. R. 2013. Identification and functional characterization of silicon transporters in soybean using comparative genomics of major intrinsic proteins in *Arabidopsis* and rice. *Plant Molecular Biology* 83(4–5), 303–15. doi:10.1007/s11103-013-0087-3.

Deshmukh, R. K., Vivancos, J., Ramakrishnan, G., Geurin, V., Carpentier, G., Sonah, H., Labbe, C., Isenring, P., Belzile, F. J. and Bélanger, R. R. 2015. A precise spacing between the NPA domains of aquaporins is essential for silicon permeability in plants. *The Plant Journal: for Cell and Molecular Biology* 83(3), 489–500. doi:10.1111/tpj.12904.

Dey, S. and Rosen, B. P. 1995. Dual mode of energy coupling by the oxyanion-translocating ArsB protein. *Journal of Bacteriology* 177(2), 385–9. doi:10.1128/jb.177.2.385-389.1995.

D'Imperio, M., Renna, M., Cardinali, A., Buttaro, D., Santamaria, P. and Serio, F. 2016. Silicon biofortification of leafy vegetables and its bioaccessibility in the edible parts. *Journal of the Science of Food and Agriculture* 96(3), 751–6. doi:10.1002/jsfa.7142.

Dorneles, A. O. S., Pereira, A. S., Rossato, L. V., Possebom, G., Sasso, V. M., Bernardy, K., Sandri, RdQ., Nicoloso, F. T., Ferreira, P. A. A. and Tabaldi, L. A. 2016. Silicon reduces aluminum content in tissues and ameliorates its toxic effects on potato plant growth. *Ciência Rural* 46(3), 506–12. doi:10.1590/0103-8478cr20150585.

Epstein, E. 1994. The anomaly of silicon in plant biology. *Proceedings of the National Academy of Sciences of the United States of America* 91(1), 11–7. doi:10.1073/pnas.91.1.11.

Epstein, E. and Bloom, A. J. 2005. *Mineral Nutrition of Plants: Principles and Perspectives* (2nd edn.). Sinaeur Associates Inc., Sunderland.

Fauteux, F., Chain, F., Belzile, F., Menzies, J. G. and Belanger, R. R. 2006. The protective role of silicon in the *Arabidopsis*-powdery mildew pathosystem. *Proceedings of the National Academy of Sciences of the United States of America* 103(46), 17554–9. doi:10.1073/pnas.0606330103.

Fekih, I. B., Zhang, C., Li, Y. P., Zhao, Y., Alwathnani, H. A., Saquib, Q., Rensing, C. and Cervantes, C. 2018. Distribution of arsenic resistance genes in prokaryotes. *Frontiers in Microbiology* 9, 2473. doi:10.3389/fmicb.2018.02473.

Fernandez, V. and Brown, P. H. 2013. From plant surface to plant metabolism: the uncertain fate of foliar-applied nutrients. *Frontiers in Plant Science* 4, 289. doi:10.3389/fpls.2013.00289.

Fernandez, V., Sotiropoulos, T. and Brown, P. 2013. *Foliar Fertilization: Scientific Principles and Field Practices*. International Fertilizer Industry Association, Paris, France.

Fleck, A. T., Schulze, S., Hinrichs, M., Specht, A., WaBmann, F., Schreiber, L. and Schenk, M. K. 2015. Silicon promotes exodermal casparian band formation in Si-accumulating and Si-excluding species by forming phenol complexes. *PLoS ONE* 10(9), e0138555. doi:10.1371/journal.pone.0138555.

Flora, C., Khandekar, S., Boldt, J. and Leisner, S. 2019. Silicon alleviates long-term copper toxicity and influences gene expression in *Nicotiana tabacum*. *Journal of Plant Nutrition* 42(8), 864–78. doi:10.1080/01904167.2019.1589508.

Franke, W. 1967. Mechanisms of foliar penetration of solutions. *Annual Review of Plant Physiology* 18(1), 281–300. doi:10.1146/annurev.pp.18.060167.001433.

Frantz, J. M., Pitchay, D. D. S., Locke, J. C., Horst, L. E. and Krause, C. R. 2005. Silicon is deposited in leaves of New Guinea impatiens. *Plant Health Progress* 6(1) doi:10.1094/PHP-2005-0217-01-RS.

Frantz, J. M., Locke, J. C., Datnoff, L., Omer, M., Widrig, A., Sturtz, D., Horst, L. and Krause, C. R. 2008. Detection, distribution and quantification of silicon in floricultural crops utilizing three distinct analytical methods. *Communications in Soil Science and Plant Analysis* 39(17–18), 2734–51. doi:10.1080/00103620802358912.

Frantz, J. M., Locke, J. C., Sturtz, D. and Leisner, S. 2010. Silicon in ornamental crops: detection, delivery, and function. In: Rodrigues, F. A. (Ed), *Silicio na agricultura: Anais do V simposio brasileiro sobre silicio na agricultura*. Universidade Federal de Vicosa, Vicosa, Brazil, pp. 111–34.

Fraysse, F., Cantais, F., Pokrovsky, O. S., Schott, J. and Meunier, J. D. 2006. Aqueous reactivity of phytoliths and plant litter: physico-chemical constraints of terrestrial biogeochemical cycle of silicon. *Journal of Geochemical Exploration* 88(1–3), 202–5. doi:10.1016/j.gexplo.2005.08.039.

French-Monar, R. D., Rodrigues, F. A., Korndorfer, G. H. and Datnoff, L. E. 2010. Silicon suppresses phytophthora blight development on bell pepper. *Journal of Phytopathology* 158(7–8), 554–60. doi:10.1111/j.439-0434.2009.01665.x.

Gengmaeo, Z., Shihui, L., Xing, S., Yizhou, W. and Zipan, C. 2015. The role of silicon in physiology of the medicinal plant (*Lonicera japonica* L.) under salt stress. *Scientific Reports* 5, 12696. doi:10.1038/srep12696.

Gowayed, M. H., Al-Zahrani, H. S. M. and Metwali, E. M. R. 2017. Improving the salinity tolerance in potato (*Solanum tuberosum*) by exogenous application of silicon dioxide nanoparticles. *International Journal of Agriculture and Biology* 19(1), 183-92. doi:10.17957/IJAB/15.0262.

Haddad, C., Arkoun, M., Jamois, F., Schwarzenberg, A., Yvin, J. C., Etienne, P. and Laine, P. 2018. Silicon promotes growth of *Brassica napus* L. and delays leaf senescence induced by nitrogen starvation. *Frontiers in Plant Science* 9, 516. doi:10.3389/fpls.2018.00516.

Haghighi, M. and Pessarakli, M. 2013. Influence of Si and nano-silicon on salinity tolerance of cherry tomatoes (*Solanum lycopersicum* L.) at early growth stage. *Scientia Horticulturae* 262, 111-7.

Haghighi, M., Afifipour, Z. and Mozafarian, M. 2012. The effect of N-Si on tomato seed germination under salinity levels. *Journal of Biology and Environmental Sciences* 6, 87-90.

Hajiboland, R., Cheraghvareh, L. and Poschenrieder, C. 2017. Improvement of drought tolerance in tobacco (*Nicotiana rustica* L.) plants by silicon. *Journal of Plant Nutrition* 40(12), 1661-76. doi:10.1080/01904167.2017.1310887.

Hill, J. F. (Trans.). 2012. *Chemical Research on Plant Growth by Theodore de Saussure*. Springer, New York, NY.

Hodson, M. J., White, P. J., Mead, A. and Broadley, M. R. 2005. Phylogenetic variation in the silicon composition of plants. *Annals of Botany* 96(6), 1027-46. doi:10.1093/aob/mci255.

Huang, C. H., Roberts, P. D. and Datnoff, L. E. 2011. Silicon suppresses fusarium crown and root rot of tomato. *Journal of Phytopathology* 159(7-8), 546-54. doi:10.1111/j.1439-0434.2011.01803.x

Isa, M., Bai, S., Yokoyama, T., Ma, J. F., Ishibashi, Y., Yuasa, T. and Iwaya-Inoue, M. 2010. Silicon enhances growth independent of silica depositions in a low-silica mutant, *lsi1*. *Plant and Soil* 331(1-2), 361-75. doi:10.1007/s11104-009-0258-9.

Jang, J. Y., Kim, D. G., Kim, Y. O., Kim, J. S. and Kang, H. 2004. An expression analysis of a gene family encoding plasma membrane aquaporins in response to abiotic stresses in *Arabidopsis thaliana*. *Plant Molecular Biology* 54(5), 713-25. doi:10.1023/B:PLAN.0000040900.61345.a6.

Kafi, M. and Rahimi, Z. 2011. Effect of salinity and silicon on root characteristics, growth, water status, proline content and ion accumulation of purslane (*Portulaca oleracea* L.). *Soil Science and Plant Nutrition* 57(2), 341-7. doi:10.1080/00380768.2011.567398.

Kalteh, M., Alipour, Z. T., Ashraf, S., Aliabadi, M. M. and Nosratabadi, A. F. 2014. Effect of silica nanoparticles on basil (*Ocimum basilicum*) under salinity stress. *Journal of Chemical Health Risks* 4, 49-55.

Kamenidou, S., Cavins, T. J. and Marek, S. 2008. Silicon supplements affect horticultural traits of greenhouse-produced ornamental sunflowers. *HortScience* 43(1), 236-9. doi:10.21273/HORTSCI.43.1.236.

Kazemi, M. 2012. Effect of cobalt, silicon, acetylsalicylic acid and sucrose as novel agents to improve vase-life of *Argyranthemum* flowers. *Trends in Applied Sciences Research* 7(7), 579-83. doi:10.3923/tasr.2012.579.583.

Kazemi, M., Asadi, M. and Aghdasi, S. 2012a. Postharvest life of cut Lisianthus flowers as affected by silicon, malic acid and acetylsalicylic acid. *Research Journal of Soil Biology* 4(1), 15-20. doi:10.3923/rjsb.2012.15.20.

Kazemi, M., Holami, M. G. and Bahmanipour, F. 2012b. Effect of silicon and acetylsalicylic acid on antioxidant activity, membrane stability and ACC-oxidase activity in relation to vase life of carnation cut flowers. *Biotechnology* 11(2), 87–90. doi:10.3923/biotech.2012.87.90.

Kieber, J. J. and Schaller, G. E. 2018. Cytokinin signaling in plant development. *Development* 145(4), dev149344. doi:10.1242/dev.149344.

Klimczyk, A., Stachura, R., Bernasowski, M. and Ledzki, A. 2012. Silicon behavior at the blast furnace process. *Journal of Achievements in Materials and Manufacturing Engineering* 55, 712–5.

Knight, C. T. G. and Kinrade, S. D. 2001. A primer on the aqueous chemistry of silicon. In: Datnoff, L. E., Snyder, G. H. and Korndörfer, G. H. (Eds), *Silicon in Agriculture, Volume 8: Studies in Plant Science*. Elsevier, Amsterdam, the Netherlands, pp. 57–84.

Korunic, Z. 1998. Diatomaceous earths, a group of natural insecticides. *Journal of Stored Products Research* 34(2–3), 87–97. doi:10.1016/S0022-474X(97)00039-8.

Kumar, S., Soukup, M. and Elbaum, R. 2017. Silicification in grasses: variation between different cell types. *Frontiers in Plant Science* 8, 438. doi:10.3389/fpls.2017.00438.

Laane, H. M. 2017. The effects of the application of foliar sprays with stabilized silicic acid: an overview of the results from 2003–2014. *Silicon* 9(6), 803–7. doi:10.1007/s12633-016-9466-0.

Laane, H. M. 2018. The effects of foliar sprays with different silicon compounds. *Plants* 7(2), 45. doi:10.3390/plants7020045.

Lang, D. Y., Fei, P. X., Cao, G. Y., Jia, X. X., Li, Y. T. and Zhang, X. H. 2019. Silicon promotes seedling growth and alters endogenous IAA, GA_3 and ABA concentrations in *Glycyrrhiza uralensis* under 100 mM NaCl stress. *The Journal of Horticultural Science and Biotechnology* 94(1), 87–93. doi:10.1080/14620316.2018.1450097.

Lee, C. W., Mahendra, S., Zodrow, K., Li, D., Tsai, Y. C., Braam, J. and Alvarez, P. J. J. 2010. Developmental phytotoxicity of metal oxide nanoparticles to *Arabidopsis thaliana*. *Environmental Toxicology and Chemistry* 29(3), 669–75. doi:10.1002/etc.58.

Lewin, J. and Reimann, B. E. F. 1969. Silicon and plant growth. *Annual Reviews of Plant Physiology* 20(1), 289–304. doi:10.1146/annurev.pp.20.060169.001445.

Li, J., Leisner, S. M. and Frantz, J. 2008. Alleviation of copper toxicity in *Arabidopsis thaliana* by silicon addition to hydroponic solutions. *Journal of the American Society for Horticultural Science* 133(5), 670–7. doi:10.21273/JASHS.133.5.670.

Li, R. Y., Ago, Y., Liu, W. J., Mitani, N., Feldmann, J., McGrath, S. P., Ma, J. F. and Zhao, F. J. 2009. The rice aquaporin *Lsi1* mediates uptake of methylated arsenic species. *Plant Physiology* 150(4), 2071–80. doi:10.1104/pp.109.140350.

Li, H., Zhu, Y., Hu, Y., Han, W. and Gong, H. 2015. Beneficial effects of silicon in alleviating salinity stress of tomato seedlings grown under sand culture. *Acta Physiologiae Plantarum* 37(4). doi:10.1007/s11738-015-1818-7.

Liang, Y., Nikolic, M., Belanger, R., Gong, H. and Song, A. 2015a. History and introduction of silicon research. In: Silicon in Agriculture: From Theory to Practice. Springer, New York, NY, pp. 1–18. doi:10.1007/978-94-017-9978-2.

Liang, X., Wang, H., Hu, Y., Mao, L., Sun, L., Dong, T., Nan, W. and Bi, Y. 2015b. Silicon does not mitigate cell death in cultured tobacco BY-2 cells subjected to salinity without ethylene emission. *Plant Cell Reports* 34(2), 331–43. doi:10.1007/s00299-014-1712-6.

Liebig, J. 1840. *Organic Chemistry in Its Application to Agriculture and Physiology*. From the manuscript of the author by Lyon Playfair. Taylor & Walton, London, UK.

Lin, Y., Sun, Z., Li, Z., Xue, R., Cui, W., Sun, S., Liu, T., Zeng, R. and Song, Y. 2019. Deficiency in silicon transporter *Lsi1* compromises inducibility of anti-herbivore defense in rice plants. *Frontiers in Plant Science* 10, 652. doi:10.3389/fpls.2019.00652.

Lins, U., Barros, C. F., da Cunha, M. and Miguens, F. C. 2002. Structure, morphology and composition of silicon biocomposites in the palms tree *Syagrus coronata* (Mart.) Beck. *Protoplasma* 220(1-2), 89-96. doi:10.1007/s00709-002-0036-5.

Liu, P., Yin, L., Wang, S., Zhang, M., Deng, X., Zhang, S. and Tanaka, K. 2015. Enhanced root hydraulic conductance by aquaporin regulation accounts for silicon alleviated salt-induced osmotic stress in *Sorghum bicolor* L. *Environmental and Experimental Botany* 111, 42-51. doi:10.1016/j.envexpbot.2014.10.006.

Ma, J. F. and Takashashi, E. 2002. *Soil, Fertilizer and Plant Silicon Research in Japan*. Elsevier Science, Amsterdam, the Netherlands.

Ma, J. F., Tamai, K., Yamaji, N., Mitani, N., Konishi, S., Katsuhara, M., Ishiguro, M., Murata, Y. and Yano, M. 2006. A silicon transporter in rice. *Nature* 440(7084), 688-91. doi:10.1038/nature04590.

Ma, J. F., Yamaji, N., Mitani, N., Tamai, K., Konishi, S., Fujiwara, T., Katsuhara, M. and Yano, M. 2007. An efflux transporter of silicon in rice. *Nature* 448(7150), 209-12. doi:10.1038/nature05964.

Ma, J. F., Yamaji, N., Mitani, N., Xu, X. Y., Su, Y. H., McGrath, S. P. and Zhao, F. J. 2008. Transporters of arenite in rice and their role in arsenic accumulation in rice grain. *Proceedings of the National Academy of Sciences* 105(29), 9931-5. doi:10.1073/pnas.0802361105.

Manivannan, A. and Ahn, Y. K. 2017. Silicon regulates potential genes involved in major physiological processes in plants to combat stress. *Frontiers in Plant Science* 8, 1346. doi:10.3389/fpls.2017.01346.

Manivannan, A., Soundararajan, P., Muneer, S., Ko, C. H. and Jeong, B. R. 2016. Silicon mitigates salinity stress by regulating the physiology, antioxidant enzyme activities and protein expression in *Capsicum annuum* 'Bugwang'. *BioMed Research International* 2016, 3076357. doi:10.1155/2016/3076357.

Manivannan, A., Soundararajan, P., Cho, Y. S., Park, J. E. and Jeong, B. R. 2018. Sources of silicon influence photosystem and redox homeostasis-related proteins during the axillary shoot multiplication of *Dianthus caryophyllus*. *Plant Biosystems - an International Journal Dealing with All Aspects of Plant Biology* 152(4), 704-10. doi:10.1080/11263504.2017.1320312.

Markovich, O., Steiner, E., Kouril, Š, Tarkowski, P., Aharoni, A. and Elbaum, R. 2017. Silicon promotes cytokinin biosynthesis and delays senescence in Arabidopsis and Sorghum. *Plant, Cell and Environment* 40(7), 1189-96. doi:10.1111/pce.12913.

Marodin, J. C., Resende, J. T., Morales, R. G., Faria, M. V., Trevisam, A. R., Figueiredo, A. S. and Dias, D. M. 2016. Tomato post-harvest durability and physiochemical quality depending on silicon sources and doses. *Horticultura Brasileira* 34(3), 361-6. doi:10.1590/S0102-05362016003009.

Matichenkov, V., Calvert, D. and Snyder, G. 1999. Silicon fertilizers for citrus in Florida. *Proceedings of the Florida State Horticultural Science* 112, 5-8.

Mditshwa, A., Bower, J. P., Bertling, I. and Mathaba, N. 2013. Investigation of the efficiency of the total antioxidants assays in silicon-treated lemon fruit (*Citrus limon*). *Acta Horticulturae* 1007(1007), 93-102. doi:10.17660/ActaHortic.2013.1007.7.

Mehrabanjoubani, P., Abdolzadeh, A., Sadeghipour, H. R. and Aghdasi, M. 2015. Silicon affects transcellular and apoplastic uptake of some nutrients in plants. *Pedosphere* 25(2), 192–201. doi:10.1016/S1002-0160(15)60004-2.

Mitani, N. N., Yamaji, N. and Ma, J. F. 2009a. Identification of maize silicon influx transporters. *Plant and Cell Physiology* 50(1), 5–12. doi:10.1093/pcp/pcn110.

Mitani, N., Chiba, Y., Yamaji, N. and Ma, J. F. 2009b. Identification and characterization of maize and barley *Lsi2*-like silicon efflux transporters reveals a distinct silicon uptake system from that in rice. *The Plant Cell* 21(7), 2133–42. doi:10.1105/tpc.109.067884.

Mitani, N., Yamaji, N., Ago, Y., Iwasaki, K. and Ma, J. F. 2011. Isolation and functional characterization of an influx silicon transporter in two pumpkin cultivars contrasting in silicon accumulation. *The Plant Journal: for Cell and Molecular Biology* 66(2), 231–40. doi:10.1111/j.1365-313X.2011.04483.x.

Montpeptit, J., Vivancos, J., Mitani-Ueno, N., Yamaji, N., Remus-Borel, W., Belzile, F., Ma, J. F. and Belanger, R. R. 2012. Cloning, functional characterization and heterologous expression of *TaLsi1*, a wheat silicon transporter gene. *Plant Molecular Biology* 79(1–2), 35–46. doi:10.1007/s11103-012-9892-3.

Moradtalab, N., Weinmann, M., Wlker, F., Hoglinger, B., Ludewig, U. and Neumann, G. 2018. Silicon improves chilling tolerance during early growth of maize by effects on micronutrient homeostasis and hormonal balances. *Frontiers in Plant Science* 9, 420. doi:10.3389/fpls.2018.00420.

Moradtalab, N., Hajiboland, R., Aliasgharzd, N., Hartmann, T. E. and Neumann, G. 2019. Silicon and the association with an arbuscular-mycorrhizal fungus (*Rhizophagus clarus*) mitigate the adverse effects of drought stress on strawberry. *Agronomy* 9(1), 41. doi:10.3390/agronomy9010041.

Neumann, D. and De Figueiredo, C. 2002. A novel mechanism of silicon uptake. *Protoplasma* 220(1–2), 59–67. doi:10.1007/s00709-002-0034-7.

Ouellette, S., Goyette, M. H., Labbe, C., Laur, J., Gaudreau, L., Gosselin, A., Dorais, M., Deshmukh, R. K. and Belanger, R. R. 2017. Silicon transporters and effects of silicon amendments in strawberry under high tunnel and field conditions. *Frontiers in Plant Science* 8, 949. doi:10.3389/fpls.2017.00949.

Pandey, C., Khan, E., Panthri, M., Tripathi, R. D. and Gupta, M. 2016. Impact of silicon on Indian mustard (*Brassica juncea* L.) root traits by regulating growth parameters, cellular antioxidants and stress modulators under arsenic stress. *Plant Physiology and Biochemistry: PPB* 104, 216–25. doi:10.1016/j.plaphy.2016.03.032.

Pavlovic, J., Samardzic, J., Maksimovic, V., Timotijevic, G., Stevic, N., Laursen, K. H., Hansen, T. H., Husted, S., Schjoerring, J. K., Liang, Y. and Nikolic, M. 2013. Silicon alleviates iron deficiency in cucumber by promoting mobilization of iron in the root apoplast. *New Phytologist* 198(4), 1096–107. doi:10.1111/nph.12213.

Pereira, H. S., Korndorfer, G. H., dos Reis, C. B. and Correa, G. F. 2007. Potencial de uso de agregados siderurgicos e outras fontes de silicio em quatro materiais de solo fase cerrado. *Bioscience Journal Uberlandia* 23, 17–31.

Powell, J. J., McNaughton, S. A., Jugdaohsingh, R., Anderson, S. H. C., Dear, J., Khot, F., Mowatt, L., Gleason, K. L., Sykes, M., Thompson, R. P. H., Bolton-Smith, C. and Hodson, M. J. 2005. A provisional database for the silicon content of foods in the United Kingdom. *British Journal of Nutrition* 94(5), 804–12. doi:10.1079/bjn20051542.

Pozza, E. A., Pozza, A. A. A. and Botelho, D. M. S. 2015. Silicon in plant disease control. *Revista Ceres* 62(3), 323–31. doi:10.1590/0034-737X201562030013.

Prychid, C. J., Rudall, P. J. and Gregory, M. 2003. Systematics and biology of silica bodies in monocotyledons. *The Botanical Review* 69(4), 377–440. doi:10.1663/0006-8101(2004)069[0377:SABOSB]2.0.CO;2.

Rangaraj, S., Gopalu, K., Rathinam, Y., Periasamy, P., Venkatachalam, R. and Narayanasamy, K. 2014. Effect of silica nanoparticles on microbial biomass and silica availability in maize rhizosphere. *Biotechnology and Applied Biochemistry* 61(6), 668–75. doi:10.1002/bab.1191.

Rezende, M. T. N. P., Komatsu, R. S., Andrade, R. B., Reis, S. L. B., Leite, C. R., Rocha, S. R., Saut, J. P. E. and Guimarães, E. C.. 2018. Economic losses due to the occurrence of cysticercosis in cattle from cities located in Minas Gerais, Brazil. *Ciência Rural* 48(12). doi:10.1590/0103-8478cr20180483.

Robberecht, H., Van Dyck, K., Bosscher, D. and Van Cauwenbergh, R. 2008. Silicon in foods: content and bioavailability. *International Journal of Food Properties* 11(3), 638–45. doi:10.1080/10942910701584252.

Robberecht, H., Van Cauwenbergh, R., Van Vlaslaer, V. and Hermans, N. 2009. Dietary silicon intake in Belgium: sources, availability from foods, and human serum levels. *Science of the Total Environment* 407(16), 4777–82. doi:10.1016/j.scitotenv.2009.05.019.

Rodrigues, F. A. and Datnoff, L. E. 2015. *Silicon and Plant Diseases*. Springer International Publishing, Switzerland. doi:10.1007/978-3-319-22930-0.

Rostkowska, C., Mota, C. M., Oliveria, T. C., Santiago, F. M., Oliveira, L. A., Korndorfer, G. H., Lana, R. M. Q., Rossi, M. L., Nogueira, N. L., Simonnet, X., Mineo, T. W. P., Silva, D. A. O. and Mineo, J. R. 2016. Si-Accumulation in *Artemisia annua* glandular trichomes increases artemisinin concentration, but does not interfere in the impairment of *Toxoplasma gondii* growth. *Frontiers in Plant Science* 7, 1430 doi:10.3389/fpls.2016.01430.

Sahebi, M., Hanafi, M. M. and Azizi, P. 2016. Application of silicon in plant tissue culture. In Vitro Cellular and Developmental Biology 52(3), 226–32. doi:10.1007/s11627-016-9757-6.

Saltikov, C. W. and Olson, B. H. 2002. Homology of *Escherichia coli* R773 *arsA*, *arsB*, and *arsC* genes in arsenic-resistant bacteria isolated from raw sewage and arsenic-enriched creek waters. *Applied and Environmental Microbiology* 68(1), 280–8. doi:10.1128/aem.68.1.280-288.2002.

Savvas, D. and Ntatsi, G. 2015. Biostimulant activity of silicon in horticulture. *Scientia Horticulturae* 196, 66–81. doi:10.1016/j.scienta.2015.09.010.

Scott, P. W., Critchley, S. R. and Wilkinson, F. C. F. 1986. The chemistry and mineralogy of some granulated and pelletized blast furnace slags. *Mineralogical Magazine* 50(355), 141–7. doi:10.1180/minmag.1986.050.355.19.

Sebastian, D., Rodrigues, H., Kinsey, C., Korndörfer, G., Pereira, H., Buck, G., Datnoff, L., Miranda, S. and Provance-Bowley, M. 2013. A 5-day method for determination of soluble silicon concentrations in nonliquid fertilizer materials using a sodium carbonate-ammonium nitrate extractant followed by visible spectroscopy with heteropoly blue analysis: single-laboratory validation. *Journal of AOAC International* 96(2), 251–9. doi:10.5740/jaoacint.12-243.

Shekari, F., Abbasi, A. and Mustafavi, S. H. 2017. Effect of silicon and selenium on enzymatic changes and productivity of dill in saline conditions. *Journal of the Saudi Society of Agricultural Sciences* 16(4), 367–74. doi:10.1016/j.jssas.2015.11.006.

Shi, Y., Zhang, Y., Han, W., Feng, R., Hu, Y., Guo, J. and Gong, H. 2016. Silicon enhances water stress tolerance by improving root hydraulic conductance in *Solanum lycopersicum* L. *Frontiers in Plant Science* 7, 196. doi:10.3389/fpls.2016.00196.

Siddiqui, M. H. and Al-Whaibi, M. H. 2014. Role of nano-SiO_2 in germination of tomato (*Lycopersicum esculentum* seeds Mill.). *Saudi Journal of Biological Sciences* 21(1), 13–7. doi:10.1016/j.sjbs.2013.04.005.

Sivanesan, I. and Park, S. W. 2014. The role of silicon in plant tissue culture. *Frontiers in Plant Science* 5, 571. doi:10.3389/fpls.2014.00571.

Soares, J. D. R., Pasqual, M., Araujo, A. Gd, Castro, E. Md, Pereira, F. J. and Braga, F. T. 2012. Leaf anatomy of orchids micropropagated with different silicon concentrations. *Acta Scientiarum. Agronomy* 34(4), 413–21. doi:10.4025/actasciagron.v34i4.15062.

Soundararajan, P., Sivanesan, I., Jana, S. and Jeong, B. R. 2014. Influence of silicon supplementation on the growth and tolerance to high temperature in *Salvia splendens*. *Horticulture, Environment, and Biotechnology* 55(4), 271–9. doi:10.1007/s13580-014-0023-8.

Soundararajan, P., Manivannan, A., Ko, C. H. and Jeong, B. R. 2018. silicon enhanced redox homeostasis and protein expression to mitigate the salinity stress in *Rosa hybrida* 'Rock Fire'. *Journal of Plant Growth Regulation* 37(1), 16–34. doi:10.1007/s00344-017-9705-7.

Sripanyakorn, S., Jugdaohsingh, R., Dissayabutr, W., Anderson, S. H. C., Thompson, R. P. H., and Powell, J. J. 2009. The comparative absorption of silicon from different foods and food supplements. British Journal of Nutrition 102(6), 825–34. doi:10.1017/S0007114509311757.

Stamatakis, A., Papadantonakis, N., Savvas, D., Lydakis-Simantiris, N. and Kefalas, P. 2003. Effects of silicon and salinity on fruit yield and quality of tomato grown hydroponically. *Acta Horticulturae* 609(609), 141–7. doi:10.17660/ActaHortic.2003.609.18.

Strigel, A. 1912. Vergleichende Untersuchungen: A. Über Mineralstoffaufnahme verschiedener Pflanzenarten aus ungedüngtem Boden. B. Über den Einfluss der botanischen Natur, der Herfunft und der Erntezeit auf die chemische Zusammensetzung von Wiesenheu. *Landwirtschaftliche Jahrbücher* 43(1), 349–71.

Struve, G. A. 1935. *De Silicia in Plantis Nonulis*. Dissertation, Berolini.

Sun, H., Guo, J., Duan, Y., Zhang, T., Huo, H. and Gong, H. 2017. Isolation and functional characterization of CsLsi1, a silicon transporter gene in *Cucumis sativus*. *Physiologia Plantarum* 159(2), 201–14. doi:10.1111/ppl.12515.

Takahashi, E., Ma, J. F. and Miyake, Y. 1990. The possibility of silicon as an essential element for higher plants. *Comments on modern chemistry. Part B, Comments on Agricultural* and *Food Chemistry* 2(2), 99–102.

Tao, Y., Ran, Z., Jing, Z., Liang, W. and Jin-Ping, F. 2014. Transgenic *Petunia hybrida* with silicon transporter proteins *OsLsi1* and *OsLsi2* genes and its drought resistance analysis. *Journal of Northeast Agricultural University* 21(3), 30–8. doi:10.1016/S1006-8104(14)60066-0.

Tripathi, D. K., Singh, V. P., Prasad, S. M., Chauhan, D. K. and Dubey, N. K. 2015. Silicon nanoparticles (SiNP) alleviate chromium (VI) phytotoxicity in *Pisum sativum* (L.) seedlings. *Plant Physiology and Biochemistry: PPB* 96, 189–98. doi:10.1016/j.plaphy.2015.07.026.

Tripathi, D. K., Shweta, Singh, S., Singh, S., Pandey, R., Singh, V. P., Sharma, N. C., Prasad, S. M., Dubey, N. K. and Chauhan, D. K. 2017. An overview of manufactured nanoparticles in plants: uptake, translocation, accumulation and phytotoxicity. *Plant Physiology and Biochemistry: PPB* 110, 2–12. doi:10.1016/j.plaphy.2016.07.030.

Tubana, B. S., Babu, T. and Datnoff, L. E. 2016. A review of silicon in soils and plants and its role in US agriculture: history and future perspectives. *Soil Science* 181, 393–411. doi:10.1097/SS.0000000000000179.

Vendrame, W. A., Palmateer, A. J., Pinares, A., More, K. A. and Datnoff, L. E. 2010. Silicon fertilization affects growth of hybrid phalaenopsis orchid liners. *HortTechnology* 20(3), 603–7. doi:10.21273/HORTTECH.20.3.603.

Wang, S., Liu, P., Chen, D., Yin, L., Li, H. and Deng, X. 2015. Silicon enhanced salt tolerance by improving the root water uptake and decreasing the ion toxicity in cucumber. *Frontiers in Plant Science* 6, 759. doi:10.3389/fpls.2015.00759.

Wang, M., Gao, L., Dong, S., Sun, Y., Shen, Q. and Guo, S. 2017. Role of silicon on plant-pathogen interactions. *Frontiers in Plant Science* 8, 701. doi:10.3389/fpls.2017.00701.

Winslow, M. D. 1992. Silicon, disease resistance and yield of rice genotypes under upland cultural conditions. *Crop Science* 32(5), 1208–13. doi:10.2135/cropsci1992.0011183X003200050030x.

Winslow, M. D., Okada, K. and Correa-Victoria, F. 1997. Silicon deficiency and the adaptation of tropical rice ecotypes. *Plant and Soil* 188(2), 239–48. doi:10.1023/A:1004298817861.

Wu, Q. S., Wan, X. Y., Su, N., Cheng, Z. J., Wang, J. K., Lei, C. L., Zhang, X., Jiang, L., Ma, J. F. and Wan, J. M. 2006. Genetic dissection of silicon uptake ability in rice (*Oryza sativa* L.). *Plant Science: an International Journal of Experimental Plant Biology* 171(4), 441–8. doi:10.1016/j.plantsci.2006.05.001.

Wu, Z., Wang, F., Liu, S., Du, Y., Li, F., Du, R., Wen, D. and Zhao, J. 2016. Comparative responses to silicon and selenium in relation to cadmium uptake, compartmentation in roots, and xylem transport in flowering Chinese cabbage (*Brassica campestris* L. ssp. *chinensis* var. *utilis*) under cadmium stress. *Environmental and Experimental Botany* 131, 173–80. doi:10.1016/j.envexpbot.2016.07.012.

Xu, C. X., Ma, Y. P. and Liu, Y. L. 2015. Effects of silicon on growth, quality and ionic homeostasis of aloe under salt stress. *South African Journal of Botany* 98, 26–36. doi:10.1016/j.sajb.2015.01.008.

Yamaji, N. and Ma, J. F. 2007. Spatial distribution and temporal variation of the rice silicon transporter *Lsi1*. *Plant Physiology* 143(3), 1306–13. doi:10.1104/pp.106.093005.

Yamaji, N., Mitani, N. and Ma, J. F. 2008. A transporter regulating silicon distribution in rice shoots. *The Plant Cell* 20(5), 1381–9. doi:10.1105/tpc.108.059311.

Yang, H. C., Fu, H. L., Lin, Y. F. and Rosen, B. P. 2012. Pathways of arsenic uptake and efflux. *Current Topics in Membranes* 69, 325–58. doi:10.1016/B978-0-12-394390-3.00012-4.

Zellner, W. L., Friedrich, R. L., Kim, S., Sturtz, D., Frantz, J., Altland, J. and Krause, C. 2015. Continuing Assessment of the 5-day sodium carbonate-ammonium nitrate extraction assay as an indicator test for silicon fertilizers. *Journal of AOAC International* 98(4), 890–5. doi:10.5740/jaoacint.14-205.

Zellner, W., Lutz, L., Khandekar, S. and Leisner, S. 2019. Identification of NtNIP2;1: an *Lsi1* silicon transporter in *Nicotiana tabacum*. *Journal of Plant Nutrition* 42(9), 1028–35. doi:10.1080/01904167.2019.1589500.

Zhang, W., Xie, Z., Wang, L., Li, M., Lang, D. and Zhang, X. 2017. Silicon alleviates salt and drought stress of *Glycyrrhiza uralensis* seedling by altering antioxidant metabolism and osmotic adjustment. *Journal of Plant Research* 130(3), 611–24. doi:10.1007/s10265-017-0927-3.

Zhang, Y., Shi, Y., Gong, H., Zhao, H., Li, H., Hu, Y. and Wang, Y. 2018. Beneficial effects of silicon on photosynthesis of tomato seedlings under water stress. *Journal of Integrative Agriculture* 17(10), 2151–9. doi:10.1016/S2095-3119(18)62038-6.

Zhao, X. Q., Mitani, N., Yamaji, N., Shen, R. F. and Ma, J. F. 2010. Involvement of silicon influx transporter OsNIP2;1 in selenite uptake in rice. *Plant Physiology* 153(4), 1871–7. doi:10.1104/pp.110.157867.

Zwack, P. J. and Rashotte, A. M. 2013. Cytokinin inhibition of leaf senescence. *Plant Signaling and Behavior* 8(7), e24737. doi:10.4161/psb.24737.

Chapter 7

Plant growth-promoting rhizobacteria (PGPR) as plant biostimulants in agriculture

Dongmei Lyu, Rachel Backer and Donald Smith, McGill University, Canada

1 Introduction

Among the most complex communities is the rhizosphere. It includes a wide array of microbes, including fungi, bacteria and many other organisms (Venturi and Keel, 2016; Mhlongo et al., 2018). A key component is the plant growth-promoting rhizobacteria (PGPR), which live both in the rhizosphere and inside of plant roots, and confer benefit with regard to plant growth (Timmusk et al., 2017). PGPR not only have significant effects on plant growth, including increased biomass production and quality improvement under optimal growing conditions, but they can also impart these benefits under stressful growing conditions by improving the resistance of plants to abiotic stresses (Backer et al., 2018).

PGPR also have the ability to promote plant growth through the production of beneficial compounds, such as phytohormones, enzymes and siderophores; improvement of nutrient absorption; nitrogen fixation; and solubilizing inorganic phosphate and iron, mineralizing organic phosphate (Backer et al., 2018; Bargaz et al., 2018). These effects have been demonstrated for various plant species including corn (Lin et al., 2018), canola (El-Howeity and Asfour, 2012), wheat (Sahin and Turan, 2014), soybean (Marinkovic et al., 2018) and

http://dx.doi.org/10.19103/AS.2020.0068.08

potato (Naqqash et al., 2016). PGPR can also improve plant stress tolerance and decrease the deleterious effects caused by the stressors, mainly through producing antibiotics and triggering induced systemic resistance (ISR) (Esitken et al., 2006). Recently, the production of specific secondary metabolites in plants has also been demonstrated as a strategy to improve plant growth under both abiotic and biotic stress; however, a limited number of studies provide insight into the precise mechanism(s) underlying these effects. Thus, gaining a complete understanding of the various mechanisms of interactions between plants and PGPR will contribute to our ability to exploit deeper and broader relationships between microbes of the phytomicrobiome and the associated plant.

2 Plant growth promotion

Plant acquisition of nutrients from soil is a very complex process and depends on several environmental and biological factors (Lakshmanan et al., 2014). Since they are efficient plant growth promoters, the application of PGPR improves plant growth and crop productivity. This can occur directly, when PGPR provide necessary plant nutrients, for example, by N fixation. Alternatively, this can occur indirectly when PGPR facilitate the solubilization and availability of Zn (Goteti et al., 2013) and P (Backer et al., 2018) and the enhancement of nitrogen fixation and siderophore production (Pii et al., 2015). Additionally, PGPR manipulate some enzymatic activities related to biomass production and quality improvement such as protein, starch and oil content accumulation (Banchio et al., 2008; Shakeri et al., 2016).

In addition, it has been demonstrated that the production of auxins, cytokinins and gibberellins can be stimulated by PGPR and that this is an important strategy for plant growth promotion (Choi et al., 2011; Duca et al., 2014; Pandya and Desai, 2014). In this way, the regulation of these hormones, affected the plant metabolism within specific pathways, is also considered to be a possible explanation for the stimulation of plant growth (Khan et al., 2019). Most importantly, the production of phytohormones by PGPR has significant effects on plant development and also affects plant defence responses, resulting in negative and positive effects under stressful growing conditions (Peres et al., 2019).

3 Plant growth-promoting rhizobacteria (PGPR) and abiotic stress

As the demand for food increases, as a result of the dramatic increase in the global population, the improvement of food production becomes a critical step in fulfilling the requirement. However, as published data shows, about 96% global land is adversely affected by environmental factors which limit crop production by approximately 70% (Boyer, 1982; Van Velthuizen, 2007). When plants grow under abiotic stress conditions, the result is decreased plant growth and yield and, under extreme conditions, plant dormancy or death. This occurs because abiotic stress conditions can inhibit the synthesis of proteins necessary for growth and alter metabolic pathways necessary for plant functions such as photosynthesis, respiration, translocation of molecules within the plant, uptake of ions from the soil, carbohydrate and nutrient metabolism and production of growth promoters (Liu and Howell, 2010). In addition to the direct effects on plant growth, abiotic stress conditions also affect the biodiversity and abundance of microorganisms in the rhizosphere (Farooq et al., 2008; Jaleel et al., 2009).

Most abiotic stresses, including heat, drought and cold, reduce plant growth and productivity (Wilhelm et al., 1999; Maestri et al., 2002; Janda et al., 2003; Fahad et al., 2016) via effects on plant dry weight and antioxidant enzyme activities, and the rate of metabolic reactions (Kang and Saltveit, 2001; Howarth, 2005; Essemine et al., 2010). Under stressful drought-growing conditions, seed germination can be reduced due to low moisture levels resulting from the high rates of evaporation or transpiration (Shao et al., 2007; Kanwal et al., 2017). In addition, severe stress also reduces root growth, which impedes the transportation of water and plant nutrients (nitrogen, phosphorus, magnesium and calcium) to the above-ground parts of the plant (Bassirirad, 2000; Garg, 2003; Wahid et al., 2007; Huang et al., 2012). Meanwhile, the rate of photosynthesis can be reduced by cold stress because chloroplasts are impacted by low temperatures. This prevents the plant from meeting its basic energy requirements (Biswal et al., 2011; Theocharis et al., 2012), which also leads to decreased chlorophyll contents (Shanmugam et al., 2013). Lastly, the activity of enzymes involved in grain filling for cereals (e.g. sucrose synthase, starch synthase, starch branching enzyme, and adenosine diphosphate glucose pyrophosphorylase) are affected by extreme growing conditions, which directly decreases the yield and quality of these crops (Ahmadi and Baker, 2001).

PGPR application to agricultural crops is a promising and efficient sustainable agricultural technology that has already been applied around the world. The application of PGPR can increase plant resistance to abiotic stresses by regulating phytohormone levels and via other mechanisms (Sergeeva et al., 2006; Kundan et al., 2015; Park et al., 2017). It has been illustrated that PGPR can promote plant growth by secreting plant hormones such as auxin, ABA, cytokinin and GA under stressful conditions (Rosier et al., 2018). For example, the application of a GA-producing PGPR increased seed germination under cold stress (Barka et al., 2006). In another experiment, wheat seeds treated with *Pseudomonas aeruginosa* (strain 2CpS1) had increased root length as a result of increased ACC-deaminase activity (Meena et al., 2015). PGPR can produce plant growth hormones, which can result in enhanced resistance to heavy metal stress (Tsukanova et al., 2017;

Fan, 2018) and alter metal translocation within plants via the secretion of acids, proteins and other chemicals (Denton, 2007; Rajkumar and Freitas, 2008) and by improving ACC deaminase activity (Dell'Amico et al., 2008).

The application of *Bacillus subtilis* GB03 to plants enhanced the chlorophyll concentration and photosynthetic efficiency by regulating the endogenous glucose content and the transformation of abscisic acid signalling in *Arabidopsis* (Zhang et al., 2008b). *P. aeruginosa* is an effective nitrogen fixer that facilitates nutrient uptake by the solubilization of phosphorus and produces phytohormones that improve the morphological and physiological aspects of plant growth, including height, root length and leaf area of wheat (Tahir et al., 2009; Meena et al., 2015). Overall, PGPR can improve plant resistance to abiotic stress by increasing chlorophyll content, photosynthetic activity and relative water content and altering mineral uptake (Turan et al., 2013; Kanwal et al., 2017).

Inoculation of PGPR can also improve plant resistance to abiotic stress by alleviating membrane injury and improving superoxide dismutase (SOD), peroxidase (POD) and catalase (CAT) activity under stressful growing conditions. Park et al. (2017) demonstrated that *B. aryabhattai* SRB02 can strengthen CAT and SOD activities, allowing plants to tolerate oxidative stress induced by H_2O_2. In another study, the application of *P. putida* strain AKMP7 prevented the production of ROS, which supported the survival of wheat plants under heat stress (Ali et al., 2011). Dehydration caused by high temperatures can also affect the accumulation of proline, which is an important amino acid, related to abiotic stress responses (Ihsan et al., 2019). Interestingly, the application of *P. putida* strain AKMP7 to wheat (Ali et al., 2011) and *Bacillus amyloliquefaciens* NBRI-SN13 to rice seedlings (Tiwari et al., 2017) enhanced the production of proline under stressful conditions. Kanwal et al. (2017) and Izanloo et al. (2008) also reported that PGPR can improve plant stress resistance by causing the plant to accumulate more soluble sugars, which is another important strategy used by plants to maintain osmolyte homeostasis.

The production of volatile organic compounds (VOCs) and exopolysaccharides (EPSs) also represents mechanisms that dramatically affect plant resistance to abiotic stresses. Plants may produce fewer metabolites under optimal growing conditions than under stressful ones in the presence of beneficial microorganisms. In particular, PGPR application increases VOC levels in plants, especially under salinity and drought stress. When plants are exposed to osmotic stresses (e.g. soil salinity), PGPR can induce the biosynthesis of choline and glycine betaine and prevent Na^+ uptake, which improves nutrient transport within the plant and plant stress tolerance (Timmusk et al., 2014). Inoculation of *Arabidopsis thaliana* with *B. subtilis* reduces the influx of Na^+ by downregulating the expression of an HKT1/K^+ transporter via the production of VOCs (Zhang et al., 2008a).

PGPR production of EPSs can lead to similar results. EPSs are organic macromolecules, mainly including carbohydrates, which assist plants under salinity and water stresses (Flemming and Wingender, 2001; Prajapati et al., 2013). Rossi and De Philippis (2015) showed that EPS production associated with hydrophyllic-biofilm formation improved plant resilience to salinity stress. Under high salt levels, the EPS produced by PGPR elicit a response similar to some VOCs and can bind Na^+ to reduce its availability for uptake by plants (Tewari and Arora, 2014; Timmusk et al., 2014; Abbas et al., 2019). The inoculation of *Helianthus annuus* with *P. aeruginosa* and *Zea mays* with *Pseudomonas* and *Rhizobium* reduced electrolyte leakage and subsequently improved salt tolerance as a result of EPS production; this led to improved plant development, growth and crop yield (Vardharajula et al., 2011; Tewari and Arora, 2014). Additionally, the presence of EPS can improve water retention and regulate organic carbon source diffusion, which protects microorganisms from desiccation under drought conditions (Sandhya et al., 2009; Naseem and Bano, 2014). In polluted soil, the EPS contributes to the transformation of heavy metals and the degradation of recalcitrant organic contaminants (Naseem and Bano, 2014), reducing their effects on crops.

PGPR also improve plant stress resistance by regulating the expression of genes related to plant defence mechanisms. For example, it was found that the response of plants to high temperatures was related to the amount of endogenous ABA production under heat stress (Mascarenhas, 1984). However, the underlying mechanism for this response was not clear until the genes related to ABA biosynthesis were identified by Park et al. (2017). Their results showed that GmZEP (Glyma01g39310) and GmNCED (Glyma05g27250) are expressed under stressful conditions, leading to high-level ABA production under heat stress conditions when soybean plants were treated with *Bacillus aryabhattai* strain SRB02 (Park et al., 2017). Interestingly, the gene expression pattern was similar to that seen as a result of exogenous ABA under drought stress. As Lv et al. (2017) reported, many *DHNs* genes are regulated by exogenous ABA under drought stress. The application of PGPR can also directly affect LEA gene expression under abiotic stress, as demonstrated in chickpea inoculated with *Pseudomonas putida* (Tiwari et al., 2016). Thus, the manipulation of defence gene expression by PGPR is a function of the type of stress and plant genotype.

4 Plant growth-promoting rhizobacteria (PGPR) against biotic stress

Biotic stresses caused by pathogenic bacteria, fungi, viruses and pests are significant limiting factors for crop yield and productivity (Cohen and Leach, 2019). The control of soil-borne pathogens is a major challenge (Choudhary et al., 2007). The chronic application of pesticides to control the spread of pathogens could lead to negative environmental consequences (Panneerselvam and Saravanamuthu 2013), such as soil and water pollution, and can eventually lead to the evolution of pesticide resistance. Among the vast genetic biodiversity of prokaryotes, PGPR play a vital role in the management of plant diseases to increase crop productivity via various mechanisms, including the production of signalling molecules, plant hormones, antibiotics and siderophores (Saraf et al., 2014). For the bacterial and fungal pathogens, the interactions between the PGPR and pathogen can be directly through competition for nutrients and/or space, secreting antibiosis metabolites and infecting the pathogen as a biocontrol agent (Pankhurst and Lynch, 2005).

4.1 Microbe-microbe interactions

4.1.1 Production of antibiotics

An antibiotic is an organic chemical substance that is produced by a plant, an animal or a microorganism and is biocidal or biostatic to another organism (Davies, 2006). In the interaction between PGPR and plants, the primary mechanism of biocontrol by PGPR involves the production of antibiotics, including volatile and non-volatile antibiotics with various antimicrobial actions against pathogens (Fernando et al., 2005; Pankhurst and Lynch, 2005). Table 1 shows some examples of antibiotics produced by PGPR that were reported in the past two decades.

The biosynthesis of antibiotics by PGPR is affected by many factors, including the genus of bacteria and growth conditions and, in particular, the nutrient composition of the growing environment, which can provide either general or specific nutrients (Duffy and Défago, 2000; Ownley et al., 2003). Interestingly, various factors such as host plant species and plant age (Smith and Goodman, 1999; Notz et al., 2001) can affect the ability of PGPR to produce antibiotics. This could be explained by the production of root exudates by plants, which are the main energy source for PGPR and dictate the type of carbon source, nitrogen, amino acids and other compounds available for microbes (Slininger and Jackson, 1992; Fiddaman and Rossall, 1994; Milner et al., 1995; Duffy and Défago, 1999; Notz et al., 2001). However, the composition of root exudates can be affected by plant genotype, health and developmental stage (Timmusk et al., 2017). For instance, root exudates from older but not younger plants induce DAPG production by bacteria (Picard et al., 2000). Evidence for the role of antibiotics produced by PGPR in the biocontrol of plant pathogens includes strong correlations between antibiotic production and biocontrol efficiency (Pankhurst and Lynch, 2005).

The effects of antibiotics are also based on the specific antibiotic(s) produced by the PGPR. It has been reported that the production of polyketides,

Table 1 Examples of antibiotics produced by biocontrol PGPR

Type	Antibiotics	Producing PGPR	Target phytopathogen
Non-volatile antibiotics	Polyketides	*Pseudomonas fluorescens*	*Pythium ultimum* and *Rhizoctonia solani* (Nowak-Thompson et al., 1994),
			Xanthomonas oryzae pv. oryzae (Velusamy et al., 2006)
	Phenazine derivatives	Fluorescent *Pseudomonas*	*Gaeumannomyces graminis* var. tritici (Thomashow and Weller, 1988)
		Bacillus spp., *Bacillus velezensis, Bacillus vallismortis*	*Botrytis cinerea* (Toral et al., 2018),
	Cyclic lipopeptides	*Pseudomonas fluorescens*	*Ralstonia solanacearum* and *Fusarium oxysporum* (Cao et al., 2018),
			Phytophthora capsica (Park et al., 2016)
			Rhizoctonia solani (Nielsen et al., 2000)
	Lipopeptides	*Bacillus licheniformis*	*Magnaporthe grisea* (Tendulkar et al., 2007)
		Bacillus amyloliquefaciens	*Magnaporthe oryzae* (Rahman et al., 2015)
Volatile antibiotics	Hydrogen cyanide	*Pseudomonas fluorescens*	*Macrophomina phaseolina* (Reetha et al., 2014)

including 2,4 diacetyl phloroglucinol, pyoluteorin and mupirocin, highly restricts the growth of plant pathogens (Fernando et al., 2005). In addition, the production of antibiotics is also dependant on the cell density of PGPR: there is a greater synthesis of antibiotics at higher cell densities, such as in the stationary phase of microbial growth. Pierson et al. (1994) reported that the biosynthesis of phenazinein by *Pseudomonas aeureofaciens* 30–84 is regulated by PhzR gene, which is affected by microbial cell density. Overall, antibiotic compound production by PGPR and the effects of those antibiotics depend on multiple factors but can be a critical indicator of the capacity for a PGPR to act as a biological control agent.

4.1.2 Competition for resources

The presence of PGPR benefits plant growth and development; in turn, the plant provides PGPR with nutrients and habitat. As a result, competition between beneficial bacteria and pathogens, to colonize the limited niche space and benefit from the nutrients in the rhizosphere, has been illustrated as a mechanism of biocontrol.

As an example, iron plays a key role in plant growth and metabolism (Rout and Sahoo, 2015). It has been illustrated that the absorption of iron with the help of microbial siderophores, a type of iron-chelating molecule, enhanced the conversion of insoluble iron to soluble iron that can be taken up by plants (Sobti et al., 2018). Siderophores can also serve as biological control agents against parasite invasion and can reduce the bioavailability of toxic metals in soil, forming stable complexes with them (Maheshwari, 2013). In PGPR–plant interactions, siderophores produced by PGPR bind iron, which restricts iron availability to competing bacterial and fungal pathogens present in the soil (Pankhurst and Lynch, 2005). This phenomenon has been reported to inhibit various plant diseases, including *Fusarium glycinia* in wheat and *Sarocladium oryzae* in both potato and rice (Sayyed et al., 2013).

4.2 Microbe-to-plant signalling

Another strategy used by PGPR for reducing the impact of pathogens on plant growth and crop yield is the induction of systemic resistance, which serves as an indirect form of pathogen control (Kumari and Srivastava, 1999). Prompted resistance can generally be divided into two pathways: systemic acquired resistance (SAR) and ISR. SAR can be triggered when the plants face virulent, avirulent and non-pathogenic microbes and is mediated by a salicylic acid (SA)-dependent process (Choudhary et al., 2007; Spoel and Dong, 2012). SAR is not triggered by the PGPR; instead, the phytohormones SA and jasmonic acid directly trigger ethylene to elicit NPR1 against the attack of pathogen and pests (Ping and Boland, 2004). In contrast, in a similar way, ISR is also activated by phytohormone signalling pathways and defence proteins, but this pathway can be stimulated by PGPR to help a plant defend itself against the pathogen attack (Pieterse et al., 2014).

PGPR-mediated ISR was discovered in *Arabidopsis thaliana* and was subsequently identified in bean, carnation, cucumber, radish, tobacco and tomato; PGPR-mediated ISR is regulated by a jasmonate (JA)- and ethylene (ET)-sensitive pathway (Van Loon et al., 1998; Spoel and Dong, 2012). During this process, IAA, GA and cytokinin production elicited by PGPR provokes strong plant responses related to defence against pathogens (Rodriguez et al., 2019), whereas pathways mediated by ET and JA can be activated without measured increases in these phytohormones (Pieterse et al., 2000). This corroborates studies by Pieterse et al. (2000) and Beneduzi et al. (2012), which suggested that the JA- and ET-dependency of WCS417r-mediated ISR is based on enhanced sensitivity to these hormones rather than on changes in their production. Provoking ISR causes plants to have more ability to convert the ACC to ethylene (Pieterse et al., 2000). However, this kind of ability is also based on the specific interaction(s) between the host plant and rhizobacteria (Beneduzi et al., 2012). ISR triggered by PGPR has also been illustrated to enhance the defence capacity of plants by priming for the potentiated expression of defence genes, including those that encode for pathogenesis-related proteins (Tjamos et al., 2005).

4.2.1 Specialized antimicrobial compound production by plants

Secondary metabolites synthesized in plants, such as alkaloids, terpenoids and phenolic compounds, are constituents that play a major role in plant defence mechanisms since they can act as antioxidants and antimicrobials in many plants (Nicholson and Hammerschmidt, 1992; Pietta, 2000; Singh et al., 2012). Among them, flavonoids and phenolic compounds can absorb free radicals, remove singlet oxygen and decompose peroxides (Havsteen, 2002; Singh et al., 2012). Moreover, the importance of flavonoids in dealing with UV-B, water deficiency and heavy metal stresses through scavenging H_2O_2 has been illustrated by Saeideh and Rashid (2014) and Michalak (2006). By contrast, flavonoids exhibit a diverse spectrum of biological functions, protecting plants against various biotic and abiotic stresses and playing an important role in the symbiotic interaction between plants and microbes (Samanta et al., 2011). Increased flavonoid production has been associated with the inoculation of rhizobacterial isolates S1p1 and S12p6, which also had beneficial traits that improved plant growth (Modi and Khanna, 2018). In addition, phenolic compounds and their derivatives can inhibit a wide range of soil-borne phytopathogens (Maurya et al., 2005, 2007). High levels of phenolic compounds have been reported for pea (*Pisum sativum*) when the seed is treated with *Pseudomonas fluorescens* (strain Pf4) and *P. aeruginosa* (*Pag*), while infection by *Erysiphe pisi* was less pronounced when pea was treated by *Pag* rather than by Pf4 strains since the production of phenolic compounds induced by *Pag* was greater than that induced by Pf4 (Singh et al., 2002).

The induction of the synthesis of terpenes by PGPR is a new possible mechanism of action against pathogens in grapevine. Maffei (2010) demonstrated that the production of VOCs by PGPR affected both the rate of terpene synthesis and the rate of terpene loss, which may inhibit the spread of the pathogen into plant tissues. Terpenes may also have a plant protective function in the face of abiotic stresses due to their antioxidant activity, which reduced free radical levels, thereby diminishing oxidative damage (Aharoni et al., 2003; Wei and Shibamoto, 2007; Salomon et al., 2014) and inducing integrity and stability of membranes (Beckett et al., 2012). It is clear that some PGPR stimulate plant growth and development by regulating the biosynthesis of plant secondary metabolites including terpenes (Salomon et al., 2016).

PGPR effects on plant secondary metabolite production are becoming an important area of interest, in addition to PGPR effects on plant growth and development. For example, in cannabis production, product quality is dictated by the levels of secondary metabolites such as THC, CBD, other cannabinoids and a variety of terpenes. However, the response of these secondary metabolites to PGPR is poorly understood because the cultivation of cannabis has been illegal for almost a century, during which our understanding of the role of PGPR was being elucidated. Interestingly, a dilute solution of CBD and tetrahydrocannabinol THC can inhibit the growth of bacteria including *S. aureus*, *B. subtilis*, *E. coli* (Schultz and Haffner, 1959), *B. cereus*, *Lactobacillus plantarum* and *Leuconostoc mesenteroides* (Gal et al., 1969; Van Klingeren and Ten Ham, 1976). In addition, the same amount of CBD has greater inhibition effects on *Alternaria alternata*, *Curvularia lunata*, *Fusarium solani* and *Trichothecium roseum* than the same amount of THC. The cannabis flowering tops, where cannabinoid concentrations are highest, are less susceptible to some fungal pathogens (McPartland, 1997). Taken together, these results indicate the antibiotic properties of these unique plant secondary metabolites. However, the induction of the synthesis of cannabinoids by PGPR as a mechanism of plant protection has been little studied.

Three quarters of volatiles around cannabis leaves can be pinenes and limonene, which are very effective insect repellents (Hood et al., 1973; Ross and ElSohly, 1996). Herbivorous insects are also repelled by methyl ketones of cannabis (Turner et al., 1980; Kashyap et al., 1991). It seems that a potentially synergistic mix of these sorts of compounds could be highly efficacious in minimizing issues related to both pathogen and insect attacks. However, the levels of specific cannabinoids can be related. Mediavilla and Steinemann (1997) reported that the cannabis plant with high THC levels also produces more limonene and pinenes than plants with lower THC levels, implying that the biosynthesis of cannabinoids could be linked to the synthesis of limonene and pinenes. Another study demonstrated that a consortium of PGPR had positive effects on the production of secondary metabolites (Pagnani et al., 2018), while PGPR strains of *Bacillus* and *Pseudomonas* have a strong potential for effects on cannabinoid profiles (Lyu et al., 2019). These effects may indirectly protect plants against plant pathogenic bacteria and fungal as well as insect pests.

It is still not fully understood how inoculation with PGPR affects the biosynthesis of secondary metabolites; however, PGPR treatment of plants can have significant effects on a variety of biosynthetic processes related to amino acids, for example, as the precursor of some important secondary metabolites related to stress responses, including indole acetate, lipid precursor, lignin and tryptophan (Alonso-Ramírez et al., 2009; Zemanová et al., 2014). Khan et al. (2019) illustrated that the phenylalanine, tyrosine and tryptophan biosynthesis pathway was upregulated in *Cicer arietinum* L. inoculated with the combination of *B. subtilis*, *Bacillus thuringiensis* and *Bacillus megaterium*. Although the effect of PGPR on the production of carbohydrates, amino acids, proteins and secondary metabolites, and on enzyme activity, related to abiotic and biotic stress has been illustrated (Ali et al., 2009; Ali et al., 2011), the broader or deeper mechanisms have not yet been elucidated using metabolomic or proteomics approaches, which will give a global view of enzyme and metabolic changes in response to stressful growing conditions.

5 Dangers of assuming the effectors or mechanisms are known

While plant growth promotion by bacteria can be associated with the mechanisms described earlier, it would be unwise to assume that these are the only mechanisms at play, particularly under field conditions. In antibiotic research, it is assumed that we have scratched only the surface of identifying antibiotics produced in nature because of the challenge related to recreating the conditions required for antibiotic production in the laboratory (Begani et al., 2018). The same may very well be true for PGPR: the growing conditions, environmental stimuli and media composition may not lead to the activation of mechanisms responsible for plant growth promotion in natural settings. It is likely that there are mechanisms for microbe-associated plant growth promotion that we currently do not understand, and we should keep this in mind as we continue to explore these complex interactions. For example, laboratory screening is limited because it often considers only one or a few microbes inoculated onto plants, while in a field setting, the interactions among microbes are vastly more complicated. This section describes how signals exchanged between plants and microbes, in the form of root exudates and secondary metabolites, and among microbes play a role in establishing plant-microbe relationships. While this is one frontier that remains to be elucidated, there are likely to be others that have not yet even been discovered.

5.1 Root exudates

Root exudates are secreted by plant roots and contain soluble organic compounds, mainly including sugars, amino acid, organic acids and enzymes (Koo et al., 2005). Those compounds have important roles both in the solubilization and the mobilization of minerals and micronutrients (Koo et al., 2005). In addition, root exudates have strong impacts on the growth of plant-associated microbes (members of the phytomicrobiome), including PGPR (Beauregard, 2015), as root exudates constitute the main reduced-carbon source for PGPR. The composition and level of root exudation can vary depending on plant growth conditions, both biotic and abiotic stress. For instance, low salinity concentrations stimulate roots to excrete more organic and amino acids, a change in exudate quality, while nutrient deficiencies could increase the overall quantity of root exudate (Koo et al., 2005). In *Macrophylla*, a salt-tolerant plant species, the concentration of salicylates, proline and/or hydroxyl-L-proline in root exudates is higher than that in root exudates produced by *Carrizo* (Vives-Peris et al., 2018). Moreover, the presence of microbes can, through release of specific secondary metabolites, affect root exudation, for instance, with regard to amino acid composition (Phillips et al., 2004). The root exudates produced by specific plant species probably have a range of functions. Li et al. (2016) showed that root exudates of maize can stimulate faba bean to fix N_2, while this was not found to be the case in barley and wheat. The variation of root exudates may interact with specific PGPR, which enables them to colonize the plant roots, and exerts control over which types become prominent in the phytomicrobiome (Czarnota et al., 2003; Rudrappa et al., 2008; Chaparro et al., 2013). Vives-Peris et al. (2018) concluded that root exudates collected from *C. macrophylla* can promote rhizobacteria *P. putida* KT2440 and *Novosphingobium* sp. HR1a growth under salt stress (60 or 90 mM NaCl) and heat stress (40°C) conditions.

Data collected from microbial ecology studies has shown that there is a core root microbiome that is influenced first by soil type and second by plant species (Yeoh et al., 2017). Plants can select for desirable microbes in the rhizosphere by secreting root exudates including sugars, poly-alcohols, organic acids, fatty acids and polycosanols (Iannucci et al., 2017). The composition of root exudates has been influenced by the domestication of crop species, and there is a high degree of heritability that influences the composition of root exudate profiles in, for instance, tetraploid wheat (Iannucci et al., 2017). There is a potential to engineer the plant production of particular root exudates to recruit beneficial microbes (Mohanram and Kumar, 2019). Root exudates can also induce antibiotic production by beneficial microbes in the rhizosphere, which allows for the successful colonization by beneficial microbes (Ogran et al., 2019).

5.2 Interspecific and intraspecific microbial interactions in the rhizosphere

Communication among microbes, not just between plant and microbes, also contributes to the establishment and maintenance of plant–microbe interactions. However, our current culture-dependent approach has focussed on the application of individual PGPR strains. Furthermore, plant growth-promoting microbial research is biased towards microbes that grow quickly in pure culture on artificial, high-nutrient media (Nai and Meyer, 2018). This limitation has been discussed in the context of antimicrobial discovery, and Molloy and

Hertweck (2017) proposed the adoption of an ecological paradigm to discover antimicrobials. This strategy could also be beneficial in the search for microbe-to-plant signal molecules. It seems it is now time to investigate how relationships between microbial species in the rhizosphere impact the production of microbe-to-plant signal molecules and, in turn, how the production of microbial secondary metabolites influences plant growth under optimal and stressed conditions.

5.2.1 Intermicrobial signalling effects on microbial metabolism

Bacteria can produce a large array of bioactive molecules, and we have only begun to scratch the surface when it comes to the identification and isolation of molecules for industrial uses (Bérdy, 2012). Because bioactive molecules are generally secondary metabolites, which are energetically expensive to produce, they are not constitutively expressed. In fact, recent research has shown that the genomes of many bacteria and fungi contain code for gene clusters that are expressed only after being activated – these are referred to as silent or cryptic gene clusters. The observation of this phenomenon led to the development of the 'one strain many compounds' (OSMAC) hypothesis, which states that one bacterial strain has the ability to produce different bioactive molecules under different growing/fermentation conditions (media composition, e.g., C, phosphate, N sources; aeration rate; temperature; pH; inclusion of inhibitors or growth enhancers) (Bode et al., 2002).

Multiple reports have promoted a paradigm shift from classical microbiology, where strains are cultivated in pure culture in favour of mixed cultivation containing two or more microbial strains (Nai and Meyer, 2018). The advantages of mixed cultures are that the consortia of microbes are able to perform more complex tasks compared to pure cultures (Brenner et al., 2008; Shank and Kolter, 2009; Bertrand et al., 2014). Within mixed cultures, members of the community are constantly excreting metabolites into the culture. These metabolites serve as intermicrobial signals, which can activate otherwise-silent gene clusters, resulting in the production of a larger array of antibiotics (Netzker et al., 2018) to allow for competition between species or can result in cooperation between species (Garcia et al., 2016). This can result from obligatorily mutualistic metabolism, also known as metabolic syntrophism (Stubbendieck et al., 2016). These interactions allow microbes to work together to access nutrients available from the environment that they would not be able to access via their own metabolic pathways while maintaining smaller genome sizes (Stubbendieck et al., 2016).

Several studies have been conducted on bacterial-bacterial co-cultures. For example, the co-culture of two bacteria, *Staphylococcus lentus* SZ2 with *Vibrio harveyi*, an aquaculture pathogen, led to increased biosurfactant production by *S. lentus* and reduced growth of *V. harveyi* planktonic cells and biofilms (Hamza et al., 2018). The authors hypothesized that the change in the growth rate of *V. harveyi* was related to the competition from *S. lentus*. *B. subtilis* was also shown to produce bacillaene, which inhibits the growth and pigment production of several *Streptomyces* spp. and contributes to defence against the soil-dwelling bacterium *Myxococcus xanthus* (Stubbendieck et al., 2016). This highlights the diverse effects of a single bacterial secondary metabolite in relation to other soil microbes.

There are also several examples in the literature of mixed bacterial–fungal cultures producing bioactive secondary metabolites that are not produced in pure culture (Rateb et al., 2013). For example, cocultures of *Sarocladium strictum* and *Fusarium oxysporum* led to *de novo* production of the toxin fusaric acid, which was involved in virulence associated with fungal infections (Bohni et al., 2016); furthermore, this effect was strain-dependent. Zhang et al. (2017) identified new molecules with potent α-glucosidase inhibitory activity produced by the co-culture of *Trichoderma* sp. 307 and *Acinetobacter johnsonii* B2; these molecules could be useful in the treatment of type 2 diabetes. Another study found that the co-culture of *Bacillus amyloliquefaciens* and *Setophoma terrestris* resulted in the production of a novel blennolide that had promising and unique effects on prostate cancer cells (Arora et al., 2018). In the previous three examples, researchers proposed that the bacteria induced bioactive molecule production by the fungal species, but did not otherwise investigate the underlying mechanism of *de novo* bioactive molecule synthesis. Netzker et al. (2015) provided a recent summary of our current understanding of mechanisms underlying *de novo* bioactive molecule biosynthesis by fungi in co-culture with other microbes, including a case study of the interaction between *Aspergillus* and *Streptomyces rapamycinicus*, and a discussion of the role of chromatin remodelling in microbial interactions. The authors highlighted the need for an improved understanding of the mechanisms underlying the biosynthesis of secondary microbial metabolites in order to benefit from the production of these molecules in the context of therapeutic agents; the same can be said for benefiting from the use of these molecules in agricultural systems.

5.2.2 Other forms of microbial interactions

Microbes use not only chemical signals to communicate, cooperate or compete but also physical signals. For example, in anaerobic methane-producing or methane-consuming communities, microbes can communicate via electron transfer over relatively long distances (μm to cm) (Lovley, 2016). This mechanism may also contribute to gut microbiome communication with host gut cells and has been used to stabilize anaerobic digestion. Most importantly, for plant scientists, this interaction has been demonstrated to occur in rice paddy soil and sediments and Arctic peat (Lovley, 2016).

To date, there are several examples of how microbial consortia can be applied in other microbiological fields. For instance, bacterial–fungal interactions have been harnessed for natural product discovery and could also prove to be useful in plant growth promotion (Haq and van Elsas, 2015). As we begin to understand how microbial consortia achieve community-level properties in natural rhizosphere environments, we can begin to use synthetic microbial ecology to achieve desirable community-level properties for agricultural applications (Dolinšek et al., 2016). Specifically, consortia could be designed to produce molecules that promote plant growth by reducing the build-up of inhibitory compounds (Santala et al., 2014; Jones and Wang, 2018). This will become easier as we begin to understand how to define the function rather than the taxonomic composition of the plant microbiome (Deines and Bosch, 2016). In addition to using microbial consortia with specific functions for plant growth promotion, there is potential to apply functional microbial consortia for bioremediation, energy production and therapeutics (Hays et al., 2015).

Bioactive molecules produced by microbes can have influences on microbial metabolism (within or between microbial species) either directly or indirectly on plants. For example, thuricin 17, a small peptide produced by *B. thuringiensis* NEB17, and lipochitooligosaccharide, produced by *Bradyrhizobium japonicum*, improved soybean germination under salt stress conditions up to 150 mM by enhancing PEP carboxylase, rubisco oxygenase large subunit, pyruvate kinase and isocitrate lyase (Subramanian et al., 2016a,b). In contrast, a molecule secreted by *Pseudomonas* sp. 23s shows in vitro antagonistic activity against the serious tomato pathogen *Clavibacter michiganensis* subspecies *michiganensis* (Yoko Takishita, personal communication). Inoculation of tomato seedlings with the same strain stimulates growth and induces systemic resistance (Takishita et al., 2018), but it remains to be determined whether this effect is associated with the same molecule that shows biocontrol activity. Only a small number of studies have reported the effects of microbial signal molecules on plant growth, and it is likely that we have identified and/or characterized only a tiny fraction of the molecules responsible for microbial effects on plant growth. Zachow et al. (2015) demonstrated that a molecule secreted by an endophytic *Pseudomonas* is involved in the suppression of a pathogenic *Rhizoctonia solani* and root colonization, which highlights that these molecules can be involved in multiple functions (Zachow et al., 2015).

6 Conclusion and future trends

Plant growth-promoting rhizobacteria have enormous potential as agricultural inputs, especially under stressful growing conditions. The potential for their impact will increase as stressful growing conditions such as heat, drought, salinity, cold, flooding and pathogens continue to increase as a result of climate change. We have a clear understanding of how bacteria can provide nutrients to plants, in bioavailable forms, and are starting to learn about the molecular mechanisms that occur in the plant in response to inoculation with PGPR. For example, phytohormone in regulation can be manipulated within the plant to result in increased growth and/or resistance to stressful growing conditions. However, as we continue to learn more about the mechanisms underlying these effects, it is becoming clear that there is even more we do not fully understand. This pertains, in particular, to signal exchange that occurs in the rhizosphere that orchestrates microbe-microbe and plant-microbe interactions. In order to begin to characterize these interactions, we will need to employ a variety of techniques to characterize the molecules that are present. In addition, these interactions should be studied in (1) a variety of soils given that the soil type is known to influence root exudates and (2) various crops given that each crop, and even each variety of a crop, can have a unique root exudate profile. Furthermore, it is important to study these interactions in the presence of indigenous soil microbial communities, given our recent advances in understanding how interactions between microbial species affect secondary metabolite biosynthesis.

7 Where to look for further information

- For a broad introduction read: Agricultural use of plant biostimulants. Another useful source can be found at: https://biostimulant.com/biostimulant-101/.
- For a deeper understanding of PGPR, researchers are referred to: https://www.frontiersin.org/articles/10.3389/fpls.2018.01473/full and https://www.frontiersin.org/articles/10.3389/fpls.2020.00634/abstract.
- For the promising application of PGPR researchers are referred to: https://www.frontiersin.org/articles/10.3389/fmicb.2019.01761/full.
- The Frontiers in Plant Science research topic: Biostimulants in Agriculture.
- For developments on the regulation, marketing requirements and legal framework around biostimulants in the EU, refer to the European Biostimulants Industry Council.
- Recommended conferences: International Plant Growth-Promoting Rhizobacteria workshop, most recently held in Victoria, Canada in 2018; miCROPe, most recently held in Vienna, Austria in 2019.

8 References

Abbas, R., Rasul, S., Aslam, K., Baber, M., Shahid, M., Mubeen, F. and Naqqash, T. 2019. Halotolerant PGPR: A hope for cultivation of saline soils. *Journal of King Saud University - Science* 31(4), 1195–201. doi:10.1016/j.jksus.2019.02.019.

Aharoni, A., Giri, A. P., Deuerlein, S., Griepink, F., de Kogel, W. J., Verstappen, F. W., Verhoeven, H. A., Jongsma, M. A., Schwab, W. and Bouwmeester, H. J. 2003. Terpenoid metabolism in wild-type and transgenic *Arabidopsis* plants. *The Plant Cell* 15(12), 2866–84. doi:10.1105/tpc.016253.

Ahmadi, A. and Baker, D. A. 2001. The effect of water stress on the activities of key regulatory enzymes of the sucrose to starch pathway in wheat. *Plant Growth Regulation* 35(1), 81–91. doi:10.1023/A:1013827600528.

Ali, S. Z., Sandhya, V., Grover, M., Kishore, N., Rao, L. V. and Venkateswarlu, B. 2009. *Pseudomonas* sp. strain AKM-P6 enhances tolerance of sorghum seedlings to elevated temperatures. *Biology and Fertility of Soils* 46(1), 45–55. doi:10.1007/s00374-009-0404-9.

Ali, S. Z., Sandhya, V., Grover, M., Linga, V. R. and Bandi, V. 2011. Effect of inoculation with a thermotolerant plant growth promoting *Pseudomonas putida* strain AKMP7 on growth of wheat (*Triticum* spp.) under heat stress. *Journal of Plant Interactions* 6(4), 239–46. doi:10.1080/17429145.2010.545147.

Alonso-Ramírez, A., Rodríguez, D., Reyes, D., Jiménez, J. A., Nicolás, G., López-Climent, M., Gómez-Cadenas, A. and Nicolás, C. 2009. Evidence for a role of gibberellins in salicylic acid- modulated early plant responses to abiotic stress in *Arabidopsis* seeds. *Plant Physiology* 150(3), 1335–44. doi:10.1104/pp.109.139352.

Arora, D., Chashoo, G., Singamaneni, V., Sharma, N., Gupta, P. and Jaglan, S. 2018. *Bacillus amyloliquefaciens* induces production of a novel blennolide K in coculture of *Setophoma terrestris*. *Journal of Applied Microbiology* 124(3), 730–9. doi:10.1111/jam.13683.

Backer, R., Rokem, J. S., Ilangumaran, G., Lamont, J., Praslickova, D., Ricci, E., Subramanian, S. and Smith, D. L. 2018. Plant growth-promoting rhizobacteria: context, mechanisms of action, and roadmap to commercialization of biostimulants for sustainable agriculture. *Frontiers in Plant Science* 9, 1473. doi:10.3389/fpls.2018.01473.

Banchio, E., Bogino, P. C., Zygadlo, J. and Giordano, W. 2008. Plant growth promoting rhizobacteria improve growth and essential oil yield in *Origanum majorana* L. *Biochemical Systematics and Ecology* 36(10), 766–71. doi:10.1016/j.bse.2008.08.006.

Bargaz, A., Lyamlouli, K., Chtouki, M., Zeroual, Y. and Dhiba, D. 2018. Soil microbial resources for improving fertilizers efficiency in an integrated plant nutrient management system. *Frontiers in Microbiology* 9, 1606. doi:10.3389/fmicb.2018.01606.

Barka, E. A., Nowak, J. and Clément, C. 2006. Enhancement of chilling resistance of inoculated grapevine plantlets with a plant growth-promoting rhizobacterium, *Burkholderia phytofirmans* strain PsJN. *Applied and Environmental Microbiology* 72(11), 7246–52. doi:10.1128/AEM.01047-06.

Bassirirad, H. 2000. Kinetics of nutrient uptake by roots: responses to global change. *The New Phytologist* 147(1), 155–69. doi:10.1046/j.1469-8137.2000.00682.x.

Beauregard, P. B. 2015. Not just sweet talkers: how roots stimulate their colonization by beneficial bacteria. *Advances in Botanical Research* 75, 1–20. doi:10.1016/bs.abr.2015.07.001.

Beckett, M., Loreto, F., Velikova, V., Brunetti, C., Di Ferdinando, M., Tattini, M., Calfapietra, C. and Farrant, J. M. 2012. Photosynthetic limitations and volatile and non-volatile isoprenoids in the poikilochlorophyllous resurrection plant *Xerophyta humilis* during dehydration and rehydration. *Plant, Cell and Environment* 35(12), 2061–74. doi:10.1111/j.1365-3040.2012.02536.x.

Begani, J., Lakhani, J. and Harwani, D. 2018. Current strategies to induce secondary metabolites from microbial biosynthetic cryptic gene clusters. *Annals of Microbiology* 68(7), 419–32. doi:10.1007/s13213-018-1351-1.

Beneduzi, A., Ambrosini, A. and Passaglia, L. M. 2012. Plant growth-promoting rhizobacteria (PGPR): their potential as antagonists and biocontrol agents. *Genetics and Molecular Biology* 35(4) ((Suppl.), 1044–51. doi:10.1590/s1415-47572012000600020.

Bérdy, J. 2012. Thoughts and facts about antibiotics: where we are now and where we are heading. *The Journal of Antibiotics* 65(8), 385–95. doi:10.1038/ja.2012.27.

Bertrand, S., Bohni, N., Schnee, S., Schumpp, O., Gindro, K. and Wolfender, J. L. 2014. Metabolite induction via microorganism co-culture: a potential way to enhance chemical diversity for drug discovery. *Biotechnology Advances* 32(6), 1180–204. doi:10.1016/j.biotechadv.2014.03.001.

Biswal, B., Joshi, P., Raval, M. and Biswal, U. 2011. Photosynthesis, a global sensor of environmental stress in green plants: stress signalling and adaptation. *Current Science*, 47–56.

Bode, H. B., Bethe, B., Höfs, R. and Zeeck, A. 2002. Big effects from small changes: possible ways to explore nature's chemical diversity. *Chembiochem: a European Journal of Chemical Biology* 3(7), 619–27. doi:10.1002/1439-7633(20020703)3:7<619::AID-CBIC619>3.0.CO;2-9.

Bohni, N., Hofstetter, V., Gindro, K., Buyck, B., Schumpp, O., Bertrand, S., Monod, M. and Wolfender, J. L. 2016. Production of fusaric acid by *Fusarium* spp. in pure culture and in solid medium co-cultures. *Molecules* 21(3), 370. doi:10.3390/molecules21030370.

Boyer, J. S. 1982. Plant productivity and environment. *Science* 218(4571), 443–8. doi:10.1126/science.218.4571.443.

Brenner, K., You, L. and Arnold, F. H. 2008. Engineering microbial consortia: a new frontier in synthetic biology. *Trends in Biotechnology* 26(9), 483–9. doi:10.1016/j.tibtech.2008.05.004.

Cao, Y., Pi, H., Chandrangsu, P., Li, Y., Wang, Y., Zhou, H., Xiong, H., Helmann, J. D. and Cai, Y. 2018. Antagonism of two plant-growth promoting *Bacillus velezensis* isolates against *Ralstonia solanacearum* and *Fusarium oxysporum*. *Scientific Reports* 8(1), 4360. doi:10.1038/s41598-018-22782-z.

Chaparro, J. M., Badri, D. V., Bakker, M. G., Sugiyama, A., Manter, D. K. and Vivanco, J. M. 2013. Root exudation of phytochemicals in *Arabidopsis* follows specific patterns that are developmentally programmed and correlate with soil microbial functions. *PLoS ONE* 8(2), e55731. doi:10.1371/journal.pone.0055731.

Choi, J., Choi, D., Lee, S., Ryu, C. M. and Hwang, I. 2011. Cytokinins and plant immunity: old foes or new friends? *Trends in Plant Science* 16(7), 388–94. doi:10.1016/j.tplants.2011.03.003.

Choudhary, D. K., Prakash, A. and Johri, B. N. 2007. Induced systemic resistance (ISR) in plants: mechanism of action. *Indian Journal of Microbiology* 47(4), 289–97. doi:10.1007/s12088-007-0054-2.

Cohen, S. P. and Leach, J. E. 2019. Abiotic and biotic stresses induce a core transcriptome response in rice. *Scientific Reports* 9(1), 6273. doi:10.1038/s41598-019-42731-8.

Czarnota, M. A., Rimando, A. M. and Weston, L. A. 2003. Evaluation of root exudates of seven sorghum accessions. *Journal of Chemical Ecology* 29(9), 2073–83. doi:10.1023/a:1025634402071.

Davies, J. 2006. Are antibiotics naturally antibiotics? *Journal of Industrial Microbiology and Biotechnology* 33(7), 496–9. doi:10.1007/s10295-006-0112-5.

Deines, P. and Bosch, T. C. G. 2016. Transitioning from microbiome composition to microbial community interactions: the potential of the metaorganism Hydra as an experimental model. *Frontiers in Microbiology* 7, 1610. doi:10.3389/fmicb.2016.01610.

Dell'Amico, E., Cavalca, L. and Andreoni, V. 2008. Improvement of *Brassica napus* growth under cadmium stress by cadmium-resistant rhizobacteria. *Soil Biology and Biochemistry* 40(1), 74–84. doi:10.1016/j.soilbio.2007.06.024.

Denton, B. 2007. Advances in phytoremediation of heavy metals using plant growth promoting bacteria and fungi. *MMG 445 Basic Biotechnology* 3, 1–5.

Dolinšek, J., Goldschmidt, F. and Johnson, D. R. 2016. Synthetic microbial ecology and the dynamic interplay between microbial genotypes. *FEMS Microbiology Reviews* 40(6), 961–79. doi:10.1093/femsre/fuw024.

Duca, D., Lorv, J., Patten, C. L., Rose, D. and Glick, B. R. 2014. Indole-3-acetic acid in plant–microbe interactions. *Antonie van Leeuwenhoek* 106(1), 85–125. doi:10.1007/s10482-013-0095-y.

Duffy, B. K. and Défago, G. 1999. Environmental factors modulating antibiotic and siderophore biosynthesis by *Pseudomonas fluorescens* biocontrol strains. *Applied and Environmental Microbiology* 65(6), 2429–38. doi:10.1128/AEM.65.6.2429-2438.1999.

Duffy, B. K. and Défago, G. 2000. Controlling instability in gacS-gacA regulatory genes during inoculant production of *Pseudomonas fluorescens* Biocontrol Strains. *Applied and Environmental Microbiology* 66(8), 3142–50. doi:10.1128/aem.66.8.3142-3150.2000.

El-Howeity, M. A. and Asfour, M. M. 2012. Response of some varieties of canola plant (*Brassica napus* L.) cultivated in a newly reclaimed desert to plant growth promoting rhizobacteria and mineral nitrogen fertilizer. *Annals of Agricultural Sciences* 57(2), 129–36. doi:10.1016/j.aoas.2012.08.006.

Esitken, A., Pirlak, L., Turan, M. and Sahin, F. 2006. Effects of floral and foliar application of plant growth promoting rhizobacteria (PGPR) on yield, growth and nutrition of sweet cherry. *Scientia Horticulturae* 110(4), 324–7. doi:10.1016/j.scienta.2006.07.023.

Essemine, J., Ammar, S. and Bouzid, S. 2010. Impact of heat stress on germination and growth in higher plants: physiological, biochemical and molecular repercussions and mechanisms of defence. *Journal of Biological Sciences* 10(6), 565–72. doi:10.3923/jbs.2010.565.572.

Fahad, S., Hussain, S., Saud, S., Hassan, S., Tanveer, M., Ihsan, M. Z., Shah, A. N., Ullah, A., Nasrullah, Khan F., Ullah, S., Alharby, H., Nasim, W., Wu, C. and Huang, J. 2016. A combined application of biochar and phosphorus alleviates heat- induced adversities on physiological, agronomical and quality attributes of rice. *Plant Physiology and Biochemistry: PPB* 103, 191–8. doi:10.1016/j.plaphy.2016.03.001.

Fan, D. 2018. Isolation of rhizobacteria in Southwestern Québec, Canada: an investigation of their impact on the growth and salinity stress alleviation in Arabidopsis thaliana and crop plants. PhD Thesis. McGill University, Montreal, Canada.

Farooq, M., Basra, S. M. A., Wahid, A., Cheema, Z. A., Cheema, M. A. and Khaliq, A. 2008. Physiological role of exogenously applied glycinebetaine to improve drought tolerance in fine grain aromatic rice (*Oryza sativa* L.). *Journal of Agronomy and Crop Science* 194(5), 325–33. doi:10.1111/j.1439-037X.2008.00323.x.

Fernando, W. D., Nakkeeran, S. and Zhang, Y. 2005. *Biosynthesis of Antibiotics by PGPR and Its Relation in Biocontrol of Plant Diseases, PGPR: Biocontrol and Biofertilization.* Springer, Dordrecht, pp. 67–109.

Fiddaman, P. and Rossall, S. 1994. Effect of Substrate on the Production of Antifungal Volatiles from *Bacillus subtilis*. *Journal of Applied Bacteriology* 76, 395–405.

Flemming, H. C. and Wingender, J. 2001. Relevance of microbial extracellular polymeric substances (EPSs)-part I: structural and ecological aspects. *Water Science and Technology: a Journal of the International Association on Water Pollution Research* 43(6), 1–8. doi:10.2166/wst.2001.0326.

Gal, I., Vajda, O. and Bekes, I. 1969. Investigation of certain properties of cannabidiolic acid from the aspect of food preservation. *Elelmiszervizsgalati Kozlem* 4, 208–16.

Garcia, E. C., Perault, A. I., Marlatt, S. A. and Cotter, P. A. 2016. Interbacterial signaling via *Burkholderia* contact-dependent growth inhibition system proteins. *Proceedings of the National Academy of Sciences of the United States of America* 113(29), 8296–301. doi:10.1073/pnas.1606323113.

Garg, B. 2003. Nutrient uptake and management under drought: nutrient-moisture interaction. *Current Agriculture* 27, 1–8.

Goteti, P. K., Emmanuel, L. D. A., Desai, S. and Shaik, M. H. A. 2013. Prospective zinc solubilising bacteria for enhanced nutrient uptake and growth promotion in maize (*Zea mays* L.). *International Journal of Microbiology* 2013, 1–7. doi:10.1155/2013/869697.

Hamza, F., Kumar, A. R. and Zinjarde, S. 2018. Coculture induced improved production of biosurfactant by *Staphylococcus lentus* SZ2: role in protecting *Artemia salina* against *Vibrio harveyi*. *Enzyme and Microbial Technology* 114, 33–9. doi:10.1016/j.enzmictec.2018.03.008.

Haq, I. U. and van Elsas, J. D. 2015. *Metagenomics and Metatranscriptomics for the Exploration of Natural Products from Soil Fungi, Biosynthesis and Molecular Genetics of Fungal Secondary Metabolites* (Vol. 2). Springer, New York, NY, pp. 67–79.

Havsteen, B. H. 2002. The biochemistry and medical significance of the flavonoids. *Pharmacology and Therapeutics* 96(2–3), 67–202. doi:10.1016/s0163-7258(02)00298-x.

Hays, S. G., Patrick, W. G., Ziesack, M., Oxman, N. and Silver, P. A. 2015. Better together: engineering and application of microbial symbioses. *Current Opinion in Biotechnology* 36, 40–9. doi:10.1016/j.copbio.2015.08.008.

Hood, L. V., Dames, M. E. and Barry, G. T. 1973. Headspace volatiles of marijuana. *Nature* 242, 402. doi:10.1038/242402a0.

Howarth, C. 2005. Genetic improvements of tolerance to high temperature. In: Ashraf, M. and Harris, P. J. C. (Eds) *Abiotic Stresses- Plant Resistance through Breeding and Molecular Approaches'*. The Haworth Press, New York, pp. 277–300.

Huang, B., Rachmilevitch, S. and Xu, J. 2012. Root carbon and protein metabolism associated with heat tolerance. *Journal of Experimental Botany* 63(9), 3455–65. doi:10.1093/jxb/ers003.

Iannucci, A., Fragasso, M., Beleggia, R., Nigro, F. and Papa, R. 2017. Evolution of the crop rhizosphere: impact of domestication on root exudates in tetraploid wheat (*Triticum turgidum* L.). *Frontiers in Plant Science* 8, 2124. doi:10.3389/fpls.2017.02124.

Ihsan, M. Z., Daur, I., Alghabari, F., Alzamanan, S., Rizwan, S., Ahmad, M., Waqas, M. and Shafqat, W. 2019. Heat stress and plant development: role of sulphur metabolites and management strategies. *Acta Agriculturae Scandinavica, Section B–Soil and Plant Science* 69(4), 332–42. doi:10.1080/09064710.2019.1569715.

Izanloo, A., Condon, A. G., Langridge, P., Tester, M. and Schnurbusch, T. 2008. Different mechanisms of adaptation to cyclic water stress in two South Australian bread wheat cultivars. *Journal of Experimental Botany* 59(12), 3327–46. doi:10.1093/jxb/ern199.

Jaleel, C. A., Manivannan, P., Wahid, A., Farooq, M., Al-Juburi, H. J., Somasundaram, R. and Panneerselvam, R. 2009. Drought stress in plants: a review on morphological characteristics and pigments composition. *International Journal of Agriculture and Biology* 11, 100–5.

Janda, T., Szalai, G., Rios-Gonzalez, K., Veisz, O. and Páldi, E. 2003. Comparative study of frost tolerance and antioxidant activity in cereals. *Plant Science* 164(2), 301–6. doi:10.1016/S0168-9452(02)00414-4.

Jones, J. A. and Wang, X. 2018. Use of bacterial co-cultures for the efficient production of chemicals. *Current Opinion in Biotechnology* 53, 33–8. doi:10.1016/j.copbio.2017.11.012.

Kang, H. M. and Saltveit, M. E. 2001. Activity of enzymatic antioxidant defense systems in chilled and heat shocked cucumber seedling radicles. *Physiologia Plantarum* 113(4), 548–56. doi:10.1034/j.1399-3054.2001.1130414.x.

Kanwal, S., Ilyas, N., Batool, N. and Arshad, M. 2017. Amelioration of drought stress in wheat by combined application of PGPR, compost, and mineral fertilizer. *Journal of Plant Nutrition* 40(9), 1250–60. doi:10.1080/01904167.2016.1263322.

Kashyap, R. K., Kennedy, G. G. and Farrar, R. R. 1991. Behavioral response of *Trichogramma pretiosum* riley and *Telenomus sphingis* (Ashmead) to trichome/methyl ketone mediated resistance in tomato. *Journal of Chemical Ecology* 17(3), 543–56. doi:10.1007/BF00982125.

Khan, N., Bano, A. and Babar, M. A. 2019. Metabolic and physiological changes induced by plant growth regulators and plant growth promoting rhizobacteria and their impact on drought tolerance in *Cicer arietinum* L. *PLOS ONE* 14(3), e0213040. doi:10.1371/journal.pone.0213040.

Koo, B., Adriano, D., Bolan, N. and Barton, C. 2005. Root exudates and microorganisms. In: Hillel, D. (Ed.) *Encyclopedia of Soils in the Environment*. Elsevier Ltd., Oxford, pp. 421–8.

Kumari, V. and Srivastava, J. S. 1999. Molecular and biochemical aspects of rhizobacterial ecology with emphasis on biological control. *World Journal of Microbiology and Biotechnology* 15(5), 535–43. doi:10.1023/A:1008958912647.

Kundan, R., Pant, G., Jadon, N. and Agrawal, P. K. 2015. Plant growth promoting rhizobacteria: mechanism and current prospective. *Journal of Fertilizers and Pesticides* 06(2), 9. doi:10.4172/2471-2728.1000155.

Lakshmanan, V., Selvaraj, G. and Bais, H. P. 2014. Functional soil microbiome: belowground solutions to an aboveground problem. *Plant Physiology* 166(2), 689–700. doi:10.1104/pp.114.245811.

Li, B., Li, Y.-Y., Wu, H.-M., Zhang, F.-F., Li, C.-J., Li, X.-X., Lambers, H. and Li, L. 2016. Root exudates drive interspecific facilitation by enhancing nodulation and N2 fixation. *Proceedings of the National Academy of Sciences* 113, 6496–501.

Lin, Y., Watts, D. B., Kloepper, J. W. and Torbert, H. A. 2018. Influence of plant growth-promoting rhizobacteria on corn growth under different fertility sources. *Communications in Soil Science and Plant Analysis* 49(10), 1239–55. doi:10.1080/00103624.2018.1457155.

Liu, J. X. and Howell, S. H. 2010. Endoplasmic reticulum protein quality control and its relationship to environmental stress responses in plants. *The Plant Cell* 22(9), 2930–42. doi:10.1105/tpc.110.078154.

Lovley, D. R. 2016. Happy together: microbial communities that hook up to swap electrons. *The ISME Journal* 11(2), 327–36. doi:10.1038/ismej.2016.136.

Lv, A., Fan, N., Xie, J., Yuan, S., An, Y. and Zhou, P. 2017. Expression of CdDHN4, a novel YSK2- type dehydrin gene from Bermudagrass, responses to drought stress through the ABA- dependent signal pathway. *Frontiers in Plant Science* 8, 748. doi:10.3389/fpls.2017.00748.

Lyu, D., Backer, R. G., Robinson, W. G. and Smith, D. L. 2019. Plant-growth promoting rhizobacteria for cannabis production: yield, cannabinoid profile and disease resistance. *Frontiers in Microbiology* 10, 1761. doi:10.3389/fmicb.2019.01761.

Maestri, E., Klueva, N., Perrotta, C., Gulli, M., Nguyen, H. T. and Marmiroli, N. 2002. Molecular Genetics of heat tolerance and heat shock proteins in cereals. *Plant Molecular Biology* 48(5–6), 667–81. doi:10.1023/a:1014826730024.

Maffei, M. E. 2010. Sites of synthesis, biochemistry and functional role of plant volatiles. *South African Journal of Botany* 76(4), 612–31. doi:10.1016/j.sajb.2010.03.003.

Maheshwari, D. K. 2013. *Bacteria in Agrobiology: Disease Management*. Springer Science & Business Media, Berlin Heidelberg.

Marinkovic, J., Bjelic, D., Tintor, B., Miladinovic, J., Dukic, V. and Dorđevic, V. 2018. Effects of soybean co-inoculation with plant growth promoting rhizobacteria in field trial. *Romanian Biotechnological Letters* 23, 13401.

Mascarenhas, J. P. 1984. *Molecular Mechanisms of Heat Stress Tolerance, Applications of Genetic Engineering to Crop Improvement*. Springer, Dordrecht, Netherlands, pp. 391–425.

Maurya, S., Singh, U., Singh, D., Singh, K. and Srivastava, J. 2005. Secondary metabolites of chickpea (*Cicer arictinum*) and their role in pathogenesis after infection by *Sclerotium rolfsii*. *Zeitschrift fur Pflanzenkrankheiten und Pflanzenschutz* 112, 118–23.

Maurya, S., Singh, R., Singh, D. P., Singh, H. B., Srivastava, J. S. and Singh, U. P. 2007. Phenolic compounds of *Sorghum vulgare* in response to *Sclerotium rolfsii* infection. *Journal of Plant Interactions* 2(1), 25–9. doi:10.1080/17429140701422504.

McPartland, J. M. 1997. Cannabis as repellent and pesticide. *Journal of the International Hemp Association* 4, 89–94.

Mediavilla, V. and Steinemann, S. 1997. Ätherische Öl-erste Prüfung einiger Hanfsorten, Biorohstoff Hanf Symposium Magazin, p. 58.

Meena, H., Ahmed, M. A. and Prakash, P. 2015. Amelioration of heat stress in wheat, *Triticum aestivum* by PGPR (*Pseudomonas aeruginosa* strain 2CpS1). *Bioscience Biotechnology Research Communications* 8, 171–4.

Mhlongo, M. I., Piater, L. A., Madala, N. E., Labuschagne, N. and Dubery, I. A. 2018. The chemistry of plant-microbe interactions in the rhizosphere and the potential for metabolomics to reveal signaling related to defense priming and induced systemic resistance. *Frontiers in Plant Science* 9, 112. doi:10.3389/fpls.2018.00112.

Michalak, A. 2006. Phenolic compounds and their antioxidant activity in plants growing under heavy metal stress. *Polish Journal of Environmental Studies* 15, 523–30.

Milner, J. L., Raffel, S. J., Lethbridge, B. J. and Handelsman, J. 1995. Culture conditions that influence accumulation of zwittermicin A by *Bacillus cereus* UW85. *Applied Microbiology and Biotechnology* 43(4), 685–91. doi:10.1007/bf00164774.

Modi, R. and Khanna, V. 2018. Evaluation of thermotolerant rhizobacteria for multiple plant growth promoting traits from pigeonpea rhizosphere. *Journal of Applied and Natural Science* 10(1), 518–21. doi:10.31018/jans.v10i1.1660.

Mohanram, S. and Kumar, P. 2019. Rhizosphere microbiome: revisiting the synergy of plant-microbe interactions. *Annals of Microbiology* 69(4), 307–20. doi:10.1007/s13213-019-01448-9.

Molloy, E. M. and Hertweck, C. 2017. Antimicrobial discovery inspired by ecological interactions. *Current Opinion in Microbiology* 39, 121–7. doi:10.1016/j.mib.2017.09.006.

Nai, C. and Meyer, V. 2018. From axenic to mixed cultures: technological advances accelerating a paradigm shift in microbiology. *Trends in Microbiology* 26(6), 538–54. doi:10.1016/j.tim.2017.11.004.

Naqqash, T., Hameed, S., Imran, A., Hanif, M. K., Majeed, A. and van Elsas, J. D. 2016. Differential response of potato toward inoculation with taxonomically diverse plant growth promoting rhizobacteria. *Frontiers in Plant Science* 7, 144. doi:10.3389/fpls.2016.00144.

Naseem, H. and Bano, A. 2014. Role of plant growth-promoting rhizobacteria and their exopolysaccharide in drought tolerance of maize. *Journal of Plant Interactions* 9(1), 689–701. doi:10.1080/17429145.2014.902125.

Netzker, T., Fischer, J., Weber, J., Mattern, D. J., König, C. C., Valiante, V., Schroeckh, V. and Brakhage, A. A. 2015. Microbial communication leading to the activation of silent fungal secondary metabolite gene clusters. *Frontiers in Microbiology* 6, 299. doi:10.3389/fmicb.2015.00299.

Netzker, T., Flak, M., Krespach, M. K. C., Stroe, M. C., Weber, J., Schroeckh, V. and Brakhage, A. A. 2018. Microbial interactions trigger the production of antibiotics. *Current Opinion in Microbiology* 45, 117–23. doi:10.1016/j.mib.2018.04.002.

Nicholson, R. L. and Hammerschmidt, R. 1992. Phenolic compounds and their role in disease resistance. *Annual Review of Phytopathology* 30(1), 369–89. doi:10.1146/annurev.py.30.090192.002101.

Nielsen, T. H., Thrane, C., Christophersen, C., Anthoni, U. and Sørensen, J. 2000. Structure, production characteristics and fungal antagonism of tensin–a new antifungal cyclic lipopeptide from *Pseudomonas fluorescens* strain 96.578. *Journal of Applied Microbiology* 89(6), 992–1001. doi:10.1046/j.1365-2672.2000.01201.x.

Notz, R., Maurhofer, M., Schnider-Keel, U., Duffy, B., Haas, D. and Défago, G. 2001. Biotic factors affecting expression of the 2, 4-diacetylphloroglucinol biosynthesis gene phlA in *Pseudomonas fluorescens* biocontrol strain CHA0 in the rhizosphere. *Phytopathology* 91(9), 873–81. doi:10.1094/PHYTO.2001.91.9.873.

Nowak-Thompson, B., Gould, S. J., Kraus, J. and Loper, J. E. 1994. Production of 2, 4-diacetylphloroglucinol by the biocontrol agent *Pseudomonas fluorescens* Pf-5. *Canadian Journal of Microbiology* 40(12), 1064–6. doi:10.1139/m94-168.

Ogran, A., Yardeni, E. H., Keren-Paz, A., Bucher, T., Jain, R., Gilhar, O. and Kolodkin-Gal, I. 2019. The plant host induces antibiotic production to select the most-beneficial colonizers. *Applied and Environmental Microbiology* 85(13), e00512-00519. doi:10.1128/AEM.00512-19.

Ownley, B. H., Duffy, B. K. and Weller, D. M. 2003. Identification and manipulation of soil properties to improve the biological control performance of phenazine-producing *Pseudomonas fluorescens*. *Applied and Environmental Microbiology* 69(6), 3333–43. doi:10.1128/aem.69.6.3333-3343.2003.

Pagnani, G., Pellegrini, M., Galieni, A., D'Egidio, S., Matteucci, F., Ricci, A., Stagnari, F., Sergi, M., Sterzo, C. L., Pisante, M. and Del Gallo, M. 2018. Plant growth-promoting rhizobacteria (PGPR) in *Cannabis sativa* 'Finola' cultivation: an alternative fertilization strategy to improve plant growth and quality characteristics. *Industrial Crops and Products* 123, 75–83. doi:10.1016/j.indcrop.2018.06.033.

Pandya, N. and Desai, P. 2014. Screening and characterization of GA3 producing *Pseudomonas monteilii* and its impact on plant growth promotion. *International Journal of Current Microbiolology and Applied Sciences* 3, 110–5.

Pankhurst, C. and Lynch, J. 2005. Biocontrol of soil-borne plant diseases. *Encyclopdia of Soil Sin the Environment*, 129–36.

Panneerselvam, A. and Saravanamuthu, R. 2013. Studies on the saprophytic survival and suppression of *Fusarium moniliforme* J. Sheld, *Helminthosporium oryzae* Breda De Haan and *Sarocladium oryzae* (Sawada) W. Gams & D. Hawksw., Causing Diseases in Paddy. In: *Microbiological Research in Agroecosystem Management*. Springer, India, pp. 297–323.

Park, K., Park, Y.-S., Ahamed, J., Dutta, S., Ryu, H., Lee, S.-H., Balaraju, K., Manir, M. and Moon, S.-S. 2016. Elicitation of induced systemic resistance of chilli pepper by iturin A analogs derived from *Bacillus vallismortis* EXTN-1. *Canadian Journal of Plant Science* 96, 564–70.

Park, Y. G., Mun, B. G., Kang, S. M., Hussain, A., Shahzad, R., Seo, C. W., Kim, A. Y., Lee, S. U., Oh, K. Y., Lee, D. Y., Lee, I. J. and Yun, B. W. 2017. *Bacillus aryabhattai* SRB02 tolerates oxidative and nitrosative stress and promotes the growth of soybean by modulating the production of phytohormones. *PLoS ONE* 12(3), e0173203. doi:10.1371/journal.pone.0173203.

Peres, A. L. G. L., Soares, J. S., Tavares, R. G., Righetto, G., Zullo, M. A. T., Mandava, N. B. and Menossi, M. 2019. Brassinosteroids, the sixth class of phytohormones: a molecular view from the discovery to hormonal interactions in plant development and stress adaptation. *International Journal of Molecular Sciences* 20(2), 331. doi:10.3390/ijms20020331.

Phillips, D. A., Fox, T. C., King, M. D., Bhuvaneswari, T. V. and Teuber, L. R. 2004. Microbial products trigger amino acid exudation from plant roots. *Plant Physiology* 136(1), 2887–94. doi:10.1104/pp.104.044222.

Picard, C., Di Cello, F., Ventura, M., Fani, R. and Guckert, A. 2000. Frequency and biodiversity of 2, 4-diacetylphloroglucinol-producing bacteria isolated from the maize rhizosphere at different stages of plant growth. *Applied and Environmental Microbiology* 66(3), 948–55. doi:10.1128/aem.66.3.948-955.2000.

Pierson, L. S., Keppenne, V. D. and Wood, D. W. 1994. Phenazine antibiotic biosynthesis in *Pseudomonas aureofaciens* 30–84 is regulated by PhzR in response to cell density. *Journal of Bacteriology* 176(13), 3966–74. doi:10.1128/jb.176.13.3966-3974.1994.

Pieterse, C. M. J., Van Pelt, J. A., Ton, J., Parchmann, S., Mueller, M. J., Buchala, A. J., Métraux, J.- P. and Van Loon, L. C. 2000. Rhizobacteria-mediated induced systemic resistance (ISR) in *Arabidopsis* requires sensitivity to jasmonate and ethylene but is not accompanied by an increase in their production. *Physiological and Molecular Plant Pathology* 57(3), 123–34. doi:10.1006/pmpp.2000.0291.

Pieterse, C. M. J., Zamioudis, C., Berendsen, R. L., Weller, D. M., Van Wees, S. C. and Bakker, P. A. 2014. Induced systemic resistance by beneficial microbes. *Annual Review of Phytopathology* 52, 347–75. doi:10.1146/annurev-phyto-082712-102340.

Pietta, P. G. 2000. Flavonoids as antioxidants. *Journal of Natural Products* 63(7), 1035–42. doi:10.1021/np9904509.

Pii, Y., Mimmo, T., Tomasi, N., Terzano, R., Cesco, S. and Crecchio, C. 2015. Microbial interactions in the rhizosphere: beneficial influences of plant growth-promoting rhizobacteria on nutrient acquisition process. A review. *Biology and Fertility of Soils* 51(4), 403–15. doi:10.1007/s00374-015-0996-1.

Ping, L. and Boland, W. 2004. Signals from the underground: bacterial volatiles promote growth in Arabidopsis. *Trends in Plant Science* 9(6), 263–6. doi:10.1016/j.tplants.2004.04.008.

Prajapati, V. D., Jani, G. K. and Khanda, S. M. 2013. Pullulan: an exopolysaccharide and its various applications. *Carbohydrate Polymers* 95(1), 540–9. doi:10.1016/j.carbpol.2013.02.082.

Rahman, A., Uddin, W. and Wenner, N. G. 2015. Induced systemic resistance responses in perennial ryegrass against *Magnaporthe oryzae* elicited by semi-purified surfactin lipopeptides and live cells of *Bacillus amyloliquefaciens*. *Molecular Plant Pathology* 16(6), 546–58. doi:10.1111/mpp.12209.

Rajkumar, M. and Freitas, H. 2008. Influence of metal resistant-plant growth-promoting bacteria on the growth of *Ricinus communis* in soil contaminated with heavy metals. *Chemosphere* 71(5), 834–42. doi:10.1016/j.chemosphere.2007.11.038.

Rateb, M. E., Hallyburton, I., Houssen, W. E., Bull, A. T., Goodfellow, M., Santhanam, R., Jaspars, M. and Ebel, R. 2013. Induction of diverse secondary metabolites in *Aspergillus fumigatus* by microbial co-culture. *RSC Advances* 3(34), 14444–50. doi:10.1039/c3ra42378f.

Reetha, A., Pavani, S. and Mohan, S. 2014. Hydrogen cyanide production ability by bacterial antagonist and their antibiotics inhibition potential on *Macrophomina phaseolina* (Tassi.) Goid. *International Journal of Current Microbiology and Applied Sciences* 3, 172–8.

Rodriguez, P. A., Rothballer, M., Chowdhury, S. P., Nussbaumer, T., Gutjahr, C. and Falter-Braun, P. 2019. Systems biology of plant microbiome interactions. *Molecular Plant* 12(6), 804–21. doi:10.1016/j.molp.2019.05.006.

Rosier, A., Medeiros, F. H. V. and Bais, H. P. 2018. Defining plant growth promoting rhizobacteria molecular and biochemical networks in beneficial plant-microbe interactions. *Plant and Soil* 428(1-2), 35–55. doi:10.1007/s11104-018-3679-5.

Ross, S. A. and ElSohly, M. A. 1996. The volatile oil composition of fresh and air-dried buds of *Cannabis sativa*. *Journal of Natural Products* 59, 49–51.

Rossi, F. and De Philippis, R. 2015. Role of cyanobacterial exopolysaccharides in phototrophic biofilms and in complex microbial mats. *Life* 5(2), 1218–38. doi:10.3390/life5021218.

Rout, G. R. and Sahoo, S. 2015. Role of iron in plant growth and metabolism. *Reviews in Agricultural Science* 3, 1–24. doi:10.7831/ras.3.1.

Rudrappa, T., Czymmek, K. J., Paré, P. W. and Bais, H. P. 2008. Root-secreted malic acid recruits beneficial soil bacteria. *Plant Physiology* 148(3), 1547–56. doi:10.1104/pp.108.127613.

Saeideh, N. and Rashid, J. 2014. Effect of silver nanoparticles and Pb (no 3) 2 On the yield and chemical composition of mung bean (*Vigna radiata*). *Journal of Stress Physiology and Biochemistry* 10.

Sahin, F. and Turan, M. 2014. The effect of Plant Growth Promoting rhizobacteria strain on wheat yield and quality parameters. *Journal of Plant Nutrition* 36, 731–48.

Salomon, M. V., Bottini, R., de Souza Filho, G. A., Cohen, A. C., Moreno, D., Gil, M. and Piccoli, P. 2014. Bacteria isolated from roots and rhizosphere of *Vitis vinifera* retard water losses, induce abscisic acid accumulation and synthesis of defense-related terpenes in in vitro cultured grapevine. *Physiologia Plantarum* 151(4), 359–74. doi:10.1111/ppl.12117.

Salomon, M. V., Purpora, R., Bottini, R. and Piccoli, P. 2016. Rhizosphere associated bacteria trigger accumulation of terpenes in leaves of *Vitis vinifera* L. cv. Malbec that protect cells against reactive oxygen species. *Plant Physiology and Biochemistry: PPB* 106, 295–304. doi:10.1016/j.plaphy.2016.05.007.

Samanta, A., Das, G. and Das, S. K. 2011. Roles of flavonoids in plants. *International Journal of Pharmacy and Pharmaceutical Sciences* 6, 12–35.

Sandhya, V., Sk. Z., A., Grover, M., Reddy, G. and Venkateswarlu, B. 2009. Alleviation of drought stress effects in sunflower seedlings by the exopolysaccharides producing *Pseudomonas putida* strain GAP-P45. *Biology and Fertility of Soils* 46(1), 17–26. doi:10.1007/s00374-009-0401-z.

Santala, S., Karp, M. and Santala, V. 2014. Rationally engineered synthetic coculture for improved biomass and product formation. *PLoS ONE* 9(12), e113786. doi:10.1371/journal.pone.0113786.

Saraf, M., Pandya, U. and Thakkar, A. 2014. Role of allelochemicals in plant growth promoting rhizobacteria for biocontrol of phytopathogens. *Microbiological Research* 169(1), 18–29. doi:10.1016/j.micres.2013.08.009.

Sayyed, R., Chincholkar, S., Reddy, M., Gangurde, N. and Patel, P. 2013. Siderophore producing PGPR for crop nutrition and phytopathogen suppression, Bacteria in Agrobiology. *Disease Management*. Springer, Heidelberg, Germany, pp. 449–71.

Schultz, O. -E. and Haffner, G. 1959. Zur Kenntnis eines sedativen und antibakteriellen Wirkstoffes aus dem Deutschen Faserhanf (*Cannabis sativa*). *Zeitschrift für Naturforschung B* 14(2), 98–100. doi:10.1515/znb-1959-0208.

Sergeeva, E., Shah, S. and Glick, B. R. 2006. Growth of transgenic canola (*Brassica napus* cv. Westar) expressing a bacterial 1-aminocyclopropane-1-carboxylate (ACC) deaminase gene on high concentrations of salt. *World Journal of Microbiology and Biotechnology* 22(3), 277–82. doi:10.1007/s11274-005-9032-1.

Shakeri, E., Modarres-Sanavy, S. A. M., Amini Dehaghi, M., Tabatabaei, S. A. and Moradi-Ghahderijani, M. 2016. Improvement of yield, yield components and oil quality in sesame (*Sesamum indicum* L.) by N-fixing bacteria fertilizers and urea. *Archives of Agronomy and Soil Science* 62(4), 547–60. doi:10.1080/03650340.2015.1064901.

Shank, E. A. and Kolter, R. 2009. New developments in microbial interspecies signaling. *Current Opinion in Microbiology* 12(2), 205–14. doi:10.1016/j.mib.2009.01.003.

Shanmugam, S., Kjaer, K. H., Ottosen, C.-O., Rosenqvist, E., Kumari Sharma, D. and Wollenweber, B. 2013. The alleviating effect of elevated CO2 on heat stress susceptibility of two wheat (*Triticum aestivum* L.) cultivars. *Journal of Agronomy and Crop Science* 199(5), 340–50. doi:10.1111/jac.12023.

Shao, H. B., Chu, L. Y., Wu, G., Zhang, J. H., Lu, Z. H. and Hu, Y.C. 2007. Changes of some anti- oxidative physiological indices under soil water deficits among 10 wheat (*Triticum aestivum* L.) genotypes at tillering stage. *Colloids and Surfaces. B, Biointerfaces* 54(2), 143–9. doi:10.1016/j.colsurfb.2006.09.004.

Singh, A., Maurya, S., Singh, R. and Singh, U. P. 2012. Antibiotic potential of plant growth promoting rhizobacteria (PGPR) against *Sclerotium rolfsii*. *Archives of Phytopathology and Plant Protection* 45(14), 1655–62. doi:10.1080/03235408.2012.702460.

Singh, U. P., Sarma, B. K., Singh, D. P. and Bahadur, A. 2002. Plant growth-promoting rhizobacteria- mediated induction of phenolics in pea (*Pisum sativum*) after infection with *Erysiphe pisi*. *Current Microbiology* 44(6), 396–400. doi:10.1007/s00284-001-0007-7.

Slininger, P. J. and Jackson, M. A. 1992. Nutritional factors regulating growth and accumulation of phenazine 1-carboxylic acid by *Pseudomonas fluorescens* 2-79. *Applied Microbiology and Biotechnology* 37(3), 388–92. doi:10.1007/BF00210998.

Smith, K. P. and Goodman, R. M. 1999. Host variation for interactions with beneficial plant-associated microbes. *Annual Review of Phytopathology* 37, 473–91. doi:10.1146/annurev.phyto.37.1.473.

Sobti, R. C., Arora, N. K. and Kothari, R. 2018. *Environmental Biotechnology: for Sustainable Future*. Springer, Singapore, 401 pp.

Spoel, S. H. and Dong, X. 2012. How do plants achieve immunity? Defence without specialized immune cells. *Nature Reviews. Immunology* 12(2), 89–100. doi:10.1038/nri3141.

Stubbendieck, R. M., Vargas-Bautista, C. and Straight, P. D. 2016. Bacterial communities: interactions to scale. *Frontiers in Microbiology* 7, 1234. doi:10.3389/fmicb.2016.01234.

Subramanian, S., Ricci, E., Souleimanov, A. and Smith, D. L. 2016a. A proteomic approach to lipo- chitooligosaccharide and thuricin 17 effects on soybean germination unstressed and salt stress. *PLoS ONE* 11(8), e0160660. doi:10.1371/journal.pone.0160660.

Subramanian, S., Souleimanov, A. and Smith, D. L. 2016b. Proteomic studies on the effects of lipo- chitooligosaccharide and thuricin 17 under unstressed and salt stressed conditions in *Arabidopsis thaliana*. *Frontiers in Plant Science* 7, 1314. doi:10.3389/fpls.2016.01314.

Tahir, I. S. A., Nakata, N., Yamaguchi, T., Nakano, J. and Mukhtar Ali, A. 2009. Physiological response of three wheat cultivars to high shoot and root temperatures during early growth stages. *Plant Production Science* 12(4), 409–19. doi:10.1626/pps.12.409.

Takishita, Y., Charron, J. B. and Smith, D. L. 2018. Biocontrol rhizobacterium *Pseudomonas* sp. 23S induces systemic resistance in tomato (*Solanum lycopersicum* L.) against bacterial canker *Clavibacter michiganensis* subsp. *michiganensis*. *Frontiers in Microbiology* 9, 2119. doi:10.3389/fmicb.2018.02119.

Tendulkar, S. R., Saikumari, Y. K., Patel, V., Raghotama, S., Munshi, T. K., Balaram, P. and Chattoo, B. B. 2007. Isolation, purification and characterization of an antifungal molecule produced by *Bacillus licheniformis* BC98, and its effect on phytopathogen *Magnaporthe grisea*. *Journal of Applied Microbiology* 103(6), 2331–9. doi:10.1111/j.1365-2672.2007.03501.x.

Tewari, S. and Arora, N. K. 2014. Multifunctional exopolysaccharides from *Pseudomonas aeruginosa* PF23 involved in plant growth stimulation, biocontrol and stress amelioration in sunflower under saline conditions. *Current Microbiology* 69(4), 484–94. doi:10.1007/s00284-014-0612-x.

Theocharis, A., Clément, C. and Barka, E. A. 2012. Physiological and molecular changes in plants grown at low temperatures. *Planta* 235(6), 1091–105. doi:10.1007/s00425-012-1641-y.

Thomashow, L. S. and Weller, D. M. 1988. Role of a phenazine antibiotic from *Pseudomonas fluorescens* in biological control of *Gaeumannomyces graminis* var. *tritici*. *Journal of Bacteriology* 170(8), 3499–508. doi:10.1128/jb.170.8.3499-3508.1988.

Timmusk, S., El-Daim, I. A. A., Copolovici, L., Tanilas, T., Kännaste, A., Behers, L., Nevo, E., Seisenbaeva, G., Stenström, E. and Niinemets, Ü. 2014. Drought-tolerance of wheat improved by rhizosphere bacteria from harsh environments: enhanced biomass production and reduced emissions of stress volatiles. *PLOS ONE* 9(5), e96086. doi:10.1371/journal.pone.0096086.

Timmusk, S., Behers, L., Muthoni, J., Muraya, A. and Aronsson, A. C. 2017. Perspectives and challenges of microbial application for crop improvement. *Frontiers in Plant Science* 8, 49. doi:10.3389/fpls.2017.00049.

Tiwari, S., Lata, C., Chauhan, P. S. and Nautiyal, C. S. 2016. *Pseudomonas putida* attunes morphophysiological, biochemical and molecular responses in *Cicer arietinum* L. during drought stress and recovery. *Plant Physiology and Biochemistry: PPB* 99, 108–17. doi:10.1016/j.plaphy.2015.11.001.

Tiwari, S., Prasad, V., Chauhan, P. S. and Lata, C. 2017. *Bacillus amyloliquefaciens* confers tolerance to various abiotic stresses and modulates plant response to phytohormones through osmoprotection and gene expression regulation in rice. *Frontiers in Plant Science* 8, 1510. doi:10.3389/fpls.2017.01510.

Tjamos, S. E., Flemetakis, E., Paplomatas, E. J. and Katinakis, P. 2005. Induction of resistance to *Verticillium dahliae* in *Arabidopsis thaliana* by the biocontrol agent K-165 and pathogenesis-related proteins gene expression. *Molecular Plant-Microbe Interactions: MPMI* 18(6), 555–61. doi:10.1094/MPMI-18-0555.

Toral, L., Rodríguez, M., Bejar, V. and Sampedro, I. 2018. Antifungal activity of lipopeptides from. *Bacillus* XT1 CECT 8661 against *Botrytis cinerea*. *Frontiers in Microbiology* 9, 1315.

Tsukanova, K. A., Chebotar, V. K, Meyer, J. J. M. and Bibikova, T. N. 2017. Effect of plant growth-promoting rhizobacteria on plant hormone homeostasis. *South African Journal of Botany* 113, 91–102. doi:10.1016/j.sajb.2017.07.007.

Turan, M., Güllüce, M., Çakmak, R. and Şahin, F. 2013. Effect of plant growth-promoting rhizobacteria strain on freezing injury and antioxidant enzyme activity of wheat and barley. *Journal of Plant Nutrition* 36(5), 731–48. doi:10.1080/01904167.2012.754038.

Turner, C. E., Elsohly, M. A. and Boeren, E. G. 1980. Constituents of *Cannabis sativa* L. XVII. A review of the natural constituents. *Journal of Natural Products* 43(2), 169–234. doi:10.1021/np50008a001.

Van Klingeren, B. and Ten Ham, M. 1976. Antibacterial activity of Δ 9-tetrahydrocannabinol and cannabidiol. *Antonie van Leeuwenhoek* 42(1–2), 9–12. doi:10.1007/bf00399444.

Van Loon, L. C., Bakker, P. A. and Pieterse, C. M. 1998. Systemic resistance induced by rhizosphere bacteria. *Annual Review of Phytopathology* 36, 453–83. doi:10.1146/annurev.phyto.36.1.453.

Van Velthuizen, H. 2007. *Mapping Biophysical Factors that Influence Agricultural Production and Rural Vulnerability*. Food & Agriculture Organization, Environment and Natural Resources Series No. 11. Rome, 96 pp.

Vardharajula, S., Zulfikar Ali, S., Grover, M., Reddy, G. and Bandi, V. 2011. Drought-tolerant plant growth promoting *Bacillus* spp.: effect on growth, osmolytes, and antioxidant status of maize under drought stress. *Journal of Plant Interactions* 6(1), 1–14. doi:10.1080/17429145.2010.535178.

Velusamy, P., Immanuel, J. E., Gnanamanickam, S. S. and Thomashow, L. 2006. Biological control of rice bacterial blight by plant-associated bacteria producing 2, 4-diacetylphloroglucinol. *Canadian Journal of Microbiology* 52(1), 56–65. doi:10.1139/w05-106.

Venturi, V. and Keel, C. 2016. Signaling in the rhizosphere. *Trends in Plant Science* 21(3), 187–98. doi:10.1016/j.tplants.2016.01.005.

Vives-Peris, V., Molina, L., Segura, A., Gómez-Cadenas, A. and Pérez-Clemente, R. M. 2018. Root exudates from citrus plants subjected to abiotic stress conditions have a positive effect on rhizobacteria. *Journal of Plant Physiology* 228, 208–17. doi:10.1016/j.jplph.2018.06.003.

Wahid, A., Gelani, S., Ashraf, M. and Foolad, M. R. 2007. Heat tolerance in plants: an overview. *Environmental and Experimental Botany* 61(3), 199–223. doi:10.1016/j.envexpbot.2007.05.011.

Wei, A. and Shibamoto, T. 2007. Antioxidant activities and volatile constituents of various essential oils. *Journal of Agricultural and Food Chemistry* 55(5), 1737–42. doi:10.1021/jf062959x.

Wilhelm, E. P., Mullen, R. E., Keeling, P. L. and Singletary, G. W. 1999. Heat stress during grain filling in maize: effects on kernel growth and metabolism. *Crop Science* 39(6), 1733–41. doi:10.2135/cropsci1999.3961733x.

Yeoh, Y. K., Dennis, P. G., Paungfoo-Lonhienne, C., Weber, L., Brackin, R., Ragan, M. A., Schmidt, S. and Hugenholtz, P. 2017. Evolutionary conservation of a core root microbiome across plant phyla along a tropical soil chronosequence. *Nature Communications* 8(1), 215. doi:10.1038/s41467-017-00262-8.

Zachow, C., Jahanshah, G., de Bruijn, I., Song, C., Ianni, F., Pataj, Z., Gerhardt, H., Pianet, I., Lämmerhofer, M., Berg, G., Gross, H. and Raaijmakers, J. M. 2015. The novel lipopeptide poaeamide of the endophyte *Pseudomonas poae* RE* 1-1-14 is involved in pathogen suppression and root colonization. *Molecular Plant-Microbe Interactions: MPMI* 28(7), 800–10. doi:10.1094/MPMI-12-14-0406-R.

Zemanová, V., Pavlík, M., Pavlíková, D. and Tlustoš, P. 2014. The significance of methionine, histidine and tryptophan in plant responses and adaptation to cadmium stress. *Plant, Soil and Environment* 60(9), 426–32. doi:10.17221/544/2014-PSE.

Zhang, H., Kim, M. S., Sun, Y., Dowd, S. E., Shi, H. and Paré, P. W. 2008a. Soil bacteria confer plant salt tolerance by tissue-specific regulation of the sodium transporter HKT1. *Molecular Plant-Microbe Interactions: MPMI* 21(6), 737–44. doi:10.1094/MPMI-21-6-0737.

Zhang, H., Xie, X., Kim, M. S., Kornyeyev, D. A., Holaday, S. and Paré, P. W. 2008b. Soil bacteria augment *Arabidopsis* photosynthesis by decreasing glucose sensing and abscisic acid levels in planta. *The Plant Journal: For Cell and Molecular Biology* 56(2), 264–73. doi:10.1111/j.1365-313X.2008.03593.x.

Zhang, L., Niaz, S. I., Khan, D., Wang, Z., Zhu, Y., Zhou, H., Lin, Y., Li, J. and Liu, L. 2017. Induction of diverse bioactive secondary metabolites from the mangrove endophytic fungus *Trichoderma* sp. (strain 307) by co-cultivation with *Acinetobacter johnsonii* (strain B2). *Marine Drugs* 15(2), 35. doi:10.3390/md15020035.

Chapter 8

Arbuscular mycorrhizal fungi as biostimulants for sustainable crop production

Michael Bitterlich, Leibniz-Institute of Vegetable and Ornamental Crops, Germany; Louis Mercy and Miguel Arato, INOQ GmbH, Germany; and Philipp Franken, Erfurt Research Centre for Horticultural Crops, University of Applied Sciences Erfurt and Institute of Microbiology, Friedrich Schiller University Jena, Germany

1 Introduction

To deal with the interrelated problems of hunger and poverty, climate change, economic and social inequity, and the need for environmental sustainability, a paradigm shift is required in agricultural production from the 'green revolution' to an 'ecological intensification' approach (UNCTAD, 2013). To achieve this, there is a need to drastically reduce the environmental impacts of conventional agriculture and broaden the scope of agroecological methods (UNCTAD, 2013). The use of biostimulants as an agroecological approach in sustainable agriculture has attracted increasing interest in recent years (du Jardin, 2015). In fact, biostimulants have found their way into the Fertilizing Products Regulations of the European Union (EU) as a distinct category therein (https://eur-lex.europa.eu/legal-content/EN/TXT/?uri=CELEX%3A32019R1009).

Symbiotic arbuscular mycorrhizal (AM) fungi are considered biostimulants (Rouphael et al., 2015; du Jardin, 2015). These members of the fungal subdivision Glomeromycotina establish mutualistic relationships with most (74%) terrestrial plant species (Van Der Heijden et al., 2015; Spatafora et al., 2016), and likely

http://dx.doi.org/10.19103/AS.2020.0068.09

participated in plant colonization of terrestrial ecosystems 410 million years ago (Kenrick and Strullu-Derrien, 2014). Many important crops, such as cereals, fruit trees, vegetables, and ornamental crops, engage in associations with AM fungi. Moreover, more than 300 types of AM fungi have been characterized to date (http://www.amf-phylogeny.com/amphylo_species.html), and this number is constantly increasing. This low host specificity, together with the large pool of potential AM fungal species, constitutes a resource that can potentially be applied in a broad range of sustainable crop production scenarios.

AM fungi are considered the 'mother' of all root endosymbioses (Parniske, 2008), and the backbone of the plant holobiont (Lee et al., 2019). The basis of the AM symbiosis is a bidirectional exchange of nutrients between the plant and the AM fungus (Smith and Read, 2008). Host plants provide a physical support and a favourable metabolic framework for the obligate biotrophic lifestyle of AM fungi. AM fungi receive carbon that is fixed by the host plant's photosynthesis in exchange for mineral nutrients that they provide to the host plant via the fungal mycelial network (Smith and Read, 2008). A frequent outcome of empirical research done on AM associations in many different settings is improved plant performance and growth (Jayne and Quigley, 2014; Lekberg and Koide, 2005; Hoeksema et al., 2010).

Successful implementation of AM fungi as tools for the transition of crop production towards sustainability requires:

- in-depth understanding of the biological functions that qualify AM fungi as biostimulants;
- knowledge of how these AM functions respond to environmental gradients and management practices;
- evaluation of which goals should be achieved by implemantation of mycorrhiza in different agricultural and societal contexts; and
- access to products that provide economic and ecological security and effectiveness for application of mycorrhiza in a broad range of settings.

Advances in science and technology have helped elucidate the fundamental physiological mechanisms involved in AM symbioses. However, there is a strong context dependency of the outcomes of the application of mycorrhizal inoculants, as well as a large variability in the responses of AM symbioses to anthropogenic interventions and environmental factors. Root colonization success by AM fungi varies, and plant responses to fungal infection range from growth depression to growth promotion, depending on the plant species and cultivar, fungal species and strain, their combinations, the soil, or even simply the site used (Azcon and Ocampo, 1981; Declerck et al., 1995; Hetrick et al., 1992; Taylor et al., 2015; Burleigh et al., 2002; Klironomos, 2003; Schläppi et al., 2017; Kokkoris et al., 2019). From an ecological point of view, this variability is not a concern and would probably be expected, but it constitutes a key challenge to developing safe and broadly effective strategies for the implementation of AM fungi in sustainable crop production. Considering the societal demand for agriculture to transition towards more sustainable methods, solutions are still needed to face the challenge of maintaining crop productivity while reducing the environmental harm caused by the intensification of agriculture. Hence, in the current chapter, we focussed on the mechanisms that prove the most fitting evidence qualifying AM fungi as biostimulants, and we identified conditions that risk inhibiting the biostimulant potential of AM fungi. We evaluated the factors needed for the successful implementation of AM fungi in crop production systems and attempted to propose a way forward to do this. Finally, we reviewed the current market situation for mycorrhizal products.

2 Functions and benefits of arbuscular mycorrhizal (AM) fungi

2.1 Bidirectional nutrient exchange

Decades ago, field and pot experiments on cereals and other crops had already revealed the fact that AM fungi are effective at stimulating growth and phosphorus (P) acquisition in plants (Jakobsen, 1986; Jakobsen and Erik Nielsen, 1983; Saif and Khan, 1977; Khan, 1975; Khan, 1972). This stimulatory effect requires that there be a hyphal network that proliferates in the soil and that is connected to the root systems of the plants. Such a hyphal network can extend several centimetres beyond the rooted zone (Li et al., 1991b; Jakobsen et al., 1992a). Due to its low mobility in most soils, P availability to plants is generally low, causing the formation of P depletion zones around roots. These P depletion zones can be bypassed by the hyphae of AM fungi (Schachtman et al., 1998), which increases the plant's P uptake efficiency (George et al., 1995). Direct delivery of P that is otherwise inaccessible to plant roots via AM hyphae has been confirmed by radioisotope tracing (Joner and Jakobsen, 1995; Jakobsen et al., 1992b; Pearson and Jakobsen, 1993). Since the existence of the membrane P transporters that enable P uptake by AM fungal hyphae from the soil has been verified (Harrison and van Buuren, 1995) and since the molecular basis for P translocation across the fungus-plant interface has been found to be conserved in several crop species (Rausch et al., 2001; Javot et al., 2007a; Javot et al., 2007b; Harrison et al., 2002), the scientific consensus is that the benefits of AM fungi are related to plant P nutrition.

Similar, albeit less extensive, evidence exists for nitrogen (N) translocation via the hyphae of AM fungi. Isotope tracer experiments verified that AM fungi were involved in the transfer of N from plant-inaccessible areas towards their roots (Johansen et al., 1994; Johansen et al., 1992). Metabolic N uptake and transport mechanisms from soils to fungal structures inside the plant by AM fungi (Jin et al., 2005; Govindarajulu et al., 2005) and across the plant-fungal interface (Guether et al., 2009b) have also been described, with the major form of N transferred to the plants in this way being ammonia (Guether et al., 2009b).

In turn, AM fungi receive a significant proportion of the photosynthetically fixed carbon from their host plants, and increase the carbon flow in the direction of the fungus (Slavíková et al., 2017; Douds et al., 1988; Wang et al., 1989). The carbon is shared with the fungi in the form of sugars and fatty acids (Pfeffer et al., 1999; Helber et al., 2011; Rich et al., 2017; Keymer et al., 2017), and fungal and mycorrhiza-specific plant ATPases guarantee that energy-dependent transport mechanisms can carry these nutrients across the root-fungal interface (Krajinski et al., 2002; Requena et al., 2003).

Not surprisingly, a large body of fundamental research has focussed on the C-N-P stoichiometry of AM symbioses. Experimental evidence suggests that a coordinated trade-off of C against P and N sustains the mutualism of the symbiosis (Fellbaum et al., 2014; Fellbaum et al., 2012; Kiers et al., 2011; Bever et al., 2009). However, plant growth and exploitation of soil nutrients or nutrient responsiveness of plants to AM fungal infection is variable, although C-N-P trade-off mechanisms are conserved across plant species and within the Glomeromycotina (Mensah et al., 2015; Cavagnaro et al., 2005; Burleigh et al., 2002; Smith et al., 2004). This species-dependent and even isolate-dependent variability of effectiveness in the same environmental setting has been termed 'functional diversity'. However, the potential benefits of AM fungi to plant nourishment and growth, among other traits, are integral agronomical features. The effectiveness of AM fungi on plant growth and bio-fertigation are emergent properties of the AM symbiosis (Janos, 2007; Johnson and Graham, 2013; Johnson et al., 1997) that depend on the suitability of their genetic properties under given environmental conditions. AM fungi have frequently been recognized as effective biostimulants for crops grown under suboptimal edaphic conditions, such as nutrient shortages, heavy metal pollution, drought, salinity stress, and adverse soil pH (Rouphael et al., 2015). In the following sections, we address soil nutrient and water limitations, as these appear to be the cases wherein the positive responsiveness and performance of AM symbioses are the most evident in many plant and fungal species.

2.2 Soil nutrient limitation

A large body of studies has shown that AM fungi promote increased shoot size and plant biomass (Smith and Read, 2008), which is accompanied by greater accumulation of nutrients, including N, P, K, Ca, Mg, Zn, S, Mn, Fe, and Cu, in many different crops depending on the fertilizer regime, plant species or cultivars, or AM fungal isolates used (e.g. Raju et al., 1990; McArthur and Knowles, 1993; Cavallazzi et al., 2007; Neumann et al., 2009; Oliveira et al.,

2011; Shao et al., 2018; Marschner and Dell, 1994; Watts-Williams et al., 2019). This is because AM fungi extend the root functions of plants, particularly nutrient uptake. They explore the soil with a dense and complex network of mycelia (2.7–20.5 m g^{-1} of soil) that bypasses the nutrient depletion zones around the roots (Pepe et al., 2018; Schachtman et al., 1998). The mycorrhizal mycelium is able to establish interconnections between compatible AM isolates via hyphal anastomosis (Giovannetti et al., 1999) and share nutritive and signalling compounds between neighbouring plants, forming a so-called common mycorrhizal network (Fellbaum et al., 2014; Barto et al., 2011).

For mycorrhizal symbioses, ecological modelling predicts that mutualism should change to parasitism with increasing soil fertility (Neuhauser and Fargione, 2004). Consistent with this view, the most consistently beneficial services of AM fungi to plant nourishment are observed under conditions of low P availability to plants (Smith and Read, 2008). Plants with mycorrhizas possess two common methods of P exploration: the direct pathway, via the roots, and the mycorrhizal pathway, wherein nutrient acquisition and translocation to the root is triggered by the mycorrhizal mycelium, which explores a larger soil volume than the roots alone (Smith et al., 2011; Smith and Smith, 2011). Moreover, it is often reported that these benefits of AM fungi are lost under conditions of high P availability and/or high plant P nutrition. This is because high P levels *in planta* or high levels of P fertilization systemically reduce root colonization by AM fungi (Menge et al., 1978; Jasper et al., 1979; Breuillin et al., 2010; Nouri et al., 2014).

The second best-studied type of nutrient flow involved in AM symbioses is N transport, and it has been shown that N transport via the mycorrhizal network can account for up to 35% of total plant N uptake (George et al., 1992). However, the contribution of AM fungi to plant N uptake is much more variable across studies (Jansa et al., 2019) because the N demand by plants is high (George et al., 1995) and the N demand of the AM fungal hyphae can also be substantial (Hodge and Fitter, 2010), which may lead to competition between partners for the N resource (Jansa et al., 2019; Püschel et al., 2016). Moreover, the efficiency of N uptake in plants with mycorrhizas varies with the available N source in the soil (Ames et al., 1983; Hodge et al., 2001), the amount of nitrogen in the soil (Mader et al., 2000; Püschel et al., 2016), the amount of plant-available P (Nouri et al., 2014), and other abiotic stressors, such as drought (Tobar et al., 1994; Augé, 2001). Variable net contributions of AM fungi to plant N nutrition may also be observed because the majority of plant N is acquired in the form of nitrate, the transfer of which towards the roots is driven by mass flow, which is highly dynamic.

Previous research has indicated that the uptake and accumulation of basically all other plant-relevant nutrients in mycorrhizal symbioses can be affected both positively and negatively, in a context-dependent manner

(Marschner and Dell, 1994; Augé, 2001; Nouri et al., 2014; Clark and Zeto, 2000; Ruiz-Lozano et al., 1995a). Increased nutrient acquisition of plants with mycorrhizas is probably restricted to situations where low-nutrient mobility in the soil limits plant nutrient uptake, as in these cases the spreading of hyphae through the soil helps them to encounter these low-mobility nutrients (Jansa et al., 2019). Hence, positive contributions of AM fungi to plant mineral nutrition can be observed for Cu, P, and Zn in most soils; for ammonium when it is the predominant N source in alkaline, clayey, or calcareous soils (Jansa et al., 2019 and references therein); for K, Ca, and Mg in more acidic soils (Clark and Zeto, 2000); or for nutrients in general if overall soil nutrient mobility is limited by low soil moisture (Tobar et al., 1994; Al-Karaki and Al-Raddad, 1997; Bethlenfalvay et al., 1988; Gholamhoseini et al., 2013; Azcón et al., 1996). The equivocal character of outcomes across studies and their strong context dependence for all other nutrients delivered by AM fungi apart from P may indicate that indirect mechanisms play the dominant role in plant nutrient acquisition, which would have mainly evolved under conditions where AM fungi alleviated nutrient stress conditions that were limiting factors for plant growth.

2.3 Soil water limitation

Earlier studies suggested that improved P fertilization of aerial tissues in plants with mycorrhiza accounts for their improved performance under drought stress (Nelsen and Safir, 1982). Further, the greater flexibility in N:P stoichiometry in plants with AM fungi may confer them resistance against drought perturbations (Mariotte et al., 2017). Nevertheless, even with the absence of net effects on nutrient accumulation, plants with mycorrhizas often grow better under drought conditions than their non-mycorrhizal counterparts, and soils frequently have to be strongly depleted of water before plants with mycorrhizas achieve a comparable physiological drought response (Augé et al., 1986; Duan et al., 1996; Ebel et al., 1997). The maintenance of the connection between the leaf and the root system in an intact plant is required for the leaf reaction to AM fungi to manifest under drought stress (Green et al., 1998).

Physiologically, plants react to declining soil moisture by closing their stomata to prevent desiccation and wilting (Tardieu and Simonneau, 1998; Tardieu et al., 1996; Tardieu and Davies, 1993); hence, stomatal conductance is a strong drought-stress indicator (Medrano et al., 2002). Stomatal conductance is often higher in plants with mycorrhizas, as has been observed in many plant and crop species, and this mycorrhizal promotion of leaf stomatal conductance increases with drought intensity (Augé et al., 2015; Augé, 2001). Several modes of actions are possibly involved in this process. The promotion of stomatal conductance was observed in the presence and absence of mycorrhizal effects on the hydraulic state of the plant; for example, changes in shoot water potentials, osmotic potentials, and water content were not required to alter stomatal conductance in plants with mycorrhizas (Ebel et al., 1997; Duan et al., 1996). Hence, non-hydraulic signalling via abscisic acid (ABA) was identified as the cause of alterations to stomatal conductance in mycorrhizal systems (Ebel et al., 1997; Duan et al., 1996; Augé et al., 1995; Augé and Duan, 1991). ABA is the compound which 'communicates' the occurrence of drought underground to the shoot, and ABA accumulation induces stomatal closure (Wilkinson and Davies, 2002). Stomatal conductance in plants with mycorrhizas is not differentially sensitive to the concentration of ABA in the leaf (Ebel et al., 1997; Duan et al., 1996), but lower ABA fluxes to the leaves were observed in these plants under conditions of low soil moisture (Duan et al., 1996). Hence, the root systems of plants with mycorrhizas either experience alleviated soil drought or sense drought differently than non-mycorrhizal roots.

It has been suggested that mycorrhizal roots are better able to scavenge water from drier soils (Augé, 2001). The root systems of plants with mycorrhizas can possess higher root hydraulic conductivities under drought stress, which is related to the altered root aquaporin expression patterns that underlies hormonal regulation and maintain root-cell water permeability under drought stress in them (Ruiz-Lozano et al., 2016; Bárzana et al., 2012; Aroca et al., 2007; Sánchez-Romera et al., 2016; Quiroga et al., 2019). These are mechanisms that potentially enable mycorrhizal roots to better scavenge water from drier soils, but these would still require sufficient water to reach the root system. Direct hyphal delivery of water to root systems likely occurs at levels too low to significantly contribute to plant water relations (George et al., 1992). More recent investigations showed that a significant water flow from hyphal compartments to plant compartments can take place, but these studies lacked evidence of direct intrahyphal water transport, such as tracking water movement inside the hyphae (Ruth et al., 2011; Khalvati et al., 2005), although it is noteworthy that matter flow in hyphae seems to be driven by plant activity (Kikuchi et al., 2016; Bitterlich and Franken, 2016).

Finally, mycorrhizal effects on the efficiency of plant water uptake under drought stress may also result indirectly from soil colonization by AM fungal hyphae (Augé et al., 2004; Augé, 2004; Augé et al., 2007). This was indicated by the observation that non-host plants growing in a mycorrhizal soil also showed enhanced stomatal conductance (Augé, 2004), and that mycorrhizal soils or substrates possess distinct water retention properties and substrate hydraulic conductivities from those of non-mycorrhizal soils or substrates (Augé et al., 2001; Daynes et al., 2013; Bitterlich et al., 2018a; Bitterlich et al., 2018c), even in equally rooted substrates (Bitterlich et al., 2018c; Augé et al., 2001) or in the absence of root ingrowth (Bitterlich et al., 2018a). These changes to the underground hydraulic environment when AM fungi colonize soils imply that roots may experience a different degree of stress under equal moisture conditions when AM fungi are present than when they are absent (Bitterlich et al., 2018b).

The different hydraulic properties of mycorrhizal soils are likely related to the effects of AM fungi on soil structure and their wetting properties (Rillig and Mummey, 2006; Leifheit et al., 2014; Rillig et al., 2010; Miller and Jastrow, 2000) because soil hydraulic properties depend on soil structure and water repellence (Durner, 1994; Hallett, 2008; Hallett et al., 2009). This leads to AM fungi's potentially conferring another important agroecological service in their positive effects on soil quality.

2.4 Soil quality

A soil of good quality provides important ecosystem services, is mechanically stable, and possesses sufficient organic matter and a pool of microorganisms that support soil structuration (Bronick and Lal, 2005; Morris et al., 2019). When the structure of a soil (i.e. its content of crumbly agglomerations of mineral and organic compounds) is favourable, this maintains soil porosity, good internal drainage and gas exchange, and resists erosion, and is thus of crucial importance not only for plant cultivation but also for the sustainability of agroecosystems (Six et al., 2004; Rillig and Mummey, 2006). AM fungi contribute to the soil organic matter pool, and are important drivers of the formation and stabilization of the soil structure (Leifheit et al., 2014; Rillig et al., 2010; Hallett et al., 2009; Rillig and Mummey, 2006; Rillig et al., 2002; Six et al., 2004; Miller and Jastrow, 2000; Tisdall et al., 1997; Piotrowski et al., 2004). AM fungi can enmesh soil particles (Clough and Sutton, 1978; Rillig and Mummey, 2006) and produce proteins that are involved in soil particle adherence (Wright and Upadhyaya, 1998; Wright and Upadhyaya, 1996). These proteins are released during hyphal turnover (Driver et al., 2005), and hyphal soil colonization can have surfactant effects that alter soil wettability (Rillig et al., 2010; Rillig, 2005). As AM fungi are biotrophic in nature, they do not directly compete with other saprobes for soil carbon (Klironomos and Kendrick, 1996), and hence are important drivers of soil carbon sequestration, the rate of which is tightly coupled to their abundance (Wilson et al., 2009). The beneficial ecosystem services that a well-structured soil provides have shifted management interest towards the maintenance of aggregated soils (Piotrowski et al., 2004).

2.5 Symptoms of arbuscular mycorrhizal (AM) associations in favourable environments

A plant that grows in high-quality soil with favourable physicochemical properties and that possesses an AM association that extends its scope for resource acquisition is expected to have a different physiological state and other metabolic capacities than a plant that does not. A complex process of plant reprogramming upon AM fungal infection was indeed demonstrated in previous studies (Schliemann et al., 2008; Schweiger and Mueller, 2015; Schweiger et al., 2014a; Schweiger et al., 2014b; Guether et al., 2009a). A literature survey revealed that a plant with mycorrhizas differs greatly metabolically from a non-mycorrhizal one. This reflects AM fungi-induced alterations to: the plant's energy metabolism and respiration (Del-Saz et al., 2017; Campos et al., 2015; Romero-Munar et al., 2017; Mercy et al., 2017); changes to basically all plant hormone levels (Ludwig-Müller, 2010; Meixner et al., 2005; Bedini et al., 2018; Cosme and Wurst, 2013; Hause and Schaarschmidt, 2009; Rivero et al., 2015; Hanlon and Coenen, 2011; Lucini et al., 2019; Sánchez-Romera et al., 2016); changes in the amounts, metabolism, and transport properties of soluble sugars (Lohse et al., 2005; Bona et al., 2017; Doidy et al., 2012; Boldt et al., 2011); changes to plant lipid metabolism (Lohse et al., 2005; Schliemann et al., 2008); altered amino acid concentrations (Lohse et al., 2005; Bücking and Kafle, 2015; Salvioli et al., 2012; Rivero et al., 2015); and different secondary metabolite levels (Scervino et al., 2007; Larose et al., 2002; Cosme et al., 2014; Latef and Chaoxing, 2011a; Zuccarini, 2007; Pinior et al., 2005; Wang et al., 2018). These metabolic alterations either function in regulating the plant-fungal interaction or are symptoms of its adjustment to particular environmental conditions. From an agronomic point of view, altered tissue accumulation of particular target compounds in AM symbioses can increase the quality of consumable plants (Giovannetti et al., 2012; Bona et al., 2017; Hart et al., 2015; Chialva et al., 2019; Schwarz et al., 2011; Sbrana et al., 2014; Zeng et al., 2013; Avio et al., 2018).

Furthermore, the altered physiological states of plants with AM fungi under favourable environmental conditions allow for them to have increased metabolic capacities. Increased levels of photosynthesis in the leaves have been commonly observed in many plants and crops with AM fungi (Aguilera-Gomez et al., 1999; Fay et al., 1996; Boldt et al., 2011; Ruiz-Lozano et al., 1995b; Black et al., 2000; Sheng et al., 2008; Mathur and Vyas, 1995; Romero-Munar et al., 2017). The photosynthetic capacities of leaves scale with their N and P levels and their protein contents (Walker et al., 2014; Niinemets et al., 1999; Yamori et al., 2010) in plants with and without mycorrhizas (Black et al., 2000; Grimoldi et al., 2006; Bitterlich et al., 2019). A mycorrhiza-induced acceleration of carbohydrate drain from photosynthetic tissues that stimulates photosynthesis has also been suggested to occur (Wright et al., 1998; Gavito et al., 2019; Kaschuk et al., 2009). The frequently observed maintenance of stomatal conductance in AM systems under drought stress (Augé et al., 2015; Augé et al., 2007; Augé, 2001; Ebel et al., 1997; Boldt et al., 2011; Ruiz-Lozano et al., 1995a) thus allows higher rates of photosynthesis to be maintained in these scenarios.

In the longer term, root growth and morphology can be altered upon AM infection (Berta et al., 1995; Hodge et al., 2009; Paszkowski and Gutjahr,

2013; Sukumar et al., 2013; Bi et al., 2018; Shao et al., 2018) by the metabolic and molecular reprogramming of plant tissues that accompanies this process (Meixner et al., 2005; Ludwig-Müller and Güther, 2007; Ludwig-Müller, 2006; Fiorilli et al., 2011; Hanlon and Coenen, 2011; Bedini et al., 2018; Liu et al., 2018). Theoretically, the properties that AM fungi confer on roots can favour plant establishment after transplanting or re-implantation in degraded soils, and thus has potential for application in horticulture, reforestation, or phytoremediation programmes (Douds et al., 1995; Scagel, 2004a; Scagel, 2004b; Fatemeh and Zaynab, 2014; Curaqueo et al., 2014). Finally, the adjustment of resource allocations by the AM fungal colonization of plants can stimulate their vegetative and generative development. For example, stimulation of the generative organs by AM fungal colonization has been shown to occur in important ornamental crops (Perner et al., 2007; Conversa et al., 2015; Abdel-Salam et al., 2018; Asrar et al., 2012) and crops producing consumable fruits (Todeschini et al., 2018; Bona et al., 2017).

Still, in spite of the above, commensalism or even parasitism in growth outcomes or the absence of a net effect of AM fungi on plant performance are frequently reported in comparative studies on the AM symbiosis (Johnson and Graham, 2013; Johnson et al., 1997; Smith et al., 2011; Ryan and Graham, 2018). Considering that the genetic properties of an organism determine under which environments it can approach its maximum growth potential, it is feasible that the variable growth effectiveness of plant-fungal combinations in the same settings is caused by the environmental conditions themselves. In the following section, we reviewed which environmental factors have been shown to have compelling detrimental effects on the effectiveness of mycorrhizal fungi.

2.6 Factors impairing mycorrhizal effectiveness

Considering the facts that the AM symbiosis has prevailed over the course of evolution for many millennia and that mycorrhizas are natural, it should be mentioned that growth depression by AM inoculation or the absence of a net mycorrhizal effect is not problematic from either an ecological or a sustainability standpoint. However, at the same time one could argue that achieving better plant growth is an economic and sustainable goal because, in this case, resources and inputs increasingly accumulate in the plant, and this increases the chance for crops attaining higher marketable yields. Therefore, the conditions that may jeopardize the effectiveness of mycorrhizal application need to be known.

Naturally, mycorrhizal benefits to plant performance are derived from the underground environment, but in most scenarios, plants grow under variable climatic conditions that can impose aboveground limitations on plant growth. However, studies on plant responses to the application of AM fungi under (suboptimal) climatic conditions like heat (Zhu et al., 2011; Cabral et al., 2016; Maya and Matsubara, 2013) and cold stress (Zhu et al., 2010; Latef and Chaoxing, 2011b; Chen et al., 2013; Charest et al., 1993), increased ozone levels (Cui et al., 2013; Ho and Trappe, 1984), or elevated atmospheric CO_2 concentrations (Alberton et al., 2005; Jongen et al., 1996; Gavito et al., 2000) are underrepresented in the literature.

As both partners in the AM symbiosis depend on plant photosynthesis to grow, climatic factors that limit photosynthesis are expected to negatively affect the positive contributions of AM fungi to plant activity and growth. Indeed, this has been observed for light deprivation in studies that compared plants with AM and non-AM plants (Son and Smith, 1988; Konvalinková and Jansa, 2016; Konvalinková et al., 2015; Augé et al., 2015; Bitterlich et al., 2019; Heinemeyer et al., 2006), as well as for conditions with limited atmospheric moisture (Bitterlich et al., 2019; Huang et al., 1985). Light availability seems to control carbon allocation to AM fungi (Heinemeyer et al., 2006), and if light becomes limiting to the photosynthetic process the potential benefits of AM fungi to leaf nutrition or the water status of the plant become superfluous because plant C acquisition becomes energy-limited (Bitterlich et al., 2019). It is also likely that such unfavourable C limitation can also occur in plants with mycorrhiza if their stomata close, which could predominantly occur due to atmospheric drought stress (Bitterlich et al., 2019; Huang et al., 1985).

Looking underground, scenarios in which soil levels of plant-available P are high are clearly a cause for concern since the intensity of the AM association seems to be under the control of the plant, and root colonization is systemically downregulated under high-P conditions (Breuillin et al., 2010). Moreover, the taxonomic richness and root colonization of AM fungi is negatively correlated with soil P fertility in agricultural fields (Verbruggen et al., 2013). Low-pH or acidic soils seem to have negative effects on the root colonization, spore germination, and soil abundances of particular AM fungus species (Carrino-Kyker et al., 2016; Svenningsen et al., 2018; Coughlan et al., 2000; Abbott and Robson, 1985; El-Kherbawy et al., 1989), which might be related to increased abundances of their competitors in acidic soils (Svenningsen et al., 2018). However, as with root activity, AM fungal hyphae are similarly capable of modulating the pH in the adjacent soil (Li et al., 1991a).

Moreover, belowground limitations that impair mycorrhizal effectiveness are also often related to human interventions and culture conditions that depart from natural scenarios. A simple example is plants in pots. Pot size matters in this case, and growth responsiveness to any treatment can become negligible in pot-bound plants (Poorter et al., 2012; Bourgault et al., 2017). Such confounds as pot size or rooting density are also likely to affect tests of mycorrhizal application scenarios (Koide, 1991), and this implies that pot studies that compare different species in the same setting may be suitable to study

AM functions across species and growth responses but have, at best, limited power to assess growth responsiveness to mycorrhiza directly. To counteract the effects of pot size as a limiting factor for growth, rooting volume should be decoupled from resource availability (i.e. nutrients, water) (Murphy et al., 2013), as is done in hydroponics. However, the more the cultivation system shifts towards hydroponics, the more difficult the root colonization becomes (Rouphael et al., 2015) and the less suitable the conditions are for beneficial mycorrhizal growth responses (see Sections 2.2 and 2.3).

Apart from limiting pot volumes, fields over-fertilized with P and N in conventional agriculture and management practices, such as unfavourable crop rotation, bare-fallow periods, and tillage methods have been shown to impede AM fungal abundance, diversity, and soil inoculation potential, and also to superimpose selection pressures that may not favour AM fungal diversity or the persistence of the most beneficial AM fungal taxa (Verbruggen et al., 2013; Kabir et al., 1999; Kabir, 2005; Castillo et al., 2006; Gosling et al., 2006; Troeh and Loynachan, 2003; Verbruggen and Toby Kiers, 2010; Jansa et al., 2006; Jansa et al., 2003; Jansa et al., 2002; Antunes et al., 2012; Rillig et al., 2019). The use of pesticides is also potentially detrimental to AM fungi, although this seems dependent on the chemical agent used (Rouphael et al., 2015). In general, it can be assumed that legal reinforcement of the ecological intensification approach and a return to integrated organic farming practices, such as pesticide reduction, the use of cover crops, and reduced tillage, can allow for the development of healthier soil microflora and increase the beneficial ecosystem services provided by AM fungi (Verbruggen et al., 2013; Verbruggen and Toby Kiers, 2010; Gosling et al., 2006; Kabir, 2005; Rillig et al., 2019).

3 Requirements for successful implementation of arbuscular mycorrhizal (AM) fungi in sustainable plant production

One of the basic services potentially provided by AM fungi in crop production in support of both sustainability and economic considerations would be better plant growth, which would result in either input-scaled yield benefits or in *de novo* yields from formerly unproductive scenarios. Against this backdrop, it has recently been questioned whether there is compelling evidence to warrant managing AM fungi in the agricultural context (Ryan et al., 2019; Ryan and Graham, 2002, 2018), and there is considerable debate on this point throughout the scientific community (Rillig et al., 2019; Zhang et al., 2019; Verbruggen et al., 2013). Without favouring one of these views, we recommend that the reader reviews the following set of relevant articles (Rillig et al., 2019; Zhang et al., 2019; Verbruggen et al., 2013; Ryan et al., 2019; Ryan and Graham, 2002, 2018) because they constitute an excellent overview of the obstacles and needs, knowledge gaps, and research priorities potentially impacting the integration of AM fungi into agricultural production systems. Critical discussions are definitely appropriate and needed, but to stay within the theme of the chapter we continued our review below, assuming that the implementation of AM fungi as biostimulants is a desired goal in the transition of agriculture towards adopting more sustainable methods.

3.1 Agronomical aspects

Successful implementation of AM fungi in sustainable plant production systems, either through management shifts towards their conservation or their application, requires that the mutualistic nature of the AM symbioses can be effectively translated into plant performance while also reducing the detrimental impacts of agriculture on the environment. From a sustainability standpoint, the targeted implementation of mycorrhiza should use safe resource inputs, reduce undesired resource losses, increase resource-use efficiency, or increase the resiliency of production systems. From an economic point of view, the implementation of AM fungi in crop production adds complexity and costs that have to be at least compensated for by reducing other costs of management, producing higher yields, or producing higher-quality consumables. This also requires that there are compelling, quantifiable, and, last but not least, predictable effects of application of mycorrhiza that can be targeted. However, the predictability of mycorrhizal effects (on yield, plant growth, or plant nourishment) across a wide variety of settings has been limited because multiple factors determine their outcomes (Ryan et al., 2019). The high variability of management practices, species used, soils, and climates imposes strong constraints on the predictability of contributions of mycorrhiza to plant production. To deal with this, a first step would be for plant-intrinsic traits involved in AM symbioses to find their way into plant breeding programmes. This is required because of the observed cultivar dependence and variable growth outcomes of AM fungal root colonization, and also because ongoing plant breeding programmes are leading to the cultivar recommendations for growers being dynamic.

Domestication of plants has probably led to 'breeding them away' from being mycorrhiza-dependent or from being able to achieve the most effective contributions of AM fungi to their performance, although domestication has also restored plant infectivity by fungi in some cases (Martín-Robles et al., 2018; Sawers et al., 2018). Forward genetic approaches could be used to identify genes that are responsible for the mycorrhizal phenotypes of plants (Sawers et al., 2018), which could then be included in breeding programmes while also accounting for other goals, such as yield component traits (Ryan et al., 2019). Nevertheless, the patterns described above as 'functional diversity' (i.e. the fact that a similar metabolic framework leads to different outcomes in AM symbioses) show that, if the ability to breed plants for these desired genetic properties is achieved, its results have to be put to the test with the appropriate tools in an array of different environmental contexts. We can certainly breed plants selectively for mycorrhiza dependency or traits that confer crops with climate, but not weather, resiliency. As plant growth and plant responsiveness to mycorrhizal infection are emergent properties (Janos, 2007; Johnson et al., 1997; Johnson et al., 2015) that develop over time and over the course of plant growth as a result of environmental variability, the contributions of AM fungi to ecosystem resiliency and crop productivity can hardly be evaluated over only one or a few growing seasons because seasonal variation is expected. Moreover, contributions of AM fungi to plant performance can depend on the developmental stage of the plant (Lehman et al., 2019), and thus within-culture monitoring is required to control this. To quantify services by mycorrhiza in agricultural systems, critical thresholds for mycorrhizal function across agroecosystems and along environmental gradients need to be identified, which requires long-term field experimentation in different settings (Rillig et al., 2019), sophisticated field experiments that comprehensively monitor influential factors (management, crop rotation, climate, and soil properties), and a large database to allow structural reviews to be performed (Rillig et al., 2019; Ryan et al., 2019). To achieve this, technologies that enable these laborious studies to be carried out over longer time frames are required. The development of mycorrhiza-defective mutants of the most important crops and growing these under field conditions would guide the assessment of mycorrhizal contributions to plant growth and ecosystem services under realistic conditions, but for many crops such mutants have not been found or produced (Rillig et al., 2019; Sawers et al., 2018). High-throughput technologies that allow for large-scale sampling and appropriate analyses of spatial heterogeneity across different scales of investigation are required. Recently, for AM fungal root colonization, foliar markers that are not restricted to a particular AM fungus-plant combination and that are applicable in field-based quantitative trait loci (QTL) mapping of AM symbiosis-related genes have been identified (Wang et al., 2018). Precision agriculture tools are available that can generate high-resolution soil texture maps to quantify spatial soil heterogeneity in the field (Boenecke et al., 2018; Meyer et al., 2019). First efforts have also been made towards being able to dynamically model P uptake, growth, and root infection by AM fungi (Schnepf et al., 2016; Schnepf and Roose, 2006). If such technologies are combined with breeding approaches, the emergence of systematic agronomy approaches with a good predictive power would become possible, and these approaches could integrate genetics, environment, and management to pave the way forward for developing future agronomic approaches (Ryan et al., 2019; Ryan and Graham, 2018) that may include managing crops with and for AM fungi.

In addition, identifying and applying signalling molecules, termed 'inducers', could be used to favour increased plant susceptibility to mycorrhiza, while also taking care to perform risk assessments of soil and food sanitary aspects and ensuring that cost-efficient application methods are available for end-users (Bedini et al., 2018).

If a comprehensive framework with good predictive power, such as a crop model with a mycorrhizal component, is developed, it will then become possible to make informed decisions about where and how to use AM fungi in agriculture. This may also allow managers to account for the different societal contexts that may lead to diverging goals that should be achieved, for example choosing to improve yields for food security over avoiding groundwater pollution, or vice versa.

3.2 Inoculum-related aspects

The application of mycorrhizal products is feasible since increasing the mycorrhizal soil potential can improve plant growth (Clapperton and Reid, 1992; Hijri, 2016; Bender et al., 2019). The ecological factors mediating the success of field inoculation have been reviewed previously (Verbruggen et al., 2013). If the application of mycorrhizal products is targeted, several challenges have to be solved. From the fungal side, classical breeding or genetic transformation to obtain AM fungi with stable, desirable traits is currently impossible due to the specific characteristics of fungal genetic systems (Chen et al., 2018). Spores of AM fungi can contain hundreds to thousands of different genetic nuclei (Marleau et al., 2011) whose compositions differ not only from parental but also from sister lines (Boon et al., 2013). Moreover, genetic exchange between compatible isolates via hyphal anastomosis adds more complexity to this system (Angelard et al., 2010). This genetic plasticity is supposed to be part of the high adaptive potential of AM fungi that lets them develop in different soil ecotypes and plant species, with different impacts on plant growth responses (Angelard et al., 2014). Specifically, it was previously shown that native AM fungi isolated from heavy metal- or salt-polluted soils could increase plant tolerance to these respective abiotic stressors compared with that achieved by other, non-native AM fungal lines (Diaz et al., 1996; Hildebrandt et al., 1999; Cantrell and Linderman, 2001). This may be turned into an advantage to application of mycorrhiza in agriculture, as one strategy could be to 'train' (i.e. cultivate) AM fungi in the presence of a specific and constant stressor to improve their survival and adaptation ability in various types of soils and situations (Feldmann and Grotkaas, 2002). Similarly to this idea, it was previously shown that better plant tolerance to high zinc concentrations was obtained after inoculation with zinc-acclimatized AM fungi that were grown in the presence of increasing zinc concentrations over several generations in a root organ culture system

(Bui and Franken, 2018). This opens up more possibilities for AM fungi to be acclimatized to other problematic environmental conditions and could guide their translation into broader-scale applications.

Moreover, the interactions between AM fungi and other microorganisms populating plant roots and their surroundings are important. On the one hand, microorganisms can modify the functions of the mycorrhizal symbiosis, while, on the other hand, AM fungi impact the compositions of these microorganisms. These interactions have been a research topic for a long time, and the implementation of this knowledge in an integrative strategy for applying AM fungi together with other beneficial microorganisms is a challenge for the future. Barbara Mosse observed in 1962 that an AM fungus was only able to colonize a root if the bacterium *Pseudomonas* spp. was added to it (Mosse, 1962). This result was confirmed in numerous subsequent studies using different combinations of AM fungal and bacterial isolates. The outcomes tested were either the development of the fungus (Mugnier and Mosse, 1987), the performance of the plant (Meyer and Linderman, 1986), or both (Azcón, 1989). The term 'mycorrhizal helper bacteria' was used for the first time in the context of such an ectomycorrhizal interaction and was later transferred to apply to AM symbioses (Garbaye, 1994; Toro et al. 1997). The approaches used to identify optimal conditions in these studies differed, depending on the underlying hypotheses. One strategy was to isolate the bacteria that were associated with mycorrhizal spores and hyphae, characterize these bacteria, and then formulate mixed inocula for applying them along with AM fungi in plant production systems (Xavier and Germida, 2003). The second strategy was based on considering the characters of known bacterial isolates. Based on the assumption that AM fungi mainly deliver phosphate to their host plants, they were combined with nitrogen-fixing bacteria to support the nutrition of plants with these two major mineral nutrients (Rao et al., 1985; Battini et al., 2017). In other studies, phosphate-solubilizing bacteria were used in an attempt to profit from a hypothesised synergy between phosphate dissolution by the bacteria and phosphate transport across depletion zones by the fungus (Raj et al., 1981). In summary, the overall positive effects observed in many studies showed that the combination of AM fungi and bacteria with particular characteristics is a meaningful and useful approach to improve plant nutrition and growth (Turrini et al., 2018). Optimization of inocula by adaptation and formulation are potential ways forward to increase the effectiveness of the application of these products.

However, production of such inoculants is still cost-intensive and, hence, is an economic concern because the obligate biotrophism of AM fungi means that they require plant cultivation for their propagation. The application of mycorrhizal fungi under field conditions requires the application of a large amount of propagules at a low cost, and will ideally be insensitive to factors that impede the proper development of mycorrhiza and can use existing machinery. However, currently no standardized protocol exists for the proper number of propagules to be applied per hectare and for particular crops, and the optimal application approach appears to be soil and crop specific (as for plant species and density). Ideally, the mycorrhizal potential of a fertile soil is between 1000 and 10 000 propagules per litre of soil, which corresponds to 3–30 billion propagules per hectare (considering a depth of 30 cm), respectively. These values and their associated costs (see Section 4) are far above the average number of propagules advised per hectare for commercial products. The most obvious strategies to efficiently increase mycorrhizal soil potentials include: (i) the application of an inoculum associated with cover/inter-cropping using strongly mycorrhiza-associated plant species for the production of AM fungi (in situ); (ii) adapting the application of mycorrhizal inocula in specific field patches lacking mycorrhizal soil potentials; (iii) formulating or co-formulating AM fungal inocula with other microorganisms (such as 'mycorrhiza helper bacteria'); and (iv) improving mycorrhizal product effectiveness by reducing the required amount applied (e.g. through seed coating).

Last but not least, all such efforts related to inoculation have to be evaluated with respect to their potential ecological risks of spreading fungal species (Schwartz et al., 2006; Verbruggen et al., 2013), as invasiveness has been documented for species of ectomycorrhizal fungi (Pringle et al., 2009; Berch et al., 2016).

4 The current market for mycorrhizal products

The market for commercial applications of AM fungi has grown considerably in recent years. The following paragraphs offer an extension of the systematic survey presented by Chen et al. (2018) with the intention of delineating the current state of the market for mycorrhizal products.

In the last few decades, the number of enterprises offering mycorrhizal products in their portfolios has increased exponentially. In Europe alone, the number of companies doing this has doubled in 10 years (Fig. 1). The European market is the dominant region for mycorrhizal products, in which Germany is the leader (Fig. 2). However, emerging markets like India and China in Asia, as well as Argentina, Brazil, Colombia, and Mexico in Latin America, have shown considerable growth in the availability of AM fungal products in the last 10 years. In fact, the European Biostimulants Council (EBIC) projects that the Asian continent will be the leading market for biostimulants in general by 2030.

In terms of company size and origin, the worldwide mycorrhizal market is mainly dominated by small- and medium-sized enterprises, with a few large

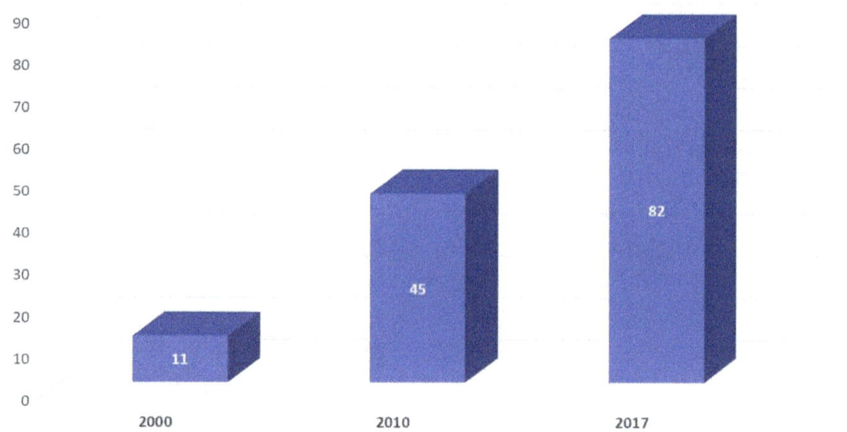

Figure 1 Number of companies selling mycorrhizal products in Europe in the last 20 years (from Chen et al., 2018).

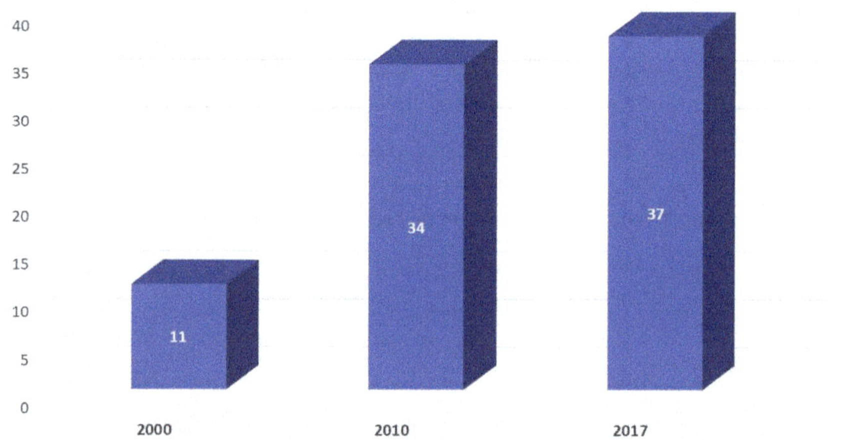

Figure 2 Number of companies selling mycorrhizal products, grouped by decade.

companies operating principally from Europe and North America. In Europe, the companies selling mycorrhizal formulations and/or mycorrhizal inocula were mainly founded in the last 10–20 years (Fig. 3), and are concentrated in Germany, Spain, Italy, the United Kingdom (UK), France, the Netherlands, the Czech Republic, Austria, and others (Fig. 2).

The main selling channels for mycorrhizal products are direct sales from producers, commercial distributors, and formulators-mixers (Fig. 4).

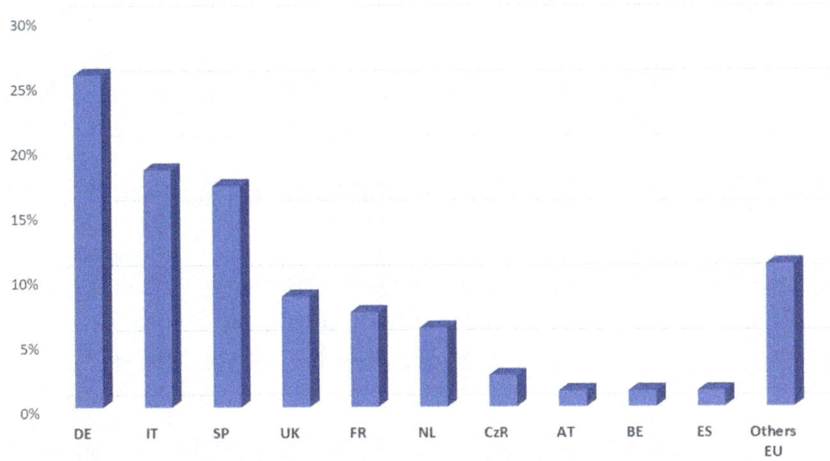

Figure 3 Enterprises selling mycorrhizal products, divided by country of origin.

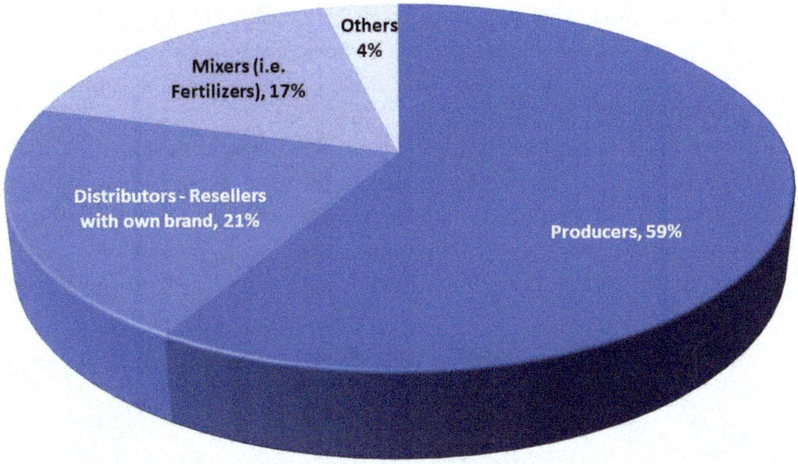

Figure 4 Percentage of selling channels of mycorrhizal products.

In the European market, around 40% of enterprises produce and sell their own mycorrhizal inocula. The remaining 60% source their inocula from a different producer and use it for developing and selling their own formulations with different substrates or fertilizers, or mix it with other products, such as humic acid, plant extracts, rhizobacteria, and others. Sourcing inocula from professional mycorrhizal producers has become a common practice among enterprises, mainly owing to the complications related to the quality control, stability, and level of expertise required to produce mycorrhizal fungi in a professional way. Enterprises opt for sourcing their inocula from other sources to avoid the risks of product inconsistency and instead concentrate on their special formulations and the market aspects of the business. Given the nature of the market, which is mainly composed of small- and medium-sized players, it is complicated to define an actual business value for this industry. The best estimations from experts indicate a global business value of biostimulants in 2016 of about $1.68 billion USD, with a forecasted value of $3 billion USD by the year 2030 (Sessitsch et al., 2018). According to the EBIC, 12% of the total biostimulants market consists of products based on mycorrhizas. Hence, the market for mycorrhizal products and mycorrhiza-based formulations could reach a potential value of up to $360 million USD by the year 2030.

Different possible means exist to produce mycorrhizal inocula, and these processes mainly include the 'on-farm' method, production in greenhouses (ex vitro), or production under monoxenic (in vitro) culture conditions (Rouphael et al., 2015). Based on our survey, the most common production process for mycorrhizal inocula reported by enterprises in Europe is production ex vitro. Less than 20% of enterprises reported using an in vitro process as their production system. Despite this latter method's potential usefulness (Ceballos et al., 2013), the reason why it is not used much might be related to the high costs and specificities of this production process, as well as there being a lack of studies that have formerly compared the inoculation rates and efficacy of in vitro and ex vitro inocula (Calvet et al., 2013).

Mycorrhizal products are found on the market in different formulations, including powder or granular formulations, or inocula mixed with a carrier material like peat, vermiculite, sand, and others (Fig. 5).

According to our survey, less than 35% of the products available for professional agriculture and horticulture are composed solely of

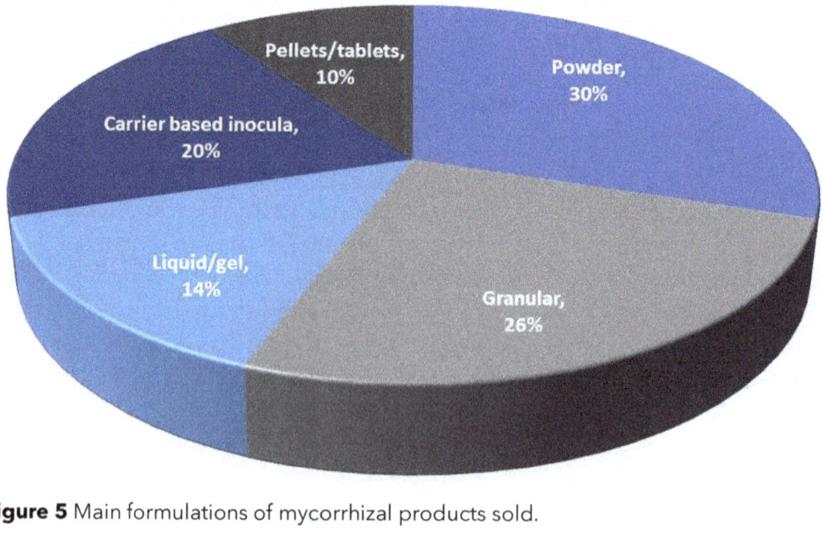

Figure 5 Main formulations of mycorrhizal products sold.

mycorrhizal fungi. Most of the products are mixtures of AM fungi with other microorganisms. Common formulations include *Bacillus subtilis*, *Lactobacillus farciminis*, *Trichoderma* spp., saprophytic fungi, and *Streptomyces avermitilis*, as well as mixtures of nitrogen, phosphorus, and potassium. The species of AM fungi commonly reported on the product labels of these formations include principally *Rhizophagus irregularis* and *Funnelliformis mosseae*, likely because these species are widespread and very common in most ecotypes (Davison et al., 2015; Savary et al., 2018). However, other species can be found, such as *Glomus etunicatum* and *Glomus fasciculatum*. Given that there are currently more than 300 characterized AM fungal species, only a very small proportion of AM fungal species are exploited as biostimulants. Some products on the market also report the use of ectomycorrhizal fungi in their formulations. However, the species and concentration are not always specified on the technical data sheets in these cases.

The mycorrhizal and mycorrhiza-based formulations offered on the market are mainly divided into two different categories: those used as biostimulants, and those used as biofertilizers along with other microorganisms. The category depends on the components included in the formulations and the applicable national regulations.

The concentration of mycorrhizal units (M.U.) that is able to generate mycorrhizas (termed mycorrhizal propagules, consisting of spores, living hyphae, and mycorrhizal root fragments containing spores and/or vesicles) in different mycorrhiza-based formulations is shown in Fig. 6. In most of the products offered on the market, both AM fungal inocula and AM fungi-based formulations include a declared concentration below 50 000 M.U. per gram.

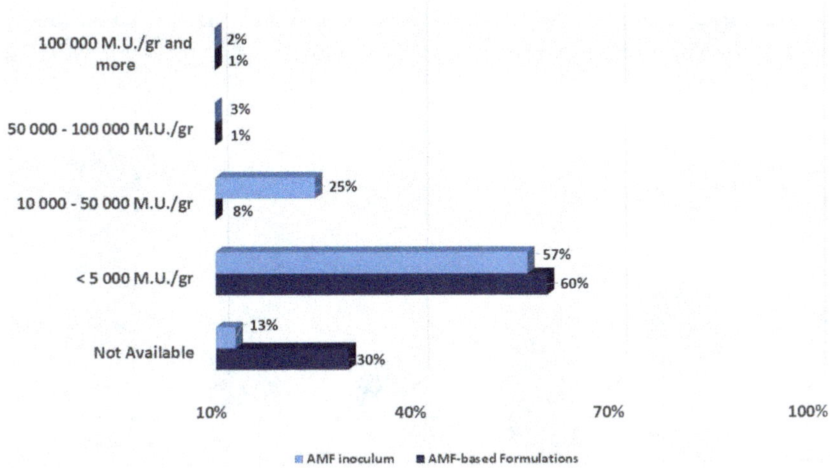

Figure 6 Percentage of mycorrhizal products with different levels of richness in mycorrhizal propagules.

Less than 5% of the AM fungal inoculum products and less than 3% of the AM fungi-based formulations offer a concentration higher than 60 000 M.U. per gram. Large variation in propagule richness among products exists, and there is also a non-negligible portion of all mycorrhizal products (21.5% of all AM fungal products) that does not mention the number of propagules, illustrating the lack of a standardized process to validate inoculum quality.

In terms of price and product application, according to our survey, the average retail price for mycorrhizal products in the European market ranges from 0.10 € to 0.50 € per plant (Fig. 7). These prices could reach higher in some cases, for instance, in hobby gardening or horticulture than those in open-culture systems (agricultural field applications). The prices are often defined based on commercial considerations and average market prices, but without any relation to the number of mycorrhizal units in each product. In agriculture and horticulture, prices of mycorrhizal and mycorrhiza-based products range between 130 € and 250 € per hectare (Chen et al., 2018; Hijri, 2016).

Regarding the sectors in which commercial mycorrhizal products are applied, these are mainly applied in gardening and landscaping, hobby and semi-professional horticulture, and forestry (Fig. 8). In professional agriculture and horticulture, despite the well-documented benefits of AM fungi to plants (e.g. Hijri, 2016), mycorrhizal products have only shown a slow market penetration in the last few decades. This slow market penetration is mainly due to the traditional practices used by farmers and food growers, as well as to inconsistencies in product performance and the poor cost competitiveness of mycorrhiza products against traditional fertilizer products (Chen et al., 2018; Vosátka et al., 2008; Berruti et al., 2016). It has only been in the last 5 to 10 years that an increasing interest in mycorrhizal products for vegetable, grain, and fruit growth has been observed in the market. This change has been principally triggered by national and European regulations

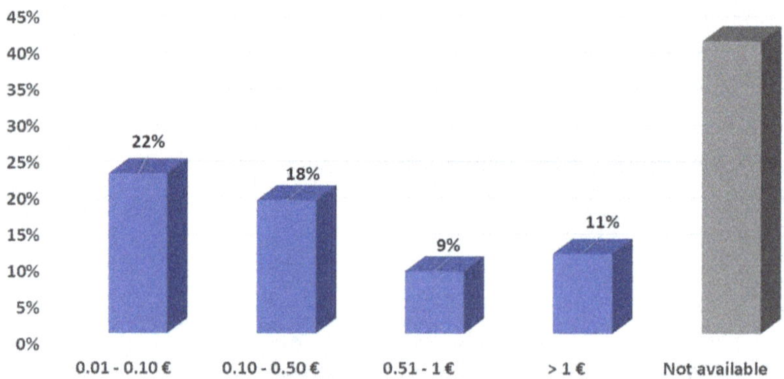

Figure 7 Price ranges per plant of mycorrhizal products.

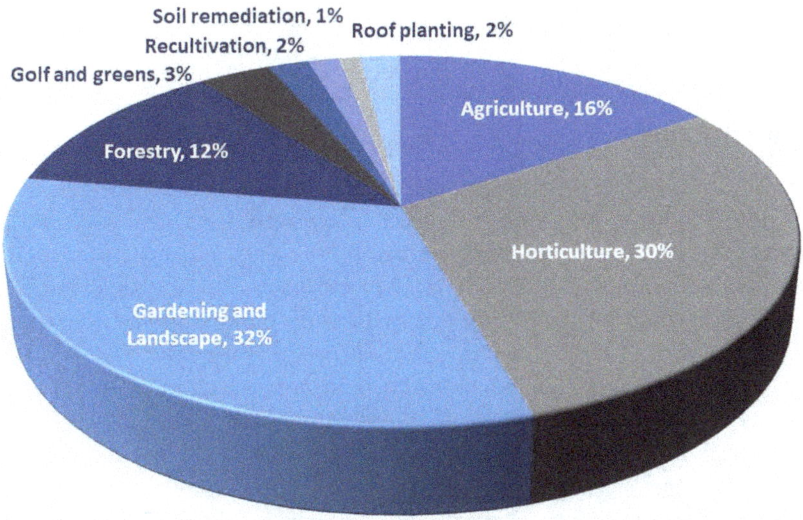

Figure 8 Sectors in which mycorrhizal fungal inocula are applied.

that have put special emphasis on promoting sustainable alternatives to traditional chemical fertilizers. In fact, the latest revisions of the European fertilizer regulations (EC No. 1069/2009 and EC No. 1107/2009) included 'Mycorrhizal fungi' as one of the four approved microorganisms for use as EC-marked biostimulants. This change in regulations will therefore present a good opportunity for business growth in the European Market (http://http://www.biostimulants.eu/2018/11/ebic-welcomes-compromise-reached-at-trilogue-meeting-on-fertilising-products-regulation-first-step-towards-eu-wide-market-creation-for-biostimulants/).

Another factor that has driven the introduction of the use of biostimulants into the food growth industry is recent changes in consumer preferences. The increasing demand for sustainable food production systems is continuously putting pressure on food growers to migrate from traditional practices towards the use of sustainable products (http://http://www.biostimulants.eu/2018/11/ebic-welcomes-compromise-reached-at-trilogue-meeting-on-fertilising-products-regulation-first-step-towards-eu-wide-market-creation-for-biostimulants/). One example is the Almeria region in Spain, where the number of organic producers and the use of sustainable products for plant growth promotion and plant protection have grown considerably in the last few years. Farmer groups, private enterprises, agronomic consultants, and research institutes are currently involved in different national and European projects searching for new sustainable inputs to large-scale horticultural production practices. Some of these projects included mycorrhiza among the biostimulants and biofertilizer products they have considered (e.g. LIFE the green link; SoilCare ID-677407; MIRA ID-722493).

5 Conclusion

Arbuscular mycorrhizal (AM) fungi have been considered as natural biostimulants for a long time. Due to their co-evolutionary history with land plants, their occurrence in most terrestrial ecosystems, and their low host specificity, it is widely suggested that they should be applied as an important element in sustainable plant production systems. Indeed, the outcome of the mycorrhizal symbiosis in many different scenarios is mostly positive in terms of plant biomass and plant health. Nevertheless, this outcome cannot be predicted, and depends on the genotypes of the plants, on the quality of the fungal inoculum, and on the environmental conditions.

Several functions of AM interactions and their underlying mechanisms constitute the ecological services provided by this symbiosis. The bidirectional nutrient exchange between the mycorrhizal partners has been relatively well characterized. This exchange is the basis for the ability of plants with mycorrhiza to overcome soil nutrient and water limitations better than their non-mycorrhizal counterparts. A further important function of AM fungi is their contribution to carbon sequestration and soil aggregation, which can result in soils having improved quality if they are pervaded by AM fungal hyphae. Plants can respond to colonization by AM fungi by increasing their primary metabolic capacities, which can increase the biomass and yield of crops. Also, the secondary metabolism of plants responds to mycorrhizal inoculation, and this has been shown to impact the quality of plant products. Beside changes in metabolism, morphological changes can also be observed after the inoculation of plants with AM fungi. Concerning the application of AM fungi, the improved rooting achieved by plants with mycorrhizas can be useful in aiding the establishment of plants after transferring them to new environments.

However, there are factors that impede the effectiveness of mycorrhizal applications. Based on the bidirectional nutrient exchange between AM partners, conditions that limit photosynthesis and cause an excess of soil nutrients are suggested to limit benefits of mycorrhiza, and this has been confirmed in numerous studies. To successfully apply the mycorrhizal symbiosis in plant production, several aspects of it need to be considered. First, it is important to determine for which sites of plant production and their environmental conditions the application of AM fungal inocula makes sense. Second, the genotype of the plant must support the positive functions of the symbiosis, which means that mycorrhizal dependency has to be considered as an important trait in future plant breeding programmes. Third, inocula should be targeted for application in particular conditions. This can be achieved by acclimatizing AM fungi during the production process, and by combining them with other beneficial soil microorganisms. Finally, should the application of AM fungi become more routine, the market for mycorrhizal and mycorrhiza-based products has to be considered. Regulations for their registration, overcoming traditional practices, and production costs will be important aspects of the AM market to consider in this case.

6 Where to look for further information

First of all, the reference list of this chapter should provide a comprehensive overview of the literature related to the biology of arbuscular mycorrhizal fungi and the scientific state of the art, without being intentionally fragmentary. To get started with the topic of mycorrhiza, the standard reference is the book *Mycorrhizal Symbiosis,* edited by Sally Smith and David Read (Smith and Read, 2008). To read further about the state of the art in mycorrhizal research *Mycorrhiza* and *The New Phytologist* are recommended journals, with a long history of reporting scientific progress in the field. The biannual 'International Conference on Mycorrhiza' (ICOM) and the 'International Molecular Mycorrhiza Meeting' (iMMM) are the main meetings where the scientific community comes together to discuss this subject area. We cannot address all aspects of mycorrhiza biology in this chapter, but we also want to acknowledge that AM fungi are also well known for their role in protecting plants against pathogens and pests (e.g. Jung et al., 2012; Pozo and Azcon-Aguilar, 2007; Schouteden et al., 2015; Vos et al., 2013). The AM fungal species characterized to date can be found at: http://www.amf-phylogeny.com/amphylo_species.html.

7 References

Abbott, L. K. and Robson, A. D. 1985. The effect of soil pH on the formation of VA mycorrhizas by two species of Glomus. *Soil Research* 23(2), 253–61. doi:10.1071/SR9850253.

Abdel-Salam, E., Alatar, A. and El-Sheikh, M. A. 2018. Inoculation with arbuscular mycorrhizal fungi alleviates harmful effects of drought stress on damask Rose. *Saudi Journal of Biological Sciences* 25(8), 1772–80. doi:10.1016/j.sjbs.2017.10.015.

Aguilera-Gomez, L., Davies Jr., F. T., Olalde-Portugal, V., Duray, S. A. and Phavaphutanon, L. 1999. Influence of phosphorus and endomycorrhiza (*Glomus intraradices*) on gas exchange and plant growth of chile ancho pepper (*Capsicum annuum L.* cv. San Luis). *Photosynthetica* 36(3), 441–9. doi:10.1023/A:1007032320951.

Alberton, O., Kuyper, T. W. and Gorissen, A. 2005. Taking mycocentrism seriously: mycorrhizal fungal and plant responses to elevated CO_2. *The New Phytologist* 167(3), 859–68. doi:10.1111/j.1469-8137.2005.01458.x.

Al-Karaki, G. N. and Al-Raddad, A. 1997. Effects of arbuscular mycorrhizal fungi and drought stress on growth and nutrient uptake of two wheat genotypes differing in drought resistance. *Mycorrhiza* 7(2), 83–8. doi:10.1007/s005720050166.

Ames, R. N., Reid, C. P. P., Porter, L. K. and Cambardella, C. 1983. Hyphal uptake and transport of nitrogen from 2 N-15-labelled sources by Glomus mosseae, a vesicular arbuscular mycorrhizal fungus. *New Phytologist* 95(3), 381–96. doi:10.1111/j.1469-8137.1983.tb03506.x.

Angelard, C., Colard, A., Niculita-Hirzel, H., Croll, D. and Sanders, I. R. 2010. Segregation in a mycorrhizal fungus alters rice growth and symbiosis-specific gene transcription. *Current Biology: CB* 20(13), 1216–21. doi:10.1016/j.cub.2010.05.031.

Angelard, C., Tanner, C. J., Fontanillas, P., Niculita-Hirzel, H., Masclaux, F. and Sanders, I. R. 2014. Rapid genotypic change and plasticity in arbuscular mycorrhizal fungi is caused by a host shift and enhanced by segregation. *The ISME Journal* 8(2), 284–94. doi:10.1038/ismej.2013.154.

Antunes, P. M., Lehmann, A., Hart, M. M., Baumecker, M. and Rillig, M. C. 2012. Long-term effects of soil nutrient deficiency on arbuscular mycorrhizal communities. *Functional Ecology* 26(2), 532–40. doi:10.1111/j.1365-2435.2011.01953.x.

Aroca, R., Porcel, R. and Ruiz-Lozano, J. M. 2007. How does arbuscular mycorrhizal symbiosis regulate root hydraulic properties and plasma membrane aquaporins in *Phaseolus vulgaris* under drought, cold or salinity stresses? *The New Phytologist* 173(4), 808–16. doi:10.1111/j.1469-8137.2006.01961.x.

Asrar, A. A., Abdel-Fattah, G. M. and Elhindi, K. M. 2012. Improving growth, flower yield, and water relations of snapdragon (*Antirrhinum majus* L.) plants grown under well-watered and water-stress conditions using arbuscular mycorrhizal fungi. *Photosynthetica* 50(2), 305–16. doi:10.1007/s11099-012-0024-8.

Augé, R. M. 2001. Water relations, drought and vesicular-arbuscular mycorrhizal symbiosis. *Mycorrhiza* 11(1), 3–42. doi:10.1007/s005720100097.

Augé, R. M. 2004. Arbuscular mycorrhizae and soil/plant water relations. *Canadian Journal of Soil Science* 84(4), 373–81. doi:10.4141/S04-002.

Augé, R. M. and Duan, X. 1991. Mycorrhizal fungis and non hydrolic root signals of soil drying. *Plant Physiology* 97(2), 821–4. doi:10.1104/pp.97.2.821.

Augé, R. M., Schekel, K. A. and Wample, R. L. 1986. Greater leaf conductance of well-watered VA mycorrhizal rose plants is not related to phosphorus nutrition. *New Phytologist* 103(1), 107–16. doi:10.1111/j.1469-8137.1986.tb00600.x.

Augé, R. M., Stodola, A. J. W., Ebel, R. C. and Duan, X. 1995. Leaf elongation and water relations of mycorrhizal sorghum in response to partial soil drying: two Glomus species at varying phosphorus fertilization. *Journal of Experimental Botany* 46(3), 297–307. doi:10.1093/jxb/46.3.297.

Augé, R. M., Stodola, A. J. W., Tims, J. E. and Saxton, A. M. 2001. Moisture retention properties of a mycorrhizal soil. *Plant and Soil* 230(1), 87–97. doi:10.1023/A:1004891210871.

Augé, R. M., Sylvia, D. M., Park, S., Buttery, B. R., Saxton, A. M., Moore, J. L. and Cho, K. 2004. Partitioning mycorrhizal influence on water relations of *Phaseolus vulgaris* into soil and plant components. *Canadian Journal of Botany* 82(4), 503–14. doi:10.1139/b04-020.

Augé, R. M., Toler, H. D., Moore, J. L., Cho, K. and Saxton, A. M. 2007. Comparing contributions of soil versus root colonization to variations in stomatal behavior and soil drying in mycorrhizal *Sorghum bicolor* and *Cucurbita pepo*. *Journal of Plant Physiology* 164(10), 1289–99. doi:10.1016/j.jplph.2006.08.005.

Augé, R. M., Toler, H. D. and Saxton, A. M. 2015. Arbuscular mycorrhizal symbiosis alters stomatal conductance of host plants more under drought than under amply watered conditions: a meta-analysis. *Mycorrhiza* 25(1), 13–24. doi:10.1007/s00572-014-0585-4.

Avio, L., Turrini, A., Giovannetti, M. and Sbrana, C. 2018. Designing the ideotype mycorrhizal symbionts for the production of healthy food. *Frontiers in Plant Science* 9, 1089. doi:10.3389/fpls.2018.01089.

Azcón, R. 1989. Selective interaction between free-living rhizosphere bacteria and vesicular arbuscular mycorrhizal fungi. *Soil Biology and Biochemistry* 21(5), 639–44. doi:10.1016/0038-0717(89)90057-6.

Azcon, R. and Ocampo, J. A. 1981. Factors affecting the vesicular-arbuscular infection and mycorrhizal dependency of thirteen wheat cultivars. *New Phytologist* 87(4), 677–85. doi:10.1111/j.1469-8137.1981.tb01702.x.

Azcón, R., Gomez, M. and Tobar, R. 1996. Physiological and nutritional responses by *Lactuca sativa* L. to nitrogen sources and mycorrhizal fungi under drought conditions. *Biology and Fertility of Soils* 22(1–2), 156–61. doi:10.1007/BF00384448.

Barto, E. K., Hilker, M., Müller, F., Mohney, B. K., Weidenhamer, J. D. and Rillig, M. C. 2011. The fungal fast lane: common mycorrhizal networks extend bioactive zones of allelochemicals in soils. *PLoS ONE* 6(11), e27195. doi:10.1371/journal.pone.0027195.

Bárzana, G., Aroca, R., Paz, J. A., Chaumont, F., Martinez-Ballesta, M. C., Carvajal, M. and Ruiz-Lozano, J. M. 2012. Arbuscular mycorrhizal symbiosis increases relative apoplastic water flow in roots of the host plant under both well-watered and drought stress conditions. *Annals of Botany* 109(5), 1009–17. doi:10.1093/aob/mcs007.

Battini, F., Grønlund, M., Agnolucci, M., Giovannetti, M. and Jakobsen, I. 2017. Facilitation of phosphorus uptake in maize plants by mycorrhizosphere bacteria. *Scientific Reports* 7(1), 4686. doi:10.1038/s41598-017-04959-0.

Bedini, A., Mercy, L., Schneider, C., Franken, P. and Lucic-Mercy, E. 2018. Unravelling the initial plant hormone signalling, metabolic mechanisms and plant defense triggering the endomycorrhizal symbiosis behavior. *Frontiers in Plant Science* 9, 1800. doi:10.3389/fpls.2018.01800.

Bender, S. F., Schlaeppi, K., Held, A. and Van Der Heijden, M. G. A. 2019. Establishment success and crop growth effects of an arbuscular mycorrhizal fungus inoculated into Swiss corn fields. *Agriculture, Ecosystems and Environment* 273, 13-24. doi:10.1016/j.agee.2018.12.003.

Berch, S. M., Kroeger, P. and Finston, T. 2016. The death cap mushroom (*Amanita phalloides*) moves to a native tree in Victoria, British Columbia. *Botany* 95(4), 435–40. doi:10.1139/cjb-2016-0183.

Berruti, A., Lumini, E., Balestrini, R. and Bianciotto, V. 2016. Arbuscular mycorrhizal fungi as natural biofertilizers: let's benefit from past successes. *Frontiers in Microbiology* 6, 1559. doi:10.3389/fmicb.2015.01559.

Berta, G., Trotta, A., Fusconi, A., Hooker, J. E., Munro, M., Atkinson, D., Giovannetti, M., Morini, S., Fortuna, P., Tisserant, Bt., Gianinazzi-Pearson, V. and Gianinazzi, S. 1995. Arbuscular mycorrhizal induced changes to plant growth and root system morphology in Prunus cerasifera. *Tree Physiology* 15(5), 281–93. doi:10.1093/treephys/15.5.281.

Bethlenfalvay, G. J., Brown, M. S., Ames, R. N. and Thomas, R. S. 1988. Effects of drought on host and endophyte development in mycorrhizal soybeans in relation to water use and phosphate uptake. *Physiologia Plantarum* 72(3), 565–71. doi:10.1111/j.1399-3054.1988.tb09166.x.

Bever, J. D., Richardson, S. C., Lawrence, B. M., Holmes, J. and Watson, M. 2009. Preferential allocation to beneficial symbiont with spatial structure maintains mycorrhizal mutualism. *Ecology Letters* 12(1), 13–21. doi:10.1111/j.1461-0248.2008.01254.x.

Bi, Y., Zhang, Y. and Zou, H. 2018. Plant growth and their root development after inoculation of arbuscular mycorrhizal fungi in coal mine subsided areas. *International Journal of Coal Science and Technology* 5(1), 47–53. doi:10.1007/s40789-018-0201-x.

Bitterlich, M. and Franken, P. 2016. Connecting polyphosphate translocation and hyphal water transport points to a key of mycorrhizal functioning. *The New Phytologist* 211(4), 1147–9. doi:10.1111/nph.14104.

Bitterlich, M., Franken, P. and Graefe, J. 2018a. Arbuscular mycorrhiza improves substrate hydraulic conductivity in the plant available moisture range under root growth exclusion. *Frontiers in Plant Science* 9, 301. doi:10.3389/fpls.2018.00301.

Bitterlich, M., Rouphael, Y., Graefe, J. and Franken, P. 2018b. Arbuscular mycorrhizas: a promising component of plant production systems provided favorable conditions for their growth. *Frontiers in Plant Science* 9, 1329. doi:10.3389/fpls.2018.01329.

Bitterlich, M., Sandmann, M. and Graefe, J. 2018c. Arbuscular mycorrhiza alleviates restrictions to substrate water flow and delays transpiration limitation to stronger drought in tomato. *Frontiers in Plant Science* 9, 154. doi:10.3389/fpls.2018.00154.

Bitterlich, M., Franken, P. and Graefe, J. 2019. Atmospheric drought and low light impede mycorrhizal effects on leaf photosynthesis--a glasshouse study on tomato under naturally fluctuating environmental conditions. *Mycorrhiza* 29(1), 13–28. doi:10.1007/s00572-018-0872-6.

Black, K. G., Mitchell, D. T. and Osborne, B. A. 2000. Effect of mycorrhizal-enhanced leaf phosphate status on carbon partitioning, translocation and photosynthesis in cucumber. *Plant, Cell and Environment* 23(8), 797–809. doi:10.1046/j.1365-3040.2000.00598.x.

Boenecke, E., Lueck, E., Ruehlmann, J., Gruendling, R. and Franko, U. 2018. Determining the within-field yield variability from seasonally changing soil conditions. *Precision Agriculture* 19(4), 750–69. doi:10.1007/s11119-017-9556-z.

Boldt, K., Pörs, Y., Haupt, B., Bitterlich, M., Kühn, C., Grimm, B. and Franken, P. 2011. Photochemical processes, carbon assimilation and RNA accumulation of sucrose transporter genes in tomato arbuscular mycorrhiza. *Journal of Plant Physiology* 168(11), 1256–63. doi:10.1016/j.jplph.2011.01.026.

Bona, E., Cantamessa, S., Massa, N., Manassero, P., Marsano, F., Copetta, A., Lingua, G., D'Agostino, G., Gamalero, E. and Berta, G. 2017. Arbuscular mycorrhizal fungi and plant growth-promoting pseudomonads improve yield, quality and nutritional value of tomato: a field study. *Mycorrhiza* 27(1), 1–11. doi:10.1007/s00572-016-0727-y.

Boon, E., Zimmerman, E., St-Arnaud, M. and Hijri, M. 2013. Allelic differences within and among sister spores of the arbuscular mycorrhizal fungus *Glomus etunicatum* suggest segregation at sporulation. *PLoS ONE* 8(12), e83301. doi:10.1371/journal.pone.0083301.

Bourgault, M., James, A. T. and Dreccer, M. F. 2017. Pot size matters revisited: does container size affect the response to elevated CO_2 and our ability to detect genotypic variability in this response in wheat? *Functional Plant Biology* 44(1), 52–61. doi:10.1071/FP16047.

Breuillin, F., Schramm, J., Hajirezaei, M., Ahkami, A., Favre, P., Druege, U., Hause, B., Bucher, M., Kretzschmar, T., Bossolini, E., Kuhlemeier, C., Martinoia, E., Franken, P., Scholz, U. and Reinhardt, D. 2010. Phosphate systemically inhibits development of arbuscular mycorrhiza in *Petunia hybrida* and represses genes involved in mycorrhizal functioning. *The Plant Journal: for Cell and Molecular Biology* 64(6), 1002–17. doi:10.1111/j.1365-313X.2010.04385.x.

Bronick, C. J. and Lal, R. 2005. Soil structure and management: a review. *Geoderma* 124(1-2), 3-22. doi:10.1016/j.geoderma.2004.03.005.

Bücking, H. and Kafle, A. 2015. Role of arbuscular mycorrhizal fungi in the nitrogen uptake of plants: current knowledge and research gaps. *Agronomy* 5(4), 587–612. doi:10.3390/agronomy5040587.

Bui, V. C. and Franken, P. 2018. Acclimatisation of *Rhizophagus irregularis* enhances Zn tolerance of the fungus and the mycorrhizal plant partner. *Frontiers in Microbiology* 9, 3156. doi:10.3389/fmicb.2018.03156.

Burleigh, S. H., Cavagnaro, T. and Jakobsen, I. 2002. Functional diversity of arbuscular mycorrhizas extends to the expression of plant genes involved in P nutrition. *Journal of Experimental Botany* 53(374), 1593–601. doi:10.1093/jxb/erf013.

Cabral, C., Ravnskov, S., Tringovska, I. and Wollenweber, B. 2016. Arbuscular mycorrhizal fungi modify nutrient allocation and composition in wheat (*Triticum aestivum* L.) subjected to heat-stress. *Plant and Soil* 408(1-2), 385–99. doi:10.1007/s11104-016-2942-x.

Calvet, C., Camprubi, A., Pérez-Hernández, A. and Lovato, P. E. 2013. Plant growth stimulation and root colonization potential of *in vivo* versus *in vitro* arbuscular mycorrhizal inocula. *HortScience* 48(7), 897–901. doi:10.21273/HORTSCI.48.7.897.

Campos, C., Cardoso, H., Nogales, A., Svensson, J., Lopez-Ráez, J. A., Pozo, M. J., Nobre, T., Schneider, C. and Arnholdt-Schmitt, B. 2015. Intra and inter-spore variability in *Rhizophagus irregularis AOX* gene. *PLoS ONE* 10(11), e0142339. doi:10.1371/journal.pone.0142339.

Cantrell, I. C. and Linderman, R. G. 2001. Preinoculation of lettuce and onion with VA mycorrhizal fungi reduces deleterious effects of soil salinity. *Plant and Soil* 233(2), 269–81. doi:10.1023/A:1010564013601.

Carrino-Kyker, S. R., Kluber, L. A., Petersen, S. M., Coyle, K. P., Hewins, C. R., DeForest, J. L., Smemo, K. A. and Burke, D. J. 2016. Mycorrhizal fungal communities respond to experimental elevation of soil pH and P availability in temperate hardwood forests. *FEMS Microbiology Ecology* 92(3), fiw024. doi:10.1093/femsec/fiw024.

Castillo, C. G., Rubio, R., Rouanet, J. L. and Borie, F. 2006. Early effects of tillage and crop rotation on arbuscular mycorrhizal fungal propagules in an Ultisol. *Biology and Fertility of Soils* 43(1), 83–92. doi:10.1007/s00374-005-0067-0.

Cavagnaro, T. R., Smith, F. A., Smith, S. E. and Jakobsen, I. 2005. Functional diversity in arbuscular mycorrhizas: exploitation of soil patches with different phosphate enrichment differs among fungal species. *Plant, Cell and Environment* 28(5), 642–50. doi:10.1111/j.1365-3040.2005.01310.x.

Cavallazzi, J. R. P., Klauberg Filho, O. K., Stürmer, S. L., Rygiewicz, P. T. and de Mendonça, M. M. 2007. Screening and selecting arbuscular mycorrhizal fungi for inoculating micropropagated apple rootstocks in acid soils. *Plant Cell, Tissue and Organ Culture* 90(2), 117–29. doi:10.1007/s11240-006-9163-6.

Ceballos, I., Ruiz, M., Fernandez, C., Pena, R., Rodriguez, A. and Sanders, I. R. 2013. The *in vitro* mass-produced model mycorrhizal fungus, *Rhizophagus irregularis*, significantly increases yields of the globally important food security crop cassava. *PLoS ONE* 8(8), e70633. doi:10.1371/journal.pone.0070633.

Charest, C., Dalpe, Y. and Brown, A. 1993. The effect of vesicular-arbuscular mycorrhizae and chilling on two hybrids of maize. *Mycorrhiza* 4(2), 89–92. doi:10.1007/BF00204064.

Chen, M., Arato, M., Borghi, L., Nouri, E. and Reinhardt, D. 2018. Beneficial services of arbuscular mycorrhizal fungi-from ecology to application. *Frontiers in Plant Science* 9, 1270. doi:10.3389/fpls.2018.01270.

Chen, S., Jin, W., Liu, A., Zhang, S., Liu, D., Wang, F., Lin, X. and He, C. 2013. Arbuscular mycorrhizal fungi (AMF) increase growth and secondary metabolism in cucumber subjected to low temperature stress. *Scientia Horticulturae* 160, 222–9. doi:10.1016/j.scienta.2013.05.039.

Chialva, M., Stefania, S., Novero, M., Bonfante, P. and Lanfranco, L. 2019. The genetics behind AM symbiosis: the case of tomato wild relatives. *Proceedings of the 4th International Molecular Mycorrhiza Meeting* (vol. 49).

Clapperton, M. J. and Reid, D. M. 1992. A relationship between plant growth and increasing VA mycorrhizal inoculum density. *New Phytologist* 120(2), 227–34. doi:10.1111/j.1469-8137.1992.tb05658.x.

Clark, R. B. and Zeto, S. K. 2000. Mineral acquisition by arbuscular mycorrhizal plants. *Journal of Plant Nutrition* 23(7), 867–902. doi:10.1080/01904160009382068.

Clough, K. S. and Sutton, J. C. 1978. Direct observation of fungal aggregates in sand dune soil. *Canadian Journal of Microbiology* 24(3), 333–5. doi:10.1139/m78-056.

Conversa, G., Bonasia, A., Lazzizera, C. and Elia, A. 2015. Influence of biochar, mycorrhizal inoculation, and fertilizer rate on growth and flowering of Pelargonium (*Pelargonium zonale* L.) plants. *Frontiers in Plant Science* 6, 429. doi:10.3389/fpls.2015.00429.

Cosme, M. and Wurst, S. 2013. Interactions between arbuscular mycorrhizal fungi, rhizobacteria, soil phosphorus and plant cytokinin deficiency change the root morphology, yield and quality of tobacco. *Soil Biology and Biochemistry* 57, 436–43. doi:10.1016/j.soilbio.2012.09.024.

Cosme, M., Franken, P., Mewis, I., Baldermann, S. and Wurst, S. 2014. Arbuscular mycorrhizal fungi affect glucosinolate and mineral element composition in leaves of *Moringa oleifera*. *Mycorrhiza* 24(7), 565–70. doi:10.1007/s00572-014-0574-7.

Coughlan, A. P., Dalpé, Y., Lapointe, L. and Piché, Y. 2000. Soil pH-induced changes in root colonization, diversity, and reproduction of symbiotic arbuscular mycorrhizal fungi from healthy and declining maple forests. *Canadian Journal of Forest Research* 30(10), 1543–54. doi:10.1139/x00-090.

Cui, X. C., Hu, J. L., Lin, X. G., Wang, F. Y., Chen, R. R., Wang, J. H. and Zhu, J. G. 2013. Arbuscular mycorrhizal fungi alleviate ozone stress on nitrogen nutrition of field wheat. *Journal of Agricultural Science and Technology* 15.

Curaqueo, G., Schoebitz, M., Borie, F., Caravaca, F. and Roldán, A. 2014. Inoculation with arbuscular mycorrhizal fungi and addition of composted olive-mill waste enhance plant establishment and soil properties in the regeneration of a heavy metal-polluted environment. *Environmental Science and Pollution Research International* 21(12), 7403–12. doi:10.1007/s11356-014-2696-z.

Davison, J., Moora, M., Öpik, M., Adholeya, A., Ainsaar, L., Bâ, A., Burla, S., Diedhiou, A. G., Hiiesalu, I., Jairus, T., Johnson, N. C., Kane, A., Koorem, K., Kochar, M., Ndiaye, C., Pärtel, M., Reier, Ü, Saks, Ü, Singh, R., Vasar, M. and Zobel, M. 2015. Global assessment of arbuscular mycorrhizal fungus diversity reveals very low endemism. *Science* 349(6251), 970–3. doi:10.1126/science.aab1161.

Daynes, C. N., Field, D. J., Saleeba, J. A., Cole, M. A. and McGee, P. A. 2013. Development and stabilisation of soil structure via interactions between organic matter, arbuscular mycorrhizal fungi and plant roots. *Soil Biology and Biochemistry* 57, 683–94. doi:10.1016/j.soilbio.2012.09.020.

Declerck, S., Plenchette, C. and Strullu, D. G. 1995. Mycorrhizal dependency of banana (*Musa acuminata*, AAA group) cultivar. *Plant and Soil* 176(1), 183–7. doi:10.1007/BF00017688.

Del-Saz, N. F., Romero-Munar, A., Alonso, D., Aroca, R., Baraza, E., Flexas, J. and Ribas-Carbo, M. 2017. Respiratory ATP cost and benefit of arbuscular mycorrhizal symbiosis with *Nicotiana tabacum* at different growth stages and under salinity. *Journal of Plant Physiology* 218, 243–8.

Diaz, G., Azcon-Aguilar, C. and Honrubia, M. 1996. Influence of arbuscular mycorrhizae on heavy metal (Zn and Pb) uptake and growth of *Lygeum spartum* and *Anthyllis cytisoides*. *Plant and Soil* 180(2), 241–9. doi:10.1007/BF00015307.

Doidy, J., Grace, E., Kuhn, C., Simon-Plas, F., Casieri, L. and Wipf, D. 2012. Sugar transporters in plants and in their interactions with fungi. *Trends in Plant Science* 17(7), 413–22. doi:10.1016/j.tplants.2012.03.009.

Douds, D. D., Johnson, C. R. and Koch, K. E. 1988. Carbon cost of the fungal symbiont relative to net leaf P accumulation in a split-root VA mycorrhizal symbiosis. *Plant Physiology* 86(2), 491–6. doi:10.1104/pp.86.2.491.

Douds, D. D., Becard, G., Pfeffer, P. E., Doner, L. W., Dymant, T. J. and Kayser, W. M. 1995. Effect of vesicular-arbuscular mycorrhizal fungi on rooting of *Sciadopitys verticillata* Sieb & Zucc. cuttings. *HortScience* 30(1), 133–4. doi:10.21273/HORTSCI.30.1.133.

Driver, J. D., Holben, W. E. and Rillig, M. C. 2005. Characterization of glomalin as a hyphal wall component of arbuscular mycorrhizal fungi. *Soil Biology and Biochemistry* 37(1), 101–6. doi:10.1016/j.soilbio.2004.06.011.

Duan, X. G., Neuman, D. S., Reiber, J. M., Green, C. D., Saxton, A. M. and Augé, R. M. 1996. Mycorrhizal influence on hydraulic and hormonal factors implicated in the control of stomatal conductance during drought. *Journal of Experimental Botany* 47(10), 1541–50. doi:10.1093/jxb/47.10.1541.

Du Jardin, P. 2015. Plant biostimulants: definition, concept, main categories and regulation. *Scientia Horticulturae* 196, 3–14. doi:10.1016/j.scienta.2015.09.021.

Durner, W. 1994. Hydraulic conductivity estimation for soils with heterogeneous pore structure. *Water Resources Research* 30(2), 211–23. doi:10.1029/93WR02676.

Ebel, R. C., Duan, X., Still, D. W. and Augé, R. M. 1997. Xylem sap abscisic acid concentration and stomatal conductance of mycorrhizal *Vigna unguiculata* in drying soil. *New Phytologist* 135(4), 755–61. doi:10.1046/j.1469-8137.1997.00674.x.

El-Kherbawy, M., Angle, J. S., Heggo, A. and Chaney, R. L. 1989. Soil pH, rhizobia and vesicular-arbuscular mycorrhiza inoculation effects on growth and heavy metal uptake of alfalfa (*Medicago sativa* L.). *Biology and Fertility of Soils* 8, 61–5.

Fatemeh, B. and Zaynab, M. 2014. Enhanced rooting of leaf bud cuttings of *Schefflera arboricola* using mycorrhizal fungi. *Annual Research and Review in Biology* 4(18), 2892–900. doi:10.9734/ARRB/2014/7914.

Fay, P., Mitchell, D. T. and Osborne, B. A. 1996. Photosynthesis and nutrient-use efficiency of barley in response to low arbuscular mycorrhizal colonization and addition of phosphorus. *The New Phytologist* 132(3), 425–33. doi:10.1111/j.1469-8137.1996.tb01862.x.

Feldmann, F. and Grotkass, C. 2002. Directed inoculum production – shall we be able to design AMF populations to achieve predictable symbiotic effectiveness? In: Gianinazzi, S., Barea, J. M. and Haselwandter, K. (Eds), *Mycorrhiza Technology in Agriculture: from Genes to Bioproducts*. Birkhäuser Verlag, Basel, pp. 261–79.

Fellbaum, C. R., Gachomo, E. W., Beesetty, Y., Choudhari, S., Strahan, G. D., Pfeffer, P. E., Kiers, E. T. and Bucking, H. 2012. Carbon availability triggers fungal nitrogen uptake and transport in arbuscular mycorrhizal symbiosis. *Proceedings of the National Academy of Sciences of the United States of America* 109(7), 2666–71. doi:10.1073/pnas.1118650109.

Fellbaum, C. R., Mensah, J. A., Cloos, A. J., Strahan, G. E., Pfeffer, P. E., Kiers, E. T. and Bücking, H. 2014. Fungal nutrient allocation in common mycorrhizal networks is regulated by the carbon source strength of individual host plants. *The New Phytologist* 203(2), 646–56. doi:10.1111/nph.12827.

Fiorilli, V., Catoni, M., Francia, D., Cardinale, F. and Lanfranco, L. 2011. The arbuscular mycorrhizal symbiosis reduces disease severity in tomato plants infected by *Botrytis cinerea*. *Journal of Plant Pathology* 93(1), 237–42.

Garbaye, J. 1994. Helper bacteria: a new dimension to the mycorrhizal symbiosis. *New Phytologist* 128(2), 197–210. doi:10.1111/j.1469-8137.1994.tb04003.x.

Gavito, M. E., Curtis, P. S., Mikkelsen, T. N. and Jakobsen, I. 2000. Atmospheric CO_2 and mycorrhiza effects on biomass allocation and nutrient uptake of nodulated pea (*Pisum sativum* L.) plants. *Journal of Experimental Botany* 51(352), 1931–8. doi:10.1093/jexbot/51.352.1931.

Gavito, M. E., Jakobsen, I., Mikkelsen, T. N. and Mora, F. 2019. Direct evidence for modulation of photosynthesis by an arbuscular mycorrhiza-induced carbon sink strength. *The New Phytologist* 223(2), 896–907. doi:10.1111/nph.15806.

George, E., Haussler, K. U., Vetterlein, D., Gorgus, E. and Marschner, H. 1992. Water and nutrient translocation by hyphae of *Glomus mosseae*. *Canadian Journal of Botany* 70(11), 2130–7. doi:10.1139/b92-265.

George, E., Marschner, H. and Jakobsen, I. 1995. Role of arbuscular mycorrhizal fungi in uptake of phosphorus and nitrogen from soils. *Critical Reviews in Biotechnology* 15(3–4), 257–70. doi:10.3109/07388559509147412.

Gholamhoseini, M., Ghalavand, A., Dolatabadian, A., Jamshidi, E. and Khodaei-Joghan, A. 2013. Effects of arbuscular mycorrhizal inoculation on growth, yield, nutrient uptake and irrigation water productivity of sunflowers grown under drought stress. *Agricultural Water Management* 117, 106–14. doi:10.1016/j.agwat.2012.11.007.

Giovannetti, M., Azzolini, D. and Citernesi, A. S. 1999. Anastomosis formation and nuclear and protoplasmic exchange in arbuscular mycorrhizal fungi. *Applied and Environmental Microbiology* 65(12), 5571-5. doi:10.1128/AEM.65.12.5571-5575.1999.

Giovannetti, M., Avio, L., Barale, R., Ceccarelli, N., Cristofani, R., Iezzi, A., Mignolli, F., Picciarelli, P., Pinto, B., Reali, D., Sbrana, C. and Scarpato, R. 2012. Nutraceutical value and safety of tomato fruits produced by mycorrhizal plants. *The British Journal of Nutrition* 107(2), 242–51. doi:10.1017/S000711451100290X.

Gosling, P., Hodge, A., Goodlass, G. and Bending, G. D. 2006. Arbuscular mycorrhizal fungi and organic farming. *Agriculture, Ecosystems and Environment* 113(1–4), 17–35. doi:10.1016/j.agee.2005.09.009.

Govindarajulu, M., Pfeffer, P. E., Jin, H. R., Abubaker, J., Douds, D. D., Allen, J. W., Bücking, H., Lammers, P. J. and Shachar-Hill, Y. 2005. Nitrogen transfer in the arbuscular mycorrhizal symbiosis. *Nature* 435(7043), 819–23. doi:10.1038/nature03610.

Green, C. D., Stodola, A. and Augé, R. M. 1998. Transpiration of detached leaves from mycorrhizal and nonmycorrhizal cowpea and rose plants given varying abscisic acid, pH, calcium, and phosphorus. *Mycorrhiza* 8(2), 93–9. doi:10.1007/s005720050218.

Grimoldi, A. A., Kavanová, M., Lattanzi, F. A., Schäufele, R. and Schnyder, H. 2006. Arbuscular mycorrhizal colonization on carbon economy in perennial ryegrass: quantification by $^{13}CO_2/^{12}CO_2$ steady-state labelling and gas exchange. *The New Phytologist* 172(3), 544–53. doi:10.1111/j.1469-8137.2006.01853.x.

Guether, M., Balestrini, R., Hannah, M., He, J., Udvardi, M. K. and Bonfante, P. 2009a. Genome-wide reprogramming of regulatory networks, transport, cell wall and membrane biogenesis during arbuscular mycorrhizal symbiosis in *Lotus japonicus*. *The New Phytologist* 182(1), 200–12. doi:10.1111/j.1469-8137.2008.02725.x.

Guether, M., Neuhauser, B., Balestrini, R., Dynowski, M., Ludewig, U. and Bonfante, P. 2009b. A mycorrhizal-specific ammonium transporter from *Lotus japonicus* acquires nitrogen released by arbuscular mycorrhizal fungi. *Plant Physiology* 150(1), 73–83. doi:10.1104/pp.109.136390.

Hallett, P. D. 2008. A brief overview of the causes, impacts and amelioration of soil water repellency-: a review. *Soil and Water Research* 3, 521–8.

Hallett, P. D., Feeney, D. S., Bengough, A. G., Rillig, M. C., Scrimgeour, C. M. and Young, I. M. 2009. Disentangling the impact of AM fungi versus roots on soil structure and water transport. *Plant and Soil* 314(1–2), 183–96. doi:10.1007/s11104-008-9717-y.

Hanlon, M. T. and Coenen, C. 2011. Genetic evidence for auxin involvement in arbuscular mycorrhiza initiation. *The New Phytologist* 189(3), 701–9. doi:10.1111/j.1469-8137.2010.03567.x.

Harrison, M. J. and van Buuren, M. L. 1995. A phosphate transporter from the mycorrhizal fungus *Glomus versiforme*. *Nature* 378(6557), 626–9. doi:10.1038/378626a0.

Harrison, M. J., Dewbre, G. R. and Liu, J. Y. 2002. A phosphate transporter from *Medicago truncatula* involved in the acquisiton of phosphate released by arbuscular mycorrhizal fungi. *The Plant Cell* 14(10), 2413–29. doi:10.1105/tpc.004861.

Hart, M., Ehret, D. L., Krumbein, A., Leung, C., Murch, S., Turi, C. and Franken, P. 2015. Inoculation with arbuscular mycorrhizal fungi improves the nutritional value of tomatoes. *Mycorrhiza* 25(5), 359–76. doi:10.1007/s00572-014-0617-0.

Hause, B. and Schaarschmidt, S. 2009. The role of jasmonates in mutualistic symbioses between plants and soil-born microorganisms. *Phytochemistry* 70(13–14), 1589–99. doi:10.1016/j.phytochem.2009.07.003.

Heinemeyer, A., Ineson, P., Ostle, N. and Fitter, A. H. 2006. Respiration of the external mycelium in the arbuscular mycorrhizal symbiosis shows strong dependence on recent photosynthates and acclimation to temperature. *The New Phytologist* 171(1), 159–70. doi:10.1111/j.1469-8137.2006.01730.x.

Helber, N., Wippel, K., Sauer, N., Schaarschmidt, S., Hause, B. and Requena, N. 2011. A versatile monosaccharide transporter that operates in the arbuscular mycorrhizal fungus *Glomus sp* is crucial for the symbiotic relationship with plants. *The Plant Cell* 23(10), 3812–23. doi:10.1105/tpc.111.089813.

Hetrick, B. A. D., Wilson, G. W. T. and Cox, T. S. 1992. Mycorrhizal dependence of modern wheat varieties, landraces, and ancestors. *Canadian Journal of Botany* 70(10), 2032–40. doi:10.1139/b92-253.

Hijri, M. 2016. Analysis of a large dataset of mycorrhiza inoculation field trials on potato shows highly significant increases in yield. *Mycorrhiza* 26(3), 209–14. doi:10.1007/s00572-015-0661-4.

Hildebrandt, U., Kaldorf, M. and Bothe, H. 1999. The zinc violet and its colonization by arbuscular mycorrhizal fungi. *Journal of Plant Physiology* 154(5-6), 709-17. doi:10.1016/S0176-1617(99)80249-1.

Ho, I. and Trappe, J. M. 1984. Effects of ozone exposure on mycorrhiza formation and growth of *Festuca arundinacea*. *Environmental and Experimental Botany* 24(1), 71-4. doi:10.1016/0098-8472(84)90062-5.

Hodge, A. and Fitter, A. H. 2010. Substantial nitrogen acquisition by arbuscular mycorrhizal fungi from organic material has implications for N cycling. *Proceedings of the National Academy of Sciences of the United States of America* 107(31), 13754-9. doi:10.1073/pnas.1005874107.

Hodge, A., Campbell, C. D. and Fitter, A. H. 2001. An arbuscular mycorrhizal fungus accelerates decomposition and acquires nitrogen directly from organic material. *Nature* 413(6853), 297-9. doi:10.1038/35095041.

Hodge, A., Berta, G., Doussan, C., Merchan, F. and Crespi, M. 2009. Plant root growth, architecture and function. *Plant and Soil* 321(1-2), 153-87. doi:10.1007/s11104-009-9929-9.

Hoeksema, J. D., Chaudhary, V. B., Gehring, C. A., Johnson, N. C., Karst, J., Koide, R. T., Pringle, A., Zabinski, C., Bever, J. D., Moore, J. C., Wilson, G. W., Klironomos, J. N. and Umbanhowar, J. 2010. A meta-analysis of context-dependency in plant response to inoculation with mycorrhizal fungi. *Ecology Letters* 13(3), 394-407. doi:10.1111/j.1461-0248.2009.01430.x.

Huang, R. U. E. Y., Smith, W. K. and Yost, R. S. 1985. Influence of vesicular-arbuscular mycorrhiza on growth, water relations, and leaf orientation in *Leucaena leucocephala* (Lam.) de Wit. *New Phytologist* 99(2), 229-43. doi:10.1111/j.1469-8137.1985.tb03652.x.

Jakobsen, I. 1986. Vesicular-arbuscular mycorrhiza in field-grown crops: III. Mycorrhizal infection and rates of phosphorus inflow in pea plants. *New Phytologist* 104(4), 573-81. doi:10.1111/j.1469-8137.1986.tb00657.x.

Jakobsen, T. and Erik Nielsen, N. 1983. Vesicular-arbuscular mycorrhiza in field-grown crops. I. Mycorrhizal infection in cereals and peas at various times and soil depths. *New Phytologist* 93(3), 401-13. doi:10.1111/j.1469-8137.1983.tb03440.x.

Jakobsen, I., Abbott, L. K. and Robson, A. D. 1992a. External hyphae of vesicular-arbuscular mycorrhizal fungi associated with *Trifolium subterraneum* L. 1. Spread of hyphae and phosphorus inflow into roots. *New Phytologist* 120(3), 371-80. doi:10.1111/j.1469-8137.1992.tb01077.x.

Jakobsen, I., Abbott, L. K. and Robson, A. D. 1992b. External hyphae of vesicular-arbuscular mycorrhizal fungi associated with *Trifolium subterraneum* L. 2. Hyphal transport of ^{32}P over defined distances. *New Phytologist* 120(4), 509-16. doi:10.1111/j.1469-8137.1992.tb01800.x.

Janos, D. P. 2007. Plant responsiveness to mycorrhizas differs from dependence upon mycorrhizas. *Mycorrhiza* 17(2), 75-91. doi:10.1007/s00572-006-0094-1.

Jansa, J., Mozafar, A., Anken, T., Ruh, R., Sanders, I. R. and Frossard, E. 2002. Diversity and structure of AMF communities as affected by tillage in a temperate soil. *Mycorrhiza* 12(5), 225-34. doi:10.1007/s00572-002-0163-z.

Jansa, J., Mozafar, A., Kuhn, G., Anken, T., Ruh, R., Sanders, I. R. and Frossard, E. 2003. Soil tillage affects the community structure of mycorrhizal fungi in maize roots. *Ecological Applications* 13(4), 1164-76. doi:10.1890/1051-0761(2003)13[1164:STATCS]2.0.CO;2.

Jansa, J., Wiemken, A. and Frossard, E. 2006. The effects of agricultural practices on arbuscular mycorrhizal fungi. *Geological Society, London, Special Publications* 266(1), 89–115. doi:10.1144/GSL.SP.2006.266.01.08.

Jansa, J., Forczek, S. T., Rozmoš, M., Püschel, D., Bukovská, P. and Hršelová, H. 2019. Arbuscular mycorrhiza and soil organic nitrogen: network of players and interactions. *Chemical and Biological Technologies in Agriculture* 6(1), 10. doi:10.1186/s40538-019-0147-2.

Jasper, D. A., Robson, A. D. and Abbott, L. K. 1979. Phosphorus and the formation of vesicular-arbuscular mycorrhizas. *Soil Biology and Biochemistry* 11(5), 501–5. doi:10.1016/0038-0717(79)90009-9.

Javot, H., Penmetsa, R. V., Terzaghi, N., Cook, D. R. and Harrison, M. J. 2007a. A *Medicago truncatula* phosphate transporter indispensable for the arbuscular mycorrhizal symbiosis. *Proceedings of the National Academy of Sciences of the United States of America* 104(5), 1720–5. doi:10.1073/pnas.0608136104.

Javot, H., Pumplin, N. and Harrison, M. J. 2007b. Phosphate in the arbuscular mycorrhizal symbiosis: transport properties and regulatory roles. *Plant, Cell and Environment* 30(3), 310–22. doi:10.1111/j.1365-3040.2006.01617.x.

Jayne, B. and Quigley, M. 2014. Influence of arbuscular mycorrhiza on growth and reproductive response of plants under water deficit: a meta-analysis. *Mycorrhiza* 24(2), 109–19. doi:10.1007/s00572-013-0515-x.

Jin, H., Pfeffer, P. E., Douds, D. D., Piotrowski, E., Lammers, P. J. and Shachar-Hill, Y. 2005. The uptake, metabolism, transport and transfer of nitrogen in an arbuscular mycorrhizal symbiosis. *The New Phytologist* 168(3), 687–96. doi:10.1111/j.1469-8137.2005.01536.x.

Johansen, A., Jakobsen, I. and Jensen, E. S. 1992. Hyphal transport of N-15-labeled nitrogen by a vesicular-arbuscular mycorrhizal fungus and its effect on depletion of inorganic soil-N. *New Phytologist* 122(2), 281–8. doi:10.1111/j.1469-8137.1992.tb04232.x.

Johansen, A., Jakobsen, I. and Jensen, E. S. 1994. Hyphal N transport by a vesicular-arbuscular mycorrhizal fungus associated with cucumber grown at 3 nitrogen levels. *Plant and Soil* 160(1), 1–9. doi:10.1007/BF00150340.

Johnson, N. C. and Graham, J. H. 2013. The continuum concept remains a useful framework for studying mycorrhizal functioning. *Plant and Soil* 363(1–2), 411–9. doi:10.1007/s11104-012-1406-1.

Johnson, N. C., Graham, J. H. and Smith, F. A. 1997. Functioning of mycorrhizal associations along the mutualism-parasitism continuum. *New Phytologist* 135(4), 575–85. doi:10.1046/j.1469-8137.1997.00729.x.

Johnson, N. C., Wilson, G. W., Wilson, J. A., Miller, R. M. and Bowker, M. A. 2015. Mycorrhizal phenotypes and the law of the minimum. *The New Phytologist* 205(4), 1473–84. doi:10.1111/nph.13172.

Joner, E. J. and Jakobsen, I. 1995. Uptake of P-32 from labeled organic-matter by mycorrhizal and nonmycorrhizal subterranean clover (*Trifolium subterraneum* L). *Plant and Soil* 172(2), 221–7. doi:10.1007/BF00011324.

Jongen, M., Fay, P. and Jones, M. B. 1996. Effects of elevated carbon dioxide and arbuscular mycorrhizal infection on Trifolium repens. *The New Phytologist* 132(3), 413–23. doi:10.1111/j.1469-8137.1996.tb01861.x.

Jung, S. C., Martinez-Medina, A., Lopez-Raez, J. A. and Pozo, M. J. 2012. Mycorrhiza-induced resistance and priming of plant defenses. *Journal of Chemical Ecology* 38(6), 651–64. doi:10.1007/s10886-012-0134-6.

Kabir, Z. 2005. Tillage or no-tillage: impact on mycorrhizae. *Canadian Journal of Plant Science* 85(1), 23–9. doi:10.4141/P03-160.

Kabir, Z., O'Halloran, I. P. and Hamel, C. 1999. Combined effects of soil disturbance and fallowing on plant and fungal components of mycorrhizal corn (*Zea mays* L.). *Soil Biology and Biochemistry* 31(2), 307–14. doi:10.1016/S0038-0717(98)00124-2.

Kaschuk, G., Kuyper, T. W., Leffelaar, P. A., Hungria, M. and Giller, K. E. 2009. Are the rates of photosynthesis stimulated by the carbon sink strength of rhizobial and arbuscular mycorrhizal symbioses? *Soil Biology and Biochemistry* 41(6), 1233–44. doi:10.1016/j.soilbio.2009.03.005.

Kenrick, P. and Strullu-Derrien, C. 2014. The origin and early evolution of roots. *Plant Physiology* 166(2), 570–80. doi:10.1104/pp.114.244517.

Keymer, A., Pimprikar, P., Wewer, V., Huber, C., Brands, M., Bucerius, S. L., Delaux, P. M., Klingl, V., von Roepenack-Lahaye, E. V., Wang, T. L., Eisenreich, W., Dörmann, P., Parniske, M. and Gutjahr, C. 2017. Lipid transfer from plants to arbuscular mycorrhiza fungi. *eLife* 6, e29107. doi:10.7554/eLife.29107.

Khalvati, M. A., Hu, Y., Mozafar, A. and Schmidhalter, U. 2005. Quantification of water uptake by arbuscular mycorrhizal hyphae and its significance for leaf growth, water relations, and gas exchange of barley subjected to drought stress. *Plant Biology* 7(6), 706–12. doi:10.1055/s-2005-872893.

Khan, A. G. 1972. The effect of vesicular-arbuscular mycorrhizal associations on growth of cereals. I. Effects on maize growth. *New Phytologist* 71(4), 613–9. doi:10.1111/j.1469-8137.1972.tb01271.x.

Khan, A. G. 1975. The effect of vesicular arbuscular mycorrhizal associations on growth of cereals: II. Effects on wheat growth. *Annals of Applied Biology* 80(1), 27–36. doi:10.1111/j.1744-7348.1975.tb01597.x.

Kiers, E. T., Duhamel, M., Beesetty, Y., Mensah, J. A., Franken, O., Verbruggen, E., Fellbaum, C. R., Kowalchuk, G. A., Hart, M. M., Bago, A., Palmer, T. M., West, S. A., Vandenkoornhuyse, P., Jansa, J. and Bücking, H. 2011. Reciprocal rewards stabilize cooperation in the mycorrhizal symbiosis. *Science* 333(6044), 880–2. doi:10.1126/science.1208473.

Kikuchi, Y., Hijikata, N., Ohtomo, R., Handa, Y., Kawaguchi, M., Saito, K., Masuta, C. and Ezawa, T. 2016. Aquaporin-mediated long-distance polyphosphate translocation directed towards the host in arbuscular mycorrhizal symbiosis: application of virus-induced gene silencing. *The New Phytologist* 211(4), 1202–8. doi:10.1111/nph.14016.

Klironomos, J. N. 2003. Variation in plant response to native and exotic arbuscular mycorrhizal fungi. *Ecology* 84(9), 2292–301. doi:10.1890/02-0413.

Klironomos, J. N. and Kendrick, W. B. 1996. Palatability of microfungi to soil arthropods in relation to the functioning of arbuscular mycorrhizae. *Biology and Fertility of Soils* 21(1–2), 43–52. doi:10.1007/BF00335992.

Koide, R. T. 1991. Density-dependent response to mycorrhizal infection in *Abutilon theophrasti* Medic. *Oecologia* 85(3), 389–95. doi:10.1007/BF00320615.

Kokkoris, V., Li, Y., Hamel, C., Hanson, K. and Hart, M. 2019. Site specificity in establishment of a commercial arbuscular mycorrhizal fungal inoculant. *The Science of the Total Environment* 660, 1135–43. doi:10.1016/j.scitotenv.2019.01.100.

Konvalinková, T. and Jansa, J. 2016. Lights off for arbuscular mycorrhiza: on its symbiotic functioning under light deprivation. *Frontiers in Plant Science* 7, 782. doi:10.3389/fpls.2016.00782.

Konvalinková, T., Püschel, D., Janouáková, M., Gryndler, M. and Jansa, J. 2015. Duration and intensity of shade differentially affects mycorrhizal growth-and phosphorus uptake responses of *Medicago truncatula*. *Frontiers in Plant Science* 6, 65.

Krajinski, F., Hause, B., Gianinazzi-Pearson, V. and Franken, P. 2002. *Mtha*1, a plasma membrane H^+-ATPase gene from *Medicago truncatula*, shows arbuscule-specific induced expression. *Plant Biology* 4(6), 754–61. doi:10.1055/s-2002-37407.

Larose, G., Chenevert, R., Moutoglis, P., Gagne, S., Piche, Y. and Vierheilig, H. 2002. Flavonoid levels in roots of *Medicago sativa* are modulated by the developmental stage of the symbiosis and the root colonizing arbuscular mycorrhizal fungus. *Journal of Plant Physiology* 159(12), 1329–39. doi:10.1078/0176-1617-00896.

Latef, A. A. H. A. and Chaoxing, H. 2011a. Effect of arbuscular mycorrhizal fungi on growth, mineral nutrition, antioxidant enzymes activity and fruit yield of tomato grown under salinity stress. *Scientia Horticulturae* 127(3), 228–33. doi:10.1016/j.scienta.2010.09.020.

Latef, A. A. H. A. and Chaoxing, H. 2011b. Arbuscular mycorrhizal influence on growth, photosynthetic pigments, osmotic adjustment and oxidative stress in tomato plants subjected to low temperature stress. *Acta Physiologiae Plantarum* 33(4), 1217–25. doi:10.1007/s11738-010-0650-3.

Lee, S. J., Morse, D. and Hijri, M. 2019. Holobiont chronobiology: mycorrhiza may be a key to linking aboveground and underground rhythms. *Mycorrhiza* 29(5), 403–12. doi:10.1007/s00572-019-00903-4.

Lehman, R. M., Osborne, S. L., Taheri, W. I., Buyer, J. S. and Chim, B. K. 2019. Comparative measurements of arbuscular mycorrhizal fungal responses to agricultural management practices. *Mycorrhiza* 29(3), 227–35. doi:10.1007/s00572-019-00884-4.

Leifheit, E. F., Veresoglou, S. D., Lehmann, A., Morris, E. K. and Rillig, M. C. 2014. Multiple factors influence the role of arbuscular mycorrhizal fungi in soil aggregation: a meta-analysis. *Plant and Soil* 374(1–2), 523–37. doi:10.1007/s11104-013-1899-2.

Lekberg, Y. and Koide, R. T. 2005. Is plant performance limited by abundance of arbuscular mycorrhizal fungi? A meta-analysis of studies published between 1988 and 2003. *The New Phytologist* 168(1), 189–204. doi:10.1111/j.1469-8137.2005.01490.x.

Li, X., Georges, E. and Marschner, H. 1991a. Phosphorus depletion and pH decrease at root-soil and hyphal-soil interfaces of VA mycorrhiza white clover fertilized with ammonium. *New Phytologist* 119(3), 397–404. doi:10.1111/j.1469-8137.1991.tb00039.x.

Li, X. L., George, E. and Marschner, H. 1991b. Extension of the phosphorus depletion zone in VA-mycorrhizal white clover in a calcareous soil. *Plant and Soil* 136(1), 41–8. doi:10.1007/BF02465218.

Liu, C. Y., Zhang, F., Zhang, D. J., Srivastava, A. K., Wu, Q. S. and Zou, Y. N. 2018. Mycorrhiza stimulates root-hair growth and IAA synthesis and transport in trifoliate orange under drought stress. *Scientific Reports* 8(1), 1978. doi:10.1038/s41598-018-20456-4.

Lohse, S., Schliemann, W., Ammer, C., Kopka, J., Strack, D. and Fester, T. 2005. Organization and metabolism of plastids and mitochondria in arbuscular mycorrhizal roots of *Medicago truncatula*. *Plant Physiology* 139(1), 329–40. doi:10.1104/pp.105.061457.

Lucini, L., Colla, G., Moreno, M. B., Bernardo, L., Cardarelli, M., Terzi, V., Bonini, P. and Rouphael, Y. 2019. Inoculation of Rhizoglomus irregulare or *Trichoderma atroviride* differentially modulates metabolite profiling of wheat root exudates. *Phytochemistry* 157, 158–67. doi:10.1016/j.phytochem.2018.10.033.

Ludwig-Müller, J. 2006. Hormonal balance in plants during colonization by mycorrhizal fungi. In: Kapulnik, Y. and Douds Jr., D. D. (Eds), *Arbuscular Mycorrhizas: Physiology and Function*. Kluwer Publishers, Dordrecht, pp. 263–85.

Ludwig-Müller, J. 2010. Hormonal responses in host plants triggered by arbuscular mycorrhizal fungi. In: Kapulnik, Y. and Douds Jr., D. D. (Eds), *Arbuscular Mycorrhizas: Physiology and Function*. Springer, Dordrecht, pp. 169–90.

Ludwig-Müller, J. and Güther, M. 2007. Auxins as signals in arbuscular mycorrhiza formation. *Plant Signaling and Behavior* 2(3), 194–6. doi:10.4161/psb.2.3.4152.

Mader, P., Vierheilig, H., Streitwolf-Engel, R., Boller, T., Frey, B., Christie, P. and Wiemken, A. 2000. Transport of N-15 from a soil compartment separated by a polytetrafluoroethylene membrane to plant roots via the hyphae of arbuscular mycorrhizal fungi. *New Phytologist* 146(1), 155–61. doi:10.1046/j.1469-8137.2000.00615.x.

Mariotte, P., Canarini, A. and Dijkstra, F. A. 2017. Stoichiometric N: P flexibility and mycorrhizal symbiosis favour plant resistance against drought. *Journal of Ecology* 105(4), 958–67. doi:10.1111/1365-2745.12731.

Marleau, J., Dalpé, Y., St-Arnaud, M. and Hijri, M. 2011. Spore development and nuclear inheritance in arbuscular mycorrhizal fungi. *BMC Evolutionary Biology* 11, 51. doi:10.1186/1471-2148-11-51.

Marschner, H. and Dell, B. 1994. Nutrient uptake in mycorrhizal symbiosis. *Plant and Soil* 159(1), 89–102. doi:10.1007/BF00000098.

Martín-Robles, N., Lehmann, A., Seco, E., Aroca, R., Rillig, M. C. and Milla, R. 2018. Impacts of domestication on the arbuscular mycorrhizal symbiosis of 27 crop species. *The New Phytologist* 218(1), 322–34. doi:10.1111/nph.14962.

Mathur, N. and Vyas, A. 1995. Influence of VA mycorrhizae on net photosynthesis and transpiration of *Ziziphus mauritiana*. *Journal of Plant Physiology* 147(3–4), 328–30. doi:10.1016/S0176-1617(11)82161-9.

Maya, M. A. and Matsubara, Y. 2013. Influence of arbuscular mycorrhiza on the growth and antioxidative activity in cyclamen under heat stress. *Mycorrhiza* 23(5), 381–90. doi:10.1007/s00572-013-0477-z.

McArthur, D. A. and Knowles, N. R. 1993. Influence of vesicular-arbuscular mycorrhizal fungi on the response of potato to phosphorus deficiency. *Plant Physiology* 101(1), 147–60. doi:10.1104/pp.101.1.147.

Medrano, H., Escalona, J. M., Bota, J., Gulías, J. and Flexas, J. 2002. Regulation of photosynthesis of C_3 plants in response to progressive drought: stomatal conductance as a reference parameter. *Annals of Botany* 89 (Spec No), 895-905. doi:10.1093/aob/mcf079.

Meixner, C., Ludwig-Muller, J., Miersch, O., Gresshoff, P., Staehelin, C. and Vierheilig, H. 2005. Lack of mycorrhizal autoregulation and phytohormonal changes in the supernodulating soybean mutant nts1007. *Planta* 222(4), 709–15. doi:10.1007/s00425-005-0003-4.

Menge, J. A., Steirle, D., Bagyaraj, D. J., Johnson, E. L. V. and Leonard, R. T. 1978. Phosphorus concentrations in plants responsible for inhibition of mycorrhizal infection. *New Phytologist* 80(3), 575–8. doi:10.1111/j.1469-8137.1978.tb01589.x.

Mensah, J. A., Koch, A. M., Antunes, P. M., Kiers, E. T., Hart, M. and Bücking, H. 2015. High functional diversity within species of arbuscular mycorrhizal fungi is associated with differences in phosphate and nitrogen uptake and fungal phosphate metabolism. *Mycorrhiza* 25(7), 533–46. doi:10.1007/s00572-015-0631-x.

Mercy, L., Lucic-Mercy, E., Nogales, A., Poghosyan, A., Schneider, C. and Arnholdt-Schmitt, B. 2017. A functional approach towards understanding the role of the mitochondrial respiratory chain in an endomycorrhizal symbiosis. *Frontiers in Plant Science* 8, 417. doi:10.3389/fpls.2017.00417.

Meyer, J. R. and Linderman, R. G. 1986. Response of subterranean clover to dual inoculation with vesicular-arbuscular fungi and a plant growth-promoting bacterium, *Pseudomonas putida*. *Soil Biology and Biochemistry* 18(2), 185–90. doi:10.1016/0038-0717(86)90025-8.

Meyer, S., Kling, C., Vogel, S., Schröter, I., Nagel, A., Kramer, E., Gebbers, R., Philipp, G., Lück, K. and Gerlach, F. 2019. Creating soil texture maps for precision liming using electrical resistivity and gamma ray mapping. In: Stafford, J. V. (Ed.), *Precision agriculture-19*. Wageningen Academic Publishers, Wageningen, pp. 539–46.

Miller, R. M. and Jastrow, J. D. 2000. Mycorrhizal fungi influence soil structure. In: Kapulnik, Y. and Douds Jr., D. D. (Eds), *Arbuscular Mycorrhizas: Physiology and Function*. Kluwer Academic Publishers, Dordrecht, pp. 3–18.

Morris, E. K., Morris, D. J. P., Vogt, S., Gleber, S. C., Bigalke, M., Wilcke, W. and Rillig, M. C. 2019. Visualizing the dynamics of soil aggregation as affected by arbuscular mycorrhizal fungi. *The ISME Journal* 13(7), 1639–46. doi:10.1038/s41396-019-0369-0.

Mosse, B. 1962. The establishment of vesicular-arbuscular mycorrhiza under aseptic conditions. *The Journal of General Microbiology* 27, 509–20. doi:10.1099/00221287-27-3-509.

Mugnier, J. and Mosse, B. 1987. Spore germination and viability of a vesicular arbuscular mycorrhizal fungus, *Glomus mosseae*. *Transactions of the British Mycological Society* 88(3), 411–3. doi:10.1016/S0007-1536(87)80018-9.

Murphy, G. P., File, A. L. and Dudley, S. A. 2013. Differentiating the effects of pot size and nutrient availability on plant biomass and allocation. *Botany* 91(11), 799–803. doi:10.1139/cjb-2013-0084.

Nelsen, C. E. and Safir, G. R. 1982. Increased drought tolerance of mycorrhizal onion plants caused by improved phosphorus nutrition. *Planta* 154(5), 407–13. doi:10.1007/BF01267807.

Neuhauser, C. and Fargione, J. E. 2004. A mutualism-parasitism continuum model and its application to plant-mycorrhizae interactions. *Ecological Modelling* 177(3–4), 337–52. doi:10.1016/j.ecolmodel.2004.02.010.

Neumann, E., Schmid, B., Römheld, V. and George, E. 2009. Extraradical development and contribution to plant performance of an arbuscular mycorrhizal symbiosis exposed to complete or partial rootzone drying. *Mycorrhiza* 20(1), 13–23. doi:10.1007/s00572-009-0259-9.

Niinemets, Ü. L. O., Tenhunen, J. D., Canta, N. R., Chaves, M. M., Faria, T., Pereira, J. S. and Reynolds, J. F. 1999. Interactive effects of nitrogen and phosphorus on the acclimation potential of foliage photosynthetic properties of cork oak, *Quercus suber*, to elevated atmospheric CO_2 concentrations. *Global Change Biology* 5(4), 455–70. doi:10.1046/j.1365-2486.1999.00241.x.

Nouri, E., Breuillin-Sessoms, F., Feller, U. and Reinhardt, D. 2014. Phosphorus and nitrogen regulate arbuscular mycorrhizal symbiosis in *Petunia hybrida*. *PLoS ONE* 9(6), e90841. doi:10.1371/journal.pone.0090841.

Oliveira, G., Nunes, A., Clemente, A. and Correia, O. 2011. Effect of substrate treatments on survival and growth of Mediterranean shrubs in a revegetated quarry: an eight-year study. *Ecological Engineering* 37(2), 255–9. doi:10.1016/j.ecoleng.2010.11.015.

Püschel, D., Janouáková, M., Hujslova, M., Slavíková, R., Gryndlerová, H. and Jansa, J. 2016. Plant-fungus competition for nitrogen erases mycorrhizal growth benefits of *Andropogon gerardii* under limited nitrogen supply. *Ecology and Evolution* 6(13), 4332–46. doi:10.1002/ece3.2207.

Parniske, M. 2008. Arbuscular mycorrhiza: the mother of plant root endosymbiosis. *Nature Reviews. Microbiology* 6(10), 763–75. doi:10.1038/nrmicro1987.

Paszkowski, U. and Gutjahr, C. 2013. Multiple control levels of root system remodeling in arbuscular mycorrhizal symbiosis. *Frontiers in Plant Science* 4, 204. doi:10.3389/fpls.2013.00204.

Pearson, J. N. and Jakobsen, I. 1993. The relative contribution of hyphae and roots to phosphorus uptake by arbuscular mycorrhizal plants, measured by dual labelling with ^{32}P and ^{33}P. *New Phytologist* 124(3), 489–94. doi:10.1111/j.1469-8137.1993.tb03840.x.

Pepe, A., Giovannetti, M. and Sbrana, C. 2018. Lifespan and functionality of mycorrhizal fungal mycelium are uncoupled from host plant lifespan. *Scientific Reports* 8(1), 10235. doi:10.1038/s41598-018-28354-5.

Perner, H., Schwarz, D., Bruns, C., Mäder, P. and George, E. 2007. Effect of arbuscular mycorrhizal colonization and two levels of compost supply on nutrient uptake and flowering of pelargonium plants. *Mycorrhiza* 17(5), 469–74. doi:10.1007/s00572-007-0116-7.

Pfeffer, P. E., Douds, D. D., Becard, G. and Shachar-Hill, Y. 1999. Carbon uptake and the metabolism and transport of lipids in an arbuscular mycorrhiza. *Plant Physiology* 120(2), 587-98. doi:10.1104/pp.120.2.587.

Pinior, A., Grunewaldt-Stocker, G., von Alten, H. and Strasser, R. J. 2005. Mycorrhizal impact on drought stress tolerance of rose plants probed by chlorophyll a fluorescence, proline content and visual scoring. *Mycorrhiza* 15(8), 596–605. doi:10.1007/s00572-005-0001-1.

Piotrowski, J. S., Denich, T., Klironomos, J. N., Graham, J. M. and Rillig, M. C. 2004. The effects of arbuscular mycorrhizas on soil aggregation depend on the interaction between plant and fungal species. *New Phytologist* 164(2), 365–73. doi:10.1111/j.1469-8137.2004.01181.x.

Poorter, H., Bühler, J., van Dusschoten, D., Climent, J. and Postma, J. A. 2012. Pot size matters: a meta-analysis of the effects of rooting volume on plant growth. *Functional Plant Biology* 39(11), 839–50. doi:10.1071/FP12049.

Pozo, M. J. and Azcon-Aguilar, C. 2007. Unraveling mycorrhiza-induced resistance. *Current Opinion in Plant Biology* 10(4), 393–8. doi:10.1016/j.pbi.2007.05.004.

Pringle, A., Adams, R. I., Cross, H. B. and Bruns, T. D. 2009. The ectomycorrhizal fungus *Amanita phalloides* was introduced and is expanding its range on the west coast of North America. *Molecular Ecology* 18(5), 817–33. doi:10.1111/j.1365-294X.2008.04030.x.

Quiroga, G., Erice, G., Ding, L., Chaumont, F., Aroca, R. and Ruiz-Lozano, J. M. 2019. The arbuscular mycorrhizal symbiosis regulates aquaporins activity and improves root cell water permeability in maize plants subjected to water stress. *Plant, Cell and Environment* 42(7), 2274–90. doi:10.1111/pce.13551.

Raj, J., Bagyaraj, D. J. and Manjunath, A. 1981. Influence of soil inoculation with vesicular-arbuscular mycorrhiza and a phosphate-dissolving bacterium on plant growth and ^{32}P-uptake. *Soil Biology and Biochemistry* 13(2), 105–8. doi:10.1016/0038-0717(81)90004-3.

Raju, P. S., Clark, R. B., Ellis, J. R. and Maranville, J. W. 1990. Effects of species of VA-mycorrhizal fungi on growth and mineral uptake of sorghum at different temperatures. *Plant and Soil* 121(2), 165–70. doi:10.1007/BF00012308.

Rao, N. S. S., Tilak, K. V. B. R. and Singh, C. S. 1985. Synergistic effect of vesicular-arbuscular mycorrhizas and *Azospirillum brasilense* on the growth of barley in pots. *Soil Biology and Biochemistry* 17(1), 119–21. doi:10.1016/0038-0717(85)90101-4.

Rausch, C., Daram, P., Brunner, S., Jansa, J., Laloi, M., Leggewie, G., Amrhein, N. and Bucher, M. 2001. A phosphate transporter expressed in arbuscule-containing cells in potato. *Nature* 414(6862), 462–70. doi:10.1038/35106601.

Requena, N., Breuninger, M., Franken, P. and Ocon, A. 2003. Symbiotic status, phosphate, and sucrose regulate the expression of two plasma membrane H^+-ATPase genes from the mycorrhizal fungus *Glomus mosseae*. *Plant Physiology* 132(3), 1540–9. doi:10.1104/pp.102.019042.

Rich, M. K., Nouri, E., Courty, P. E. and Reinhardt, D. 2017Diet of arbuscular mycorrhizal fungi: bread and butter? *Trends in Plant Science* 22(8), 652–60. doi:10.1016/j.tplants.2017.05.008.

Rillig, M. C. 2005. A connection between fungal hydrophobins and soil water repellency? *Pedobiologia* 49(5), 395–9. doi:10.1016/j.pedobi.2005.04.004.

Rillig, M. C. and Mummey, D. L. 2006. Mycorrhizas and soil structure. *The New Phytologist* 171(1), 41–53. doi:10.1111/j.1469-8137.2006.01750.x.

Rillig, M. C., Wright, S. F. and Eviner, V. T. 2002. The role of arbuscular mycorrhizal fungi and glomalin in soil aggregation: comparing effects of five plant species. *Plant and Soil* 238(2), 325–33. doi:10.1023/A:1014483303813.

Rillig, M. C., Mardatin, N. F., Leifheit, E. F. and Antunes, P. M. 2010. Mycelium of arbuscular mycorrhizal fungi increases soil water repellency and is sufficient to maintain water-stable soil aggregates. *Soil Biology and Biochemistry* 42(7), 1189–91. doi:10.1016/j.soilbio.2010.03.027.

Rillig, M. C., Aguilar-Trigueros, C. A., Camenzind, T., Cavagnaro, T. R., Degrune, F., Hohmann, P., Lammel, D. R., Mansour, I., Roy, J., van der Heijden, M. G. A. and Yang, G. 2019. Why farmers should manage the arbuscular mycorrhizal symbiosis. *New Phytologist* 222(3), 1171–5. doi:10.1111/nph.15602.

Rivero, J., Gamir, J., Aroca, R., Pozo, M. J. and Flors, V. 2015. Metabolic transition in mycorrhizal tomato roots. *Frontiers in Microbiology* 6, 598. doi:10.3389/fmicb.2015.00598.

Romero-Munar, A., Del-Saz, N. F., Ribas-Carbó, M., Flexas, J., Baraza, E., Florez-Sarasa, I., Fernie, A. R. and Gulías, J. 2017. Arbuscular mycorrhizal symbiosis with *Arundo donax* decreases root respiration and increases both photosynthesis and plant biomass accumulation. *Plant, Cell and Environment* 40, 1115–26.

Rouphael, Y., Franken, P., Schneider, C., Schwarz, D., Giovannetti, M., Agnolucci, M., De Pascale, S. D., Bonini, P. and Colla, G. 2015. Arbuscular mycorrhizal fungi act as biostimulants in horticultural crops. *Scientia Horticulturae* 196, 91–108. doi:10.1016/j.scienta.2015.09.002.

Ruiz-Lozano, J. M., Azcon, R. and Gomez, M. C. 1995a. Effects of arbuscular-mycorrhizal *Glomus* species on drought tolerance - physiological and nutritional plant-responses. *Applied and Environmental Microbiology* 61(2), 456–60. doi:10.1128/AEM.61.2.456-460.1995.

Ruiz-Lozano, J. M., Gomez, M. and Azcon, R. 1995b. Influence of different Glomus species on the time-course of physiological plant-responses of lettuce to progressive drought stress periods. *Plant Science* 110(1), 37–44. doi:10.1016/0168-9452(95)04184-V.

Ruiz-Lozano, J. M., Aroca, R., Zamarreño, ÁM., Molina, S., Andreo-Jiménez, B., Porcel, R., García-Mina, J. M., Ruyter-Spira, C. and López-Ráez, J. A. 2016. Arbuscular mycorrhizal symbiosis induces strigolactone biosynthesis under drought and improves drought tolerance in lettuce and tomato. *Plant, Cell and Environment* 39(2), 441–52. doi:10.1111/pce.12631.

Ruth, B., Khalvati, M. and Schmidhalter, U. 2011. Quantification of mycorrhizal water uptake via high-resolution on-line water content sensors. *Plant and Soil* 342(1–2), 459–68. doi:10.1007/s11104-010-0709-3.

Ryan, M. H. and Graham, J. H. 2002. Is there a role for arbuscular mycorrhizal fungi in production agriculture? *Plant and Soil* 244, 263–71.

Ryan, M. H. and Graham, J. H. 2018. Little evidence that farmers should consider abundance or diversity of arbuscular mycorrhizal fungi when managing crops. *The New Phytologist* 220(4), 1092–107. doi:10.1111/nph.15308.

Ryan, M. H., Graham, J. H., Morton, J. B. and Kirkegaard, J. A. 2019. Research must use a systems agronomy approach if management of the arbuscular mycorrhizal symbiosis is to contribute to sustainable intensification. *The New Phytologist* 222(3), 1176–8. doi:10.1111/nph.15600.

Saif, S. R. and Khan, A. G. 1977. The effect of vesicular-arbuscular mycorrhizal associations on growth of cereals. *Plant and Soil* 47(1), 17–26. doi:10.1007/BF00010364.

Salvioli, A., Zouari, I., Chalot, M. and Bonfante, P. 2012. The arbuscular mycorrhizal status has an impact on the transcriptome profile and amino acid composition of tomato fruit. *BMC Plant Biology* 12, 44. doi:10.1186/1471-2229-12-44.

Sánchez-Romera, B., Ruiz-Lozano, J. M., Zamarreño, ÁM., García-Mina, J. M. and Aroca, R. 2016. Arbuscular mycorrhizal symbiosis and methyl jasmonate avoid the inhibition of root hydraulic conductivity caused by drought. *Mycorrhiza* 26(2), 111–22. doi:10.1007/s00572-015-0650-7.

Savary, R., Masclaux, F. G., Wyss, T., Droh, G., Corella, J. C., Machado, A. P., Morton, J. B. and Sanders, I. R. 2018. A population genomics approach shows widespread geographical distribution of cryptic genomic forms of the symbiotic fungus *Rhizophagus irregularis*. *The ISME Journal* 12, 17.

Sawers, R. J. H., Ramírez-Flores, M. R., Olalde-Portugal, V. and Paszkowski, U. 2018. The impact of domestication and crop improvement on arbuscular mycorrhizal symbiosis in cereals: insights from genetics and genomics. *The New Phytologist* 220(4), 1135–40. doi:10.1111/nph.15152.

Sbrana, C., Avio, L. and Giovannetti, M. 2014. Beneficial mycorrhizal symbionts affecting the production of health-promoting phytochemicals. *Electrophoresis* 35(11), 1535–46. doi:10.1002/elps.201300568.

Scagel, C. F. 2004a. Inoculation with vesicular-arbuscular mycorrhizal fungi and rhizobacteria alters nutrient allocation and flowering of harlequin flower. *Horttechnology* 14(1), 39–48. doi:10.21273/HORTTECH.14.1.0039.

Scagel, C. F. 2004b. Mycorrhiza-induced changes in partitioning and composition alters flower and vegetative production of floral geophytes. *HortScience* 39, 767C–767.

Scervino, J. M., Ponce, M. A., Erra-Bassells, R., Bornpadre, J., Vierheilig, H., Ocampo, J. A. and Godeas, A. 2007. The effect of flavones and flavonols on colonization of tomato plants by arbuscular mycorrhizal fungi of the genera *Gigaspora* and *Glomus*. *Canadian Journal of Microbiology* 53(6), 702–9. doi:10.1139/W07-036.

Schachtman, D. P., Reid, R. J. and Ayling, S. M. 1998. Phosphorus uptake by plants - from soil to cell. *Plant Physiology* 116(2), 447–53. doi:10.1104/pp.116.2.447.

Schläppi, K., Köhl, L., Bender, F., Held, A., Mascher, F. and van der Heijden, M. 2017. Teamwork im Untergrund: Mykorrhizapilze zur Förderung des Pflanzenwachstums. *Agrarforschung Schweiz* 8, 96–101.

Schliemann, W., Ammer, C. and Strack, D. 2008. Metabolite profiling of mycorrhizal roots of *Medicago truncatula*. *Phytochemistry* 69(1), 112–46. doi:10.1016/j.phytochem.2007.06.032.

Schnepf, A. and Roose, T. 2006. Modelling the contribution of arbuscular mycorrhizal fungi to plant phosphate uptake. *The New Phytologist* 171(3), 669–82. doi:10.1111/j.1469-8137.2006.01771.x.

Schnepf, A., Leitner, D., Schweiger, P. F., Scholl, P. and Jansa, J. 2016. L-System model for the growth of arbuscular mycorrhizal fungi, both within and outside of their host roots. *Journal of the Royal Society, Interface* 13(117). doi:10.1098/rsif.2016.0129.

Schouteden, N., De Waele, D., Panis, B. and Vos, C. M. 2015. Arbuscular mycorrhizal fungi for the biocontrol of plant-parasitic nematodes: a review of the mechanisms involved. *Frontiers in Microbiology* 6, 1280. doi:10.3389/fmicb.2015.01280.

Schwartz, M. W., Hoeksema, J. D., Gehring, C. A., Johnson, N. C., Klironomos, J. N., Abbott, L. K. and Pringle, A. 2006. The promise and the potential consequences of the global transport of mycorrhizal fungal inoculum. *Ecology Letters* 9(5), 501–15. doi:10.1111/j.1461-0248.2006.00910.x.

Schwarz, D., Welter, S., George, E., Franken, P., Lehmann, K., Weckwerth, W., Dolle, S. and Worm, M. 2011. Impact of arbuscular mycorrhizal fungi on the allergenic potential of tomato. *Mycorrhiza* 21(5), 341–9. doi:10.1007/s00572-010-0345-z.

Schweiger, R. and Mueller, C. 2015. Leaf metabolome in arbuscular mycorrhizal symbiosis. *Current Opinion in Plant Biology* 26, 120–6. doi:10.1016/j.pbi.2015.06.009.

Schweiger, R., Baier, M. C. and Müller, C. 2014a. Arbuscular mycorrhiza-induced shifts in foliar metabolism and photosynthesis mirror the developmental stage of the symbiosis and are only partly driven by improved phosphate uptake. *Molecular Plant-Microbe Interactions : MPMI* 27(12), 1403–12. doi:10.1094/MPMI-05-14-0126-R.

Schweiger, R., Baier, M. C., Persicke, M. and Müller, C. 2014b. High specificity in plant leaf metabolic responses to arbuscular mycorrhiza. *Nature Communications* 5, 3886. doi:10.1038/ncomms4886.

Sessitsch, A., Brader, G., Pfaffenbichler, N., Gusenbauer, D. and Mitter, B. 2018. The contribution of plant microbiota to economy growth. *Microbial Biotechnology* 11(5), 801–5. doi:10.1111/1751-7915.13290.

Shao, Y.-D., Zhang, D.-J., Hu, X.-C., Wu, Q.-S., Jiang, C.-J., Xia, T.-J., Gao, X.-B. and Kuca, K 2018. Mycorrhiza-induced changes in root growth and nutrient absorption of tea plants. *Plant, Soil and Environment* 64(6), 283–9. doi:10.17221/126/2018-PSE.

Sheng, M., Tang, M., Chen, H., Yang, B. W., Zhang, F. F. and Huang, Y. H. 2008. Influence of arbuscular mycorrhizae on photosynthesis and water status of maize plants under salt stress. *Mycorrhiza* 18(6–7), 287–96. doi:10.1007/s00572-008-0180-7.

Six, J., Bossuyt, H., Degryze, S. and Denef, K. 2004. A history of research on the link between (micro) aggregates, soil biota, and soil organic matter dynamics. *Soil and Tillage Research* 79(1), 7–31. doi:10.1016/j.still.2004.03.008.

Slavíková, R., Püschel, D., Janouáková, M., Hujslová, M., Konvalinková, T., Gryndlerová, H., Gryndler, M., Weiser, M. and Jansa, J. 2017. Monitoring CO_2 emissions to gain a dynamic view of carbon allocation to arbuscular mycorrhizal fungi. *Mycorrhiza* 27(1), 35–51. doi:10.1007/s00572-016-0731-2.

Smith, S. E. and Read, D. J. 2008. *Mycorrhizal Symbiosis*. Academic Press, London, UK.

Smith, F. A. and Smith, S. E. 2011. What is the significance of the arbuscular mycorrhizal colonisation of many economically important crop plants? *Plant and Soil* 348(1–2), 63–79. doi:10.1007/s11104-011-0865-0.

Smith, S. E., Smith, F. A. and Jakobsen, I. 2004. Functional diversity in arbuscular mycorrhizal (AM) symbioses: the contribution of the mycorrhizal P uptake pathway is not correlated with mycorrhizal responses in growth or total P uptake. *New Phytologist* 162(2), 511–24. doi:10.1111/j.1469-8137.2004.01039.x.

Smith, S. E., Jakobsen, I., Gronlund, M. and Smith, F. A. 2011. Roles of arbuscular mycorrhizas in plant phosphorus nutrition: interactions between pathways of phosphorus uptake in arbuscular mycorrhizal roots have important implications for understanding and manipulating plant phosphorus acquisition. *Plant Physiology* 156(3), 1050–7. doi:10.1104/pp.111.174581.

Son, C. L. and Smith, S. E. 1988. Mycorrhizal growth responses: interactions between photon irradiance and phosphorus nutrition. *New Phytologist* 108(3), 305–14. doi:10.1111/j.1469-8137.1988.tb04167.x.

Spatafora, J. W., Chang, Y., Benny, G. L., Lazarus, K., Smith, M. E., Berbee, M. L., Bonito, G., Corradi, N., Grigoriev, I., Gryganskyi, A., James, T. Y., O'Donnell, K., Roberson, R. W., Taylor, T. N., Uehling, J., Vilgalys, R., White, M. M. and Stajich, J. E. 2016. A phylum-level phylogenetic classification of zygomycete fungi based on genome-scale data. *Mycologia* 108(5), 1028–46. doi:10.3852/16-042.

Sukumar, P., Legue, V., Vayssieres, A., Martin, F., Tuskan, G. A. and Kalluri, U. C. 2013. Involvement of auxin pathways in modulating root architecture during beneficial plant-microorganism interactions. *Plant, Cell and Environment* 36(5), 909–19. doi:10.1111/pce.12036.

Svenningsen, N. B., Watts-Williams, S. J., Joner, E. J., Battini, F., Efthymiou, A., Cruz-Paredes, C., Nybroe, O. and Jakobsen, I. 2018. Suppression of the activity of arbuscular mycorrhizal fungi by the soil microbiota. *The ISME Journal* 12(5), 1296–307. doi:10.1038/s41396-018-0059-3.

Tardieu, F. and Davies, W. J. 1993. Integration of hydraulic and chemical signalling in the control of stomatal conductance and water status of droughted plants. *Plant, Cell and Environment* 16(4), 341–9. doi:10.1111/j.1365-3040.1993.tb00880.x.

Tardieu, F. and Simonneau, T. 1998. Variability among species of stomatal control under fluctuating soil water status and evaporative demand: modelling isohydric and anisohydric behaviours. *Journal of Experimental Botany* 49(Special), 419–32. doi:10.1093/jxb/49.Special_Issue.419.

Tardieu, F., Lafarge, T. and Simonneau, T. 1996. Stomatal control by fed or endogenous xylem ABA in sunflower: interpretation of correlations between leaf water potential and stomatal conductance in anisohydric species. *Plant, Cell and Environment* 19(1), 75–84. doi:10.1111/j.1365-3040.1996.tb00228.x.

Taylor, A., Pereira, N., Thomas, B., Pink, D. A. C., Jones, J. E. and Bending, G. D. 2015. Growth and nutritional responses to arbuscular mycorrhizal fungi are dependent on onion genotype and fungal species. *Biology and Fertility of Soils* 51(7), 801–13. doi:10.1007/s00374-015-1027-y.

Tisdall, J. M., Smith, S. E. and Rengasamy, P. 1997. Aggregation of soil by fungal hyphae. *Soil Research* 35(1), 55–60. doi:10.1071/S96065.

Tobar, R., Azcon, R. and Barea, J. M. 1994. Improved nitrogen uptake and transport from N-15-labeled nitrate by external hyphae of arbuscular mycorrhiza under water-stressed conditions. *New Phytologist* 126(1), 119–22. doi:10.1111/j.1469-8137.1994.tb07536.x.

Todeschini, V., AitLahmidi, N., Mazzucco, E., Marsano, F., Gosetti, F., Robotti, E., Bona, E., Massa, N., Bonneau, L., Marengo, E., Wipf, D., Berta, G. and Lingua, G. 2018. Impact of beneficial microorganisms on strawberry growth, fruit production, nutritional quality and volatilome. *Frontiers in Plant Science* 9, 1611. doi:10.3389/fpls.2018.01611.

Toro, M., Azcon, R. and Barea, J. 1997. Improvement of arbuscular mycorrhiza development by inoculation of soil with phosphate-solubilizing rhizobacteria to improve rock phosphate bioavailability (^{32}P) and nutrient cycling. *Applied and Environmental Microbiology* 63(11), 4408–12. doi:10.1128/AEM.63.11.4408-4412.1997.

Troeh, Z. I. and Loynachan, T. E. 2003. Endomycorrhizal fungal survival in continuous corn, soybean, and fallow. *Agronomy Journal* 95(1), 224–30. doi:10.2134/agronj2003.0224.

Turrini, A., Avio, L., Giovannetti, M. and Agnolucci, M. 2018. Functional complementarity of arbuscular mycorrhizal fungi and associated microbiota: the challenge of translational research. *Frontiers in Plant Science* 9, 1407. doi:10.3389/fpls.2018.01407.

UNCTAD (United Nations Conference on Trade and Development). 2013. *Trade and Environment Review 2013*. UNCTAD/DITC/TED/2012/3. ISSN 1810-5432.

Van Der Heijden, M. G., Martin, F. M., Selosse, M. A. and Sanders, I. R. 2015. Mycorrhizal ecology and evolution: the past, the present, and the future. *The New Phytologist* 205(4), 1406–23. doi:10.1111/nph.13288.

Verbruggen, E. and Toby Kiers, E. T. 2010. Evolutionary ecology of mycorrhizal functional diversity in agricultural systems. *Evolutionary Applications* 3(5–6), 547–60. doi:10.1111/j.1752-4571.2010.00145.x.

Verbruggen, E., Van Der Heijden, M. G., Rillig, M. C. and Kiers, E. T. 2013. Mycorrhizal fungal establishment in agricultural soils: factors determining inoculation success. *The New Phytologist* 197(4), 1104–9. doi:10.1111/j.1469-8137.2012.04348.x.

Vos, C., Schouteden, N., Van Tuinen, D., Chatagnier, O., Elsen, A., De Waele, D., Panis, B. and Gianinazzi-Pearson, V. 2013. Mycorrhiza-induced resistance against the root-knot nematode *Meloidogyne incognita* involves priming of defense gene responses in tomato. *Soil Biology and Biochemistry* 60, 45–54. doi:10.1016/j.soilbio.2013.01.013.

Vosátka, M., Albrechtová, J. and Patten, R. 2008. The international market development for mycorrhizal technology. In: Varma, A. (Ed.), *Mycorrhiza*. Springer, Berlin Heidelberg, pp. 419–38.

Walker, A. P., Beckerman, A. P., Gu, L., Kattge, J., Cernusak, L. A., Domingues, T. F., Scales, J. C., Wohlfahrt, G., Wullschleger, S. D. and Woodward, F. I. 2014. The relationship of leaf photosynthetic traits-V_{cmax} and J_{max}-to leaf nitrogen, leaf phosphorus, and specific leaf area: a meta-analysis and modeling study. *Ecology and Evolution* 4(16), 3218–35. doi:10.1002/ece3.1173.

Wang, G. M., Coleman, D. C., Freckman, D. W., Dyer, M. I., McNaughton, S. J., Agra, M. A. and Goeschl, J. D. 1989. Carbon partitioning patterns of mycorrhizal versus non-mycorrhizal plants: real-time dynamic measurements using $^{11}CO_2$. *The New Phytologist* 112(4), 489–93. doi:10.1111/j.1469-8137.1989.tb00342.x.

Wang, M., Schäfer, M., Li, D., Halitschke, R., Dong, C., McGale, E., Paetz, C., Song, Y., Li, S., Dong, J., Heiling, S., Groten, K., Franken, P., Bitterlich, M., Harrison, M. J., Paszkowski,

U. and Baldwin, I. T. 2018. Blumenols as shoot markers of root symbiosis with arbuscular mycorrhizal fungi. *eLife* 7, e37093. doi:10.7554/eLife.37093.

Watts-Williams, S. J., Smith, F. A. and Jakobsen, I. 2019. Soil phosphorus availability is a driver of the responses of maize (*Zea mays*) to elevated CO_2 concentration and arbuscular mycorrhizal colonisation. *Symbiosis* 77(1), 73–82. doi:10.1007/s13199-018-0573-0.

Wilkinson, S. and Davies, W. J. 2002. ABA-based chemical signalling: the co-ordination of responses to stress in plants. *Plant, Cell and Environment* 25(2), 195–210. doi:10.1046/j.0016-8025.2001.00824.x.

Wilson, G. W., Rice, C. W., Rillig, M. C., Springer, A. and Hartnett, D. C. 2009. Soil aggregation and carbon sequestration are tightly correlated with the abundance of arbuscular mycorrhizal fungi: results from long-term field experiments. *Ecology Letters* 12(5), 452–61. doi:10.1111/j.1461-0248.2009.01303.x.

Wright, S. F. and Upadhyaya, A. 1996. Extraction of an abundant and unusual protein from soil and comparison with hyphal protein of arbuscular mycorrhizal fungi. *Soil Science* 161(9), 575–86. doi:10.1097/00010694-199609000-00003.

Wright, S. F. and Upadhyaya, A. 1998. A survey of soils for aggregate stability and glomalin, a glycoprotein produced by hyphae of arbuscular mycorrhizal fungi. *Plant and Soil* 198(1), 97–107. doi:10.1023/A:1004347701584.

Wright, D. P., Read, D. J. and Scholes, J. D. 1998. Mycorrhizal sink strength influences whole plant carbon balance of *Trifolium repens* L. *Plant, Cell and Environment* 21(9), 881–91. doi:10.1046/j.1365-3040.1998.00351.x.

Xavier, L. J. C. and Germida, J. J. 2003. Bacteria associated with *Glomus clarum* spores influence mycorrhizal activity. *Soil Biology and Biochemistry* 35(3), 471–8. doi:10.1016/S0038-0717(03)00003-8.

Yamori, W., Evans, J. R. and Von Caemmerer, S. 2010. Effects of growth and measurement light intensities on temperature dependence of CO_2 assimilation rate in tobacco leaves. *Plant, Cell and Environment* 33(3), 332–43. doi:10.1111/j.1365-3040.2009.02067.x.

Zeng, Y., Guo, L. P., Chen, B. D., Hao, Z. P., Wang, J. Y., Huang, L. Q., Yang, G., Cui, X. M., Yang, L., Wu, Z. X., Chen, M. L. and Zhang, Y. 2013. Arbuscular mycorrhizal symbiosis and active ingredients of medicinal plants: current research status and prospectives. *Mycorrhiza* 23(4), 253–65. doi:10.1007/s00572-013-0484-0.

Zhang, S., Lehmann, A., Zheng, W., You, Z. and Rillig, M. C. 2019. Arbuscular mycorrhizal fungi increase grain yields: a meta-analysis. *The New Phytologist* 222(1), 543–55. doi:10.1111/nph.15570.

Zhu, X. C., Song, F. B. and Xu, H. W. 2010. Arbuscular mycorrhizae improves low temperature stress in maize via alterations in host water status and photosynthesis. *Plant and Soil* 331(1–2), 129–37. doi:10.1007/s11104-009-0239-z.

Zhu, X. C., Song, F. B., Liu, S. Q. and Liu, T. D. 2011. Effects of arbuscular mycorrhizal fungus on photosynthesis and water status of maize under high temperature stress. *Plant and Soil* 346(1–2), 189–99. doi:10.1007/s11104-011-0809-8.

Zuccarini, P. 2007. Mycorrhizal infection ameliorates chlorophyll content and nutrient uptake of lettuce exposed to saline irrigation. *Plant, Soil and Environment* 53(7), 283–9. doi:10.17221/2209-PSE.

Part 3

Innovation and practical applications

Chapter 9

Designing and formulating microbial and non-microbial biostimulants

Paolo Bonini, NGAlab, Spain; Veronica Cirino, Atens Agrotecnologias Naturales S.L., Spain; Helene Reynaud, Italpollina USA, USA; Youssef Rouphael, University of Naples Federico II, Italy; Mariateresa Cardarelli, CREA, Centro di ricerca Orticoltura e Florovivaismo, Italy; and Giuseppe Colla, University of Tuscia, Italy

1 Introduction

There is an urgent need to find new and increasingly sustainable means of production to increase plant productivity and ensure food security while preserving soil quality and providing business opportunities for farmers. Biostimulant product manufacturers have developed innovative products targeting specific agronomic needs, attracting the attention of industry stakeholders, including policymakers, scientists and crop producers. The importance of plant biostimulants is rising in global agriculture due to the growing pressure to shift toward increasingly sustainable, environmentally friendly and organic production systems. Plant biostimulants have the advantages of improving nutrient use, uptake and utilization efficiency (Calvo et al., 2014), reducing the need for synthetic fertilizers and increasing the quality of products. Moreover, plant biostimulants can also enhance the tolerance of agricultural and horticultural crops to environmental stressors such as drought, salinity and extreme temperatures (Colla and Rouphael, 2015).

Thousands of plant biostimulants derived from natural raw material, chemical synthesis or beneficial microorganisms are available on the market (Du Jardin, 2015; Yakhin et al., 2017). In 2019, this market had an estimated size

http://dx.doi.org/10.19103/AS.2020.0068.11

of USD 2.6 billion, and it is expected to reach USD 4.9 billion in 2025 (Markets and Markets, 2019). The main drivers of market growth are:

1 product efficacy (yield increase of approximately 5–15%; quality improvement in staple foods and horticultural products; less use of agrochemicals (approximately 10–15%));
2 innovation (development of new biostimulant products and application strategies; involvement of multinational companies);
3 market expansion (increased number of biostimulant-treated crops such as broadacre crops; new roles for plant biostimulants, such as soil and seed applications); and
4 sustainability (environmental constraints reducing fertilizer and pesticide input; regulations promoting sustainable agriculture practices).

Every year, new enterprises enter the biostimulant market attracted by opportunities for high profits; however, some of them have little knowledge of the biostimulant sector and inadequate investments in internal research and development (R&D) programs. Small and medium enterprises are often used to cooperate with public research centers and universities in developing new and potential plant biostimulants to increase their R&D output. Partnerships between large corporations and academia are also common. The lack of a harmonized regulatory framework in Europe and worldwide results in limited flexibility in terms of innovation, market fragmentation and compromised competition among companies.

For the above reasons, plant-biostimulant production enterprises recently founded associations, such as the European Biostimulants Industry Council (EBIC) and the Biological Products Industry Alliance (BPIA), to promote a newly harmonized regulation beyond fertilizers and/or pesticide products in the European Union and the United States. On 25 June 2019, the first EU Fertilising Products Regulation (2019/1009), including regulation of plant biostimulants, was published in the *Official Journal of the European Union* and is expected to be applied in the member states on 16 July 2022. In the new EU regulation, a plant biostimulant is defined as a product stimulating plant nutrition processes independently of the product's nutrient content, with the sole aim of improving one or more of the following characteristics of the plant or the plant rhizosphere: nutrient use efficiency, tolerance to abiotic stress, quality traits, and availability of confined nutrients in soil or rhizosphere.

In the EU Fertilising Products Regulation, plant biostimulants have been divided into two categories: non-microbial and microbial. Several microorganisms have been included in microbial biostimulants such as mycorrhizal fungi, *Rhizobium* spp., *Azotobacter* spp. and *Azospirillum* spp.

In the United States, the Agriculture Improvement Act of 2018, also known as the Farm Bill, was signed into law on 20 December 2018, and it provides the first statutory language about plant biostimulants in any federal law. This is the first law in the United States to recognize the importance of plant biostimulants as an emerging technology for agriculture. The 2018 Farm Bill describes a plant biostimulant as 'a substance or micro-organism that, when applied to seeds, plants, or the rhizosphere, stimulates natural processes to enhance or benefit nutrient uptake, nutrient efficiency, tolerance to abiotic stress, or crop quality and yield'.

Considering the biostimulant definitions in the EU Regulation and the 2018 Farm Bill, it is clear that plant biostimulants are defined mainly by what they do instead of what they are. Therefore, it is pivotal to develop appropriate testing procedures that are able to define the biostimulant activity of substances and microorganisms. The development of new plant biostimulants is a long process that involves not only rigorous testing procedures but also quality control, optimization of raw material selection, production processes and product formulation. This scientific approach can assure the development of innovative products, enhancing the credibility of the plant biostimulant industry.

Three main bottlenecks hamper market introduction and growers in reaping the full benefits of biostimulants:

1 high variation in product quality, which leads to difficulty in standardization and unreliable performance of biostimulant products;
2 lack of knowledge of the physiological and molecular mode of actions of biostimulant products; and
3 lack of farmer awareness and knowledge of the specific benefits of biostimulant solutions and the farming strategies that would synergize biostimulant effects to have maximum impact at optimal costs.

This chapter provides the first overview of the optimal design and formulation of microbial and non-microbial biostimulants to address these challenges. In this chapter, the production steps for a plant biostimulant are first presented and discussed, and then two successful case studies of plant biostimulant development are reported in detail.

2 The biostimulant development process

The historical approach in biostimulant development was based on testing samples of substances or microorganisms to see if they might have some biological function on plants and can be scaled; this approach has been largely used for many products created from available wastes that needed to be disposed of (e.g. collagen from leather industry), or loosely based upon prior observations of utility (e.g. seaweeds). Products developed on this rationale are not targeted to specific biological challenges/stresses and as such are 'products looking for a purpose'. In recent years, many companies have started to develop advanced products that are 'purpose built', based upon a dissection of metabolic pathways and biological principles. These biostimulant products are tailored for a specific purpose, such as cold-stress tolerance, stimulation of root development and nutrient absorption, reduction of fruit cracking, and so on.

The development process for a plant biostimulant is shown in Fig. 1. The process is composed of seven steps, each of which affects the subsequent step (feedforward) but can also contribute to the process development step (feedback). The plant-biostimulant development process starts with the generation of new biostimulant product ideas which are also selected considering their potential market opportunities. The next step is to develop the production process for generating samples of the product prototypes. Accurate testing of product samples in a controlled environment is necessary to evaluate biostimulant activities and understand their physiological and molecular modes of action. Quality control during the production process and

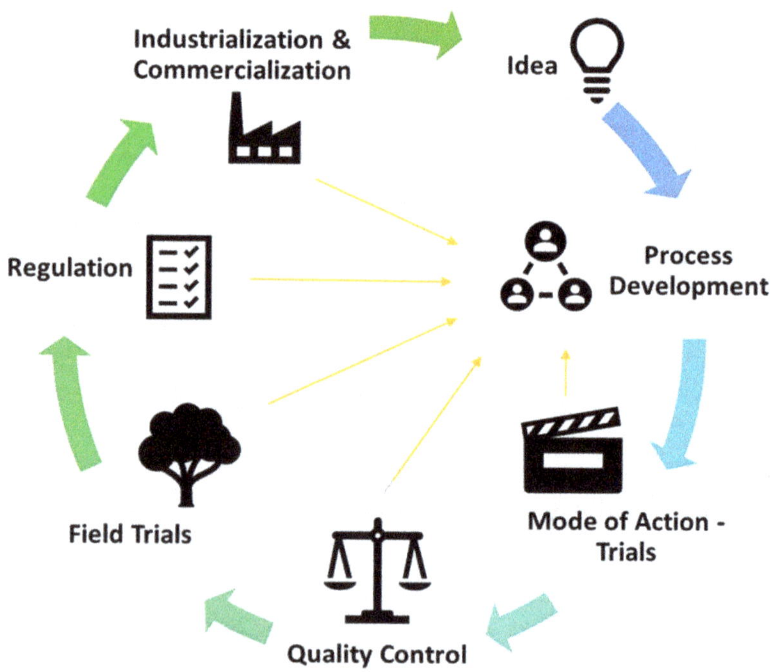

Figure 1 Circular workflow for the development of plant biostimulants.

of the finished product is also an important step to guarantee consistency of biostimulant performances. Greenhouse and field trials are then carried out together with molecular and biochemical tests to optimize the rate, timing and application method for products in different cropping systems. The best-performing products are then registered with the relevant national regulatory authority as plant biostimulants. Finally, a production scale-up is needed to support commercialization. Organization and management of the different steps is crucial for the success of the whole process. This is the reason why well-established biostimulant production companies can achieve an average success rate of 65% in biostimulant development, whereas this value can drop to 10% for new companies having no experience in biostimulant production (Castellion and Markham, 2013).

2.1 Generation of product ideas and preliminary assessments

The development of a biostimulant starts from brainstorming sections to produce ideas for new products for solving specific problems; in this first phase, it is fundamental to establish a continuous interaction between R&D staff and the field advisors and marketing team of a company for defining the most promising market opportunities for new biostimulant products. Once a draft idea for a new product has been defined, a scientific review of published papers is advised to determine the strengths and weaknesses of the idea (Brown and Saa, 2015). Scientific papers can also be a great source of new ideas. There is much research conducted in the biostimulant sector, and new microbial strains, innovative raw materials or active compounds can be found by examining the scientific literature (Rouphael and Colla, 2018). Biostimulant companies should be the link between public researchers and farmers to promote the implementation of ideas that otherwise would remain relegated to the scientific community (Rouphael et al., 2018). Published studies of efficacy, production process or active compound identification should be indexed and carefully annotated.

In developing microbial consortia-based biostimulants, it is very important to consider the compatibility of different microbial strains in the same product. Microbial consortia with incompatible microorganisms must be avoided (i.e. arbuscular mycorrhizal fungi and *Trichoderma harzianum*). Microbial strains taxonomically related to human or plant pathogens must be avoided even if they present biostimulant activity (i.e. *Fusarium fujikuroi, Agrobacterium tumefaciens* and *Escherichia coli*). The toxicity of candidate biostimulant substances to humans and the environment must be considered, as must their ability to cause allergic reactions in the human body.

Because regulation is one of the main limitations in bringing new biostimulant products in the market, the current regulatory framework must be carefully considered at this stage with reference to the target market of the new product (Caradonia et al., 2019). Each country has its own regulation that in many cases lacks any reference to biostimulant products. In the absence of a legislative framework for plant biostimulants, companies could consider registering the product as fertilizer, eventually enriching the biostimulant with plant nutrients. Evolution of regulatory frameworks must also be considered in biostimulant development.

Raw material selection is important for setting the cost of the new biostimulant or at least for trying to predict it. Very expensive raw materials should be avoided, and increased attention should be paid to agro-industrial wastes following the circular economy approach (Xu and Geelen, 2018). However, agro-industrial waste composition may vary significantly, making it difficult to guarantee a biostimulant product with a constant composition. The production of biomass such as microalgae cells or crop tissues (shoots, seeds) using specific protocols and growing conditions can limit the variability in the composition of raw material used for biostimulant production.

Regardless of the origin of the raw material, it would be important that adequate amounts of target bioactive molecules are provided to assure biostimulant activity in the final product. However, this approach cannot be followed for many biostimulants where bioactive molecules have not unequivocally determined, and the mode of function is not known. Moreover, it is important to consider that some biostimulants may indeed only function when provided as a consortium of constituents (so called 'emergent' properties – Yakhin et al., 2017). Considering that EU/US regulatory frameworks define biostumulants by what they do and not what they contain, this lack of identification of 'target bioactive molecules' is not problematic for regulatory point of view. In-depth market analysis must also be performed to identify the target selling price of the new biostimulants by considering the advantages of the product over its competitors.

2.2 Process development

The development of the production process requires a continuous interaction among experts in different fields: biology, chemistry, microbiology, engineering and business/economics. The production process needs to maximize the content of the target bioactive compounds/active fractions or propagules in non-microbe-based and microbe-based biostimulants, respectively. For non-microbe-based biostimulants, several procedures have been developed to maximize the extraction or production of bioactive compounds from different raw materials. For instance, the seaweed production industry employs a wide range of proprietary extraction processes to disrupt the cells of seaweeds and release bioactive compounds (polysaccharides, sugar alcohols, betaines, phenolic compounds, vitamins, phytohormones, etc.) into an extract. Some of the processes involve alkali extraction, acid extraction, cold cell-lysis technology, supercritical CO_2 extraction and microwave-assisted extraction.

The chemical composition of the extract largely depends on the method of extraction and on the chemical products used during the production process. Therefore, the biological activities of extracts of the same seaweed raw material obtained by different extraction processes may be considerably different (Battacharyya et al., 2015). Similarly, industrial protein hydrolysate production uses different hydrolysis methods (chemical hydrolysis under acidic or alkaline conditions, enzymatic hydrolysis, thermal hydrolysis or a combination of multiple methods) to produce bioactive compounds (amino acids and peptides) from protein-rich raw materials. Therefore, the chemical composition of protein hydrolysates can be extremely variable depending on the production process. Peptide concentrations are usually higher in enzymatically derived protein hydrolysates than in chemically derived hydrolysates, whereas free amino-acid concentrations follow the opposite trend (Colla et al., 2015a). Chemical hydrolysis can alter an amino-acid composition by destroying several amino acids (e.g., tryptophan) and also reduce the biological activity of free amino acids by converting them from the L-form to the D-form (racemization). Since the amino acids in proteins of living organisms are present only in L-form, plants cannot directly use D-amino acids in their metabolism, making protein hydrolysates with a high racemization degree ineffective or even potentially toxic for plants (Colla et al., 2015a).

The maintenance of product quality, safety and efficacy throughout the shelf life - so-called product stability - is considered a prerequisite for the acceptance and approval of any biostimulant product. Stability can be difficult to achieve in non-microbial biostimulants containing organic substances due to the complex nature of the products. The physical stability of liquid products is mandatory if the product is to be used in foliar sprays or soil-drench applications. Polymerization and precipitation of substances can occur in liquid products via a variety of reaction mechanisms that are affected by chemical composition, pH and temperature. It is important to prevent these processes through the removal of critical substances (i.e. alginates in seaweed extracts), adjusting the pH, and storing the product at appropriate temperature (10–35°C). Loss of biological activity through oxidation must also be prevented by adding antioxidants (i.e. ascorbic acid) to the liquid-biostimulant products.

Biological stability is critical in organic-based products. Microorganisms can alter the physical and chemical characteristics of a product by using organic compounds as a source of energy. Therefore, microbial activity needs to be stopped in liquid formulations by the generation of a hostile environment that prevents the growth of microorganisms (i.e. through acidification, increased salinity or addition of antimicrobial preservatives). For microbe-based biostimulants, several processes have been developed to maximize the production of highly viable propagules to be used as inoculants. Using production processes combining microbial inoculants with the co-formulates used in preparing the final product can affect the efficacy of a product by changing the survival rate of propagules during the storage period (shelf life) and re-growth after product application on crops.

Mycorrhizal fungal propagules can be produced using three different processes:

1 co-culture of mycorrhizal fungi and host plants under field conditions;
2 co-culture of mycorrhizal fungi and host plants in sterilized inert substrate under greenhouse conditions; and
3 in vitro co-culture of mycorrhizal fungi and hairy roots of carrots.

The inoculum quality of the same mycorrhizal fungal strain obtained by different production processes differs considerably (Rouphael et al., 2015). Field-produced inoculum is less expensive than lab-reared inocula but presents many disadvantages, such as variability in propagule concentration and a high risk of contamination with plant pathogens. Greenhouse-produced inoculum is pathogen free and contains a high concentration of mycorrhizal spores with a great potential effectiveness of mycorrhization over time (Calvet et al., 2013). In vitro-produced inoculum is pathogen free and contains a high concentration of mycorrhizal spores. However, in vitro-produced spores are smaller with low reserve substances and thin cell walls making the in vitro-produced inoculum less effective in comparison with the greenhouse-produced inoculum (Calvet et al., 2013).

Spore harvesting, processing and packaging are critical points of the production process. During harvesting and spore processing, it is important to prevent any damage to the spores and to mix the spores with co-formulations, which allows easy application of the inoculum and preservation of mycorrhization effectiveness over time (long shelf life of the inoculum). Packaging needs to increase the shelf life of the product, limiting UV radiation activity and humidification of the solid inocula. After the most appropriate production process is selected, prototype samples of the new biostimulant products are generated on a laboratory scale for further testing of activity.

2.3 Screening products for biostimulant activity and understanding their mode of action

Prototype samples of new biostimulant products need to be evaluated in trials under a controlled environment. Moreover, knowledge of the mode of action of the prototype samples is a key factor in optimizing biostimulant activity. Laboratory bioassays are used to evaluate the hormone-like activity of prototype samples (Gvulru and Heszxy, 1994). The auxin-like activity of biostimulant substances can be evaluated by determining the corn coleoptile elongation rate and/or rooting of tomato cuttings, while assessing the dwarf pea's stem-elongation rate and/or lettuce's hypocotyl elongation can be used to determine gibberellin-like activity (Ertani et al., 2009; Colla et al., 2014). Many tests are also available for measuring the cytokinin-like activity of biostimulant products (Gvulru and Heszxy, 1994). Preliminary tests conducted by Cardarelli et al. (unpublished data) showed positive results by testing wheat leaf chlorophyll retention to quantify the cytokinin-like activity of protein hydrolysates.

Knowing the dose-response relationship for different hormones enables the hormone-like activity of product samples to be expressed as equivalent units of a reference hormone (i.e. indole-3-acetic acid for auxin-like activity, GA3 for gibberellin-like activity and kinetin for cytokinin-like activity). The evaluation of hormone-like activity of biostimulant products can also be performed using the model plant *Arabidopsis thaliana*. Dozens of mutants that show specific morphological responses to certain types of hormones have been isolated and characterized for their involvement in plant hormone actions (Peng et al., 2009). A web interface of the Arabidopsis hormone database (http://ahd.cbi.pku.edu.cn) has been set up for quick access to information about hormone-related genes, including sequences, functional categories, mutant information, phenotypic descriptions, microarray data and linked publications. The web interface can be used to select Arabidopsis mutants that are used for screening product samples for the different hormone-like activities and for gaining information on their mode of action (i.e. hormone biosynthesis, metabolism, transport, perception and signal transduction).

Controlled-environment tests can also be used to screen product samples for improvement in abiotic stress tolerance of model crops (i.e. *A. thaliana* and Micro-Tom). Single or multiple stresses can be applied together with different biostimulant treatments (types, timing, rate, method of application) to seeds/seedlings of Arabidopsis grown in multiwell plates or on seeds/young plants of model crops grown in small pots, and morpho-physiological traits (i.e. seed germination, rosette size, shoot biomass, root biomass, leaf area, leaf chlorophyll content, chlorophyll fluorescence, stomata conductance, leaf temperature and leaf gas exchange) can be measured to evaluate the efficacy of biostimulant treatment in improving the crop response to stress factors (Lucini et al., 2015; De Diego et al., 2017).

A large number of product samples can be screened when bioassays are performed on high-throughput plant phenotyping platforms. This technology provides opportunities to screen for automated, fast scoring of several plant growth and developmental traits, following them over time using noninvasive sensors (De Diego et al., 2017). Other 'omics' technologies, such as transcriptomics, proteomics and metabolomics, can be applied to the best-performing products to determine their mode of action on plants. An approach based on the use of high-throughput phenotyping technologies and metabolomics was successfully tested on tomato crops to identify new plant-derived protein hydrolysates with biostimulant activity and to study the effects of protein hydrolysates on plants at the metabolic level (Paul et al., 2019a). The same approach was also successfully used to elucidate the biostimulant action of a vegetal-derived protein hydrolysate application method (foliar spray or substrate drench) on morpho-physiological traits and the metabolic profile of tomato grown under limited water availability (Paul et al., 2019b).

2.4 Quality control and safety

Standardization of the production process is impossible without proper quality control. By its very variable nature, biostimulant production is a delicate and demanding process, requiring precise attention to detail and quality standards to yield the best results. While a comprehensive body of legal texts regulating fertilizer production and quality control exists in most countries, a legislative framework for plant biostimulants is lacking. The very diverse nature of plant biostimulants also hinders the development of a general methodology for control for quality and safety.

In the new EU Fertilising Products Regulation (2019/1009), safety parameters for plant biostimulants have been introduced, such as limits on the amounts of heavy metals (Cr(VI), Pb, Hg, Ni, inorganic As, Cu, Zn), phosphonates, human pathogenic microorganisms (*Salmonella* spp., *E. coli*, *Listeria monocytogenes*, *Vibrio* spp., *Shigella* spp., *Staphylococcus aureus*, *Enterococcaceae*), anaerobic bacteria (unless the microbial biostimulant is an aerobic bacterium) and yeasts and moulds (unless the microbial biostimulant is a fungus).

Precise control of the raw material and the production process is necessary to prevent any contamination of the resulting biostimulant products with undesirable compounds and microorganisms. Cleaning and disinfection of production rooms, equipment and utensils must be regularly scheduled together with checks for staff hygiene. A hazard analysis and risk assessment of biostimulant production should be performed to improve programs for cleaning and disinfection operations. Quality control also needs to consider other physical, chemical and biological parameters according to the type of biostimulant product. For non-microbial plant biostimulants, quality control should include a normal appearance, water solubility, pH, organic matter/carbon content, moisture, density, and mineral composition and depending upon the product characteristics, levels of specific carbohydrates (for seaweed extracts), amino acids, peptides and degree of hydrolysis (for protein hydrolysates), humic and fulvic acids (for humic substances), and growth regulators (for seaweeds, protein hydrolysates and vegetal-derived extracts). Due to the very complex nature of non-microbial plant biostimulants, which include many bioactive compounds that are not always well known, characterizing and monitoring the product quality with a chemical fingerprint (metabolomic analysis), as recently reported for food authentication and origin certification, is more appropriate than the other methods. This approach can be used for all types of plant biostimulants independent of their origin and without the limitation of knowing all the involved bioactive compounds. For microbial plant biostimulants, quality control includes the genetic identity of the microbial strains, concentration of propagules and their viability, and content of undesirable compounds (i.e., heavy metals) and microorganisms (i.e., pathogens for plants and humans).

2.5 Field trials

Once the best product formulations are chosen from screening trials (bioassays), field trials need to be performed to verify the biostimulant activity under real growing conditions and to understand the biostimulant × genotype × environment interaction. Field trials are also necessary to define the right dose, application mode and timing for the different tested crops. Because multiple crop cycles per year can be performed under protected cultivation, greenhouse crops are appropriate subjects to undergo several trials in a short time, allowing a rapid evaluation of biostimulant products. The best-performing products under greenhouse conditions can also be tested on fruit trees and extensive field crops (i.e. cereals, soybean, sunflower, sugar beet and sugar cane) with the limitation of one trial per year due to the long growing cycle of these crops.

The use of high-throughput plant phenotyping platforms allows a tremendous increase in the capability of assessing biostimulant activity in the field, with nondestructive, repeated, objective observations and without the requirement of an extensive labor force. Several high-throughput plant phenotyping platforms for use in the field are available on a commercial scale, such as phenomobiles, autonomous mobile-sensing platforms, and unmanned aerial vehicles. These platforms can be equipped with multifunctional sensors (high-resolution visible, chlorophyll fluorescence, thermal infrared, hyperspectral and 3D laser sensors) for monitoring nondestructive morpho-physiological crop traits (Rouphael et al., 2018). The study of phenotypic traits should be integrated with omics analysis, such as metabolomics, proteomics or transcriptomics, to study the product mode of action.

2.6 Regulation and market positioning

All the data compiled before have to be synthesized to allow the marketing department of the company to design the strategy and the positioning of the product in the global market. The biostimulant product should be registered in the countries of interest according to national regulations. However, in many countries, regulations for biostimulants are lacking, and the term 'biostimulant' cannot be used on the label, which prevents companies from making legitimate, data-supported claims and thus limits how a company describes and differentiates their product. Consequently, companies must choose to either (1) limit the actual plant benefits they claim and register the products as nutrients, plant inoculants or soil amendments or (2) extend their claims and register the product as a plant protection product, even though the claims may not scientifically qualify the product as a pesticide or plant-growth regulator (Biological Products Industry Alliance, 2020).

2.7 Industrialization and commercialization

Industrial scale-up of biostimulant production is the final step of the process. It may be necessary to design new machines, processes and production facilities. All previous laboratory experiences from the process development have to be transferred to the engineering department to avoid unnecessary costs or bias. Quality control should be performed to ensure that the desired product can be produced by the new industrial process. To increase the success ratio, the operations department should be involved starting at the beginning of the design of new products, allowing them to identify possible problems and solutions. Once the biostimulant product is manufactured and its performance is supported by scientific data and efficacy trials, the final step is strategically to position it in the market. A positioning strategy requires a set of actions and processes to improve the image and visibility of the brand, company and product.

First, it is fundamental to understand that the consumer decision-making process is based on five steps:

1 recognition of the need for biostimulant products;
2 search for products that can be used to satisfy the need;
3 evaluation of all products and brands that satisfy the need;
4 purchase decision; and
5 post-purchase evaluation of the decision made.

Appropriate marketing communication and technical support are necessary to create the customer's need for biostimulant products and to differentiate the product from its competitors. For successful communication, it is important to understand how the potential customer wants to perceive the biostimulant product and to consider that potential customers want to become involved in the reasons behind the product development and the company vision.

It must be taken into account that purchase decisions are impacted not only by rational factors such as economic return (costs of the biostimulant application versus benefits in terms of crop yield increase or quality improvement) and utility (solution of farmer technical problems) but also by emotional aspects. Therefore, building trust and credibility with potential customers is crucial for the commercial success of the biostimulant product. This is especially true in the biostimulant market because there are many players that try to affirm their products by copying technical advances of other products and competing at lower price. In this scenario, product-marketing success may be largely due to emotional factors.

3 Industrial case study 1: mycorrhizal inoculants

Arbuscular mycorrhizal fungi are obligate root symbionts that provide a wide spectrum of benefits to their hosts, such as improved nutrient uptake and stress tolerance (Rouphael et al., 2015). Mycorrhizal inoculants have been studied for many years with limited successful applications at the farm level. This is due to the reduced effectiveness of mycorrhizal fungal inoculants in providing crop yield increases and due to the difficulties of delivering inoculum to crops. To address these problems, Atens, a Spanish company based in Tarragona, established the production of inoculants based on the arbuscular mycorrhizal fungus, *Rhizoglomus irregulare* BEG72 (Estaun et al., 1994), in 1994, and a few years later, it also produced some inoculants based on the arbuscular mycorrhizal fungus *Funneliformis mosseae* BEG234. These two mycorrhizal fungal strains were selected due to their capability to establish symbiosis for a wide range of host plants and environmental conditions and provide benefits to many crops in terms of growth, yield and produce quality. Atens developed an advanced in vivo mycorrhizal production process that includes the following steps:

1 inoculation of leek seedlings grown in trays filled with vermiculite under greenhouse conditions with selected mycorrhizal fungal strains;
2 transplantation of mycorrhized seedlings in greenhouse benches filled with vermiculite;
3 cultivation of mycorrhized plants for 4 months using biological crop management;
4 harvest of propagules from roots/substrates at the end of the leaf senescence phase of leek plants; and
5 processing of raw inoculum for producing the final product.

The use of an in vivo production process of mycorrhizal fungal propagules under a controlled environment guarantees the achievement of powder-based products that provide high effectiveness of mycorrhizal inoculation, long shelf life (over 5 years) and absence of undesirable microorganisms (i.e. plant and human pathogens) and compounds (i.e. heavy metals).

The in vivo propagation technology ensures a high presence of mycorrhiza helper bacteria (i.e. *Bacillus megaterium* MHBM77 and *B. megaterium* MHBM06) in the inoculum (Agnolucci et al., 2018). These bacteria are associated with mycorrhizal spores and hyphae, and they increase the effectiveness of mycorrhizal inoculation by:

1 promoting mycorrhiza spore germination, root colonization and mycelial growth;
2 enhancing the efficiency of mycorrhiza in nutrient uptake;
3 producing phytostimulation compounds; and
4 exhibiting antagonistic activity against many plant pathogens.

In these ways they enhance the functionality of the AMF product.

Rigorous quality control of the production process is performed continuously by monitoring physical, chemical and biological factors such as the greenhouse environmental conditions; composition of the substrate, irrigation water and fertilizers; morpho-physiological traits of the host crop; phytosanitary status; and mycorrhizal helper bacterial population and mycorrhizal root colonization. Quality control also includes specific analysis of the resulting product for verifying the genetic identity of mycorrhizal strains, the mycorrhizal spore concentration and spore vitality, the mycorrhizal helper bacteria propagules, and the absence of harmful microorganisms and undesirable compounds (product safety).

Mycorrhizal fungal inoculum can also be enriched with conidia of selected strains of saprophytic fungi *Trichoderma atroviride* AT10 or *Trichoderma koningii* TK7. These *Trichoderma* strains have been selected due to their ability to grow quickly over a wide range of temperatures (10–30°C) and soil conditions (pH from 5.5 to 8.0, up to 3% NaCl in the medium) and to exert biostimulant action on crops by producing auxin-like compounds (which stimulate root growth) and siderophores (which participate in the mobilization of iron and other plant nutrients) and activating crop secondary metabolism (which improves nutritional quality of product and crop stress resistance). *Trichoderma* propagules are produced by Atens using an advanced process based on the use of solid-state fermenters filled with sterile wheat bran-based substrate. The use of this technology allows the production of solid inoculum containing mainly spores with thick walls and a high content of reserve substances (the inoculum has a long shelf life), whereas liquid culture production of *Trichoderma* propagules is not adopted because this technology leads to liquid inoculum containing mainly mycelium and, to a lesser extent, spores with fine and incomplete walls (short shelf life of inoculum). The two *Trichoderma* strains are highly compatible with arbuscular mycorrhizal fungi, unlike other *Trichoderma* strains belonging to other Trichoderma species (i.e. *T. harzianum*), which can reduce the activity of mycorrhizal fungi, especially through mycoparasitism and competition for root-infection sites (De Jaeger et al., 2010). Microbial consortia containing arbuscular mycorrhizal fungi, *Trichoderma* spp., and mycorrhizal helper bacteria have been developed to promote synergistic interactions among microorganisms, enhancing the biostimulant activity of inoculants.

One of the major constraints limiting the use of microbial inoculants is the difficulty of delivering beneficial microorganisms to crops. To address this issue, Atens started from powder-based inoculum to develop many formulations for seed treatment, soil/substrate application, fertigation and inoculation of dormant bare root vines. These products resulted from intense research activity, which is also evident from the number of scientific articles found in a search in Google Scholar (https://scholar.google.com/) using the key words 'Aegis + Glomus' (i.e. as a demonstration for the product 'Aegis'); the search indeed resulted in 176 scientific papers.

In 2014, a microbial-based tablet containing a mixture of fertilizers, organic matter, mycorrhizal fungi, *Trichoderma* spp. and beneficial bacteria was developed; the tablet allows simplification of the inoculation of horticultural crops by placing the product directly into planting holes at transplanting.

The idea of the tablet product occurred in Almeria, the Spanish region where most greenhouses are located. It was observed that the application of arbuscular mycorrhizal fungi alone had an effect only after the symbiosis process had been established and farmers had to wait up to 1–2 months from the inoculation to see any visual effect on crops.

To make the effect of microbial inoculation on crops immediately visible, fertilizers and organic substances were added to the mycorrhizal inoculum (Lucini et al., 2018) to promote fast root and shoot growth and nutrient uptake. *Trichoderma* propagules were also included in the formulation to further enhance root growth and nutrient uptake and to improve crop resistance to environmental stress (Fiorentino et al., 2018). Several formulations have been tested in greenhouse trials using mixture design and multivariate studies with the aim of optimizing tablet composition. Special attention was paid to reducing the salinity index of the formulation to avoid root burn and phytotoxicity and to increasing the rate at which the fungal spores survive the compression required for making the tablet. All these preliminary trials were developed internally in the company laboratories and greenhouses.

Process development was performed with the support of the engineering department to select the mixing equipment and the right machine to make the tablets. The technology used by pharmaceutical industries to produce drugs in tablet form achieved the best quality of the microbial-based tablet. More than one hundred tests were performed to optimize the mixing of powdered ingredients (microbial spores, fertilizers, organic substances) and to define the compression force suitable for making the tablet and avoiding damage to the spores. Quality control of mycorrhizal spores in the tablet was developed by separating the spores from the matrix to allow counting with a microscope. When the production process was set up at the industrial scale and a high-quality batch of the microbial-based tablets was produced, field trials were performed internally and externally.

Independent institutions such as universities and research institutions demonstrated that tablet application at transplanting significantly increased the shoot dry weight by 167% in lettuce, 56% in melon, 115% in pepper, 68% in tomato and 58% in zucchini plants after 35 days of growth under greenhouse conditions in comparison with this weight in untreated plants (Colla et al., 2015b); moreover, field trials at Tuscia University confirmed the positive effects of tablet application on fresh shoot biomass (+70%) and fruit yield (+15%) of lettuce and zucchini, respectively.

The improved crop performance of inoculated lettuce and zucchini plants was attributed to the phytostimulation effect of beneficial microorganisms and to improved plant nutritional status (increased N, P, Fe, Mn, Zn and B concentrations for lettuce and increased P, Fe, Zn and B concentrations for leaf zucchini) of crops (Colla et al., 2015b). Similar results were also obtained in other locations for greenhouse-vegetable crops and fruit trees, especially after the second year from tree planting. In 2018, a new formulation of the microbial-based tablet was developed for enhancing water and nutrient availability to the transplants. These effects were due to the addition of hydrophilic vegetal fiber tablets that are able to increase water absorption and reduce nutrient leaching (International patent number WO2018134465).

Registration of the microbial-based tablet was difficult. In 2015, the product was registered in Italy as a mycorrhizal fungus-based inoculant according to the Legislative Decree 75/2010 and successive amendments. In Spain, the registration of the 'Asir®' tablet as a microorganism-based biostimulant product was achieved in the middle of 2019 (four years later from product development) following the procedures reported in Real Decreto 999/2017.

4 Industrial case study 2: vegetal-based protein hydrolysates

Protein hydrolysates are plant biostimulants containing mixtures of mainly polypeptides, oligopeptides and amino acids that are manufactured from protein sources using partial hydrolysis (Colla et al., 2015a). There has been growing interest in vegetal-derived protein hydrolysates during recent years due to their positive effects on crop performances, especially under environmental stress, and the lack of restrictions on their use in food crops for organic markets or for people requesting food without contamination by animal-derived products (i.e., vegetarians). The company Italpollina developed an advanced technology system for enzymatic hydrolysis (LISIVEG®) of legume seed proteins that leads to commercial vegetal-derived protein hydrolysates.

Product development started in 2002 in Nixe Laboratoires, France. The original idea was to produce a low-cost liquid fertilizer rich in organic nitrogen starting from plant biomass to offer an alternative to synthetic nitrogen fertilizers. After an extensive review of the literature, vegetal-derived proteins were selected as the best source for producing soluble organic nitrogen compounds. Enzymatic hydrolysis was preferred to chemical hydrolysis because proteolytic enzymes, which are highly selective for the cleavage of peptide bonds, allow the production of a continuous and consistent supply of the desired protein hydrolysate. Several approaches have been tested for hydrolysing vegetal proteins, from the use of fungi or bacteria grown directly on seed substrates to the application of commercial proteolytic enzymes. It took many years to produce a commercial product that met all quality criteria (high solubility, stability, and nitrogen content and a low salinity index). Feasibility and reproductivity have been the driving criteria in selecting the production process. The economic profit resulting from the candidate production processes was also considered.

The selection of the vegetal raw material as a source of proteins was changed several times. Initially, legume seeds from native plants of Venezuela, where Italpollina was present from the 1990s with a production plant for vegetal extracts and liquid fertilizers, were chosen. However, political difficulties discouraged the development of a new manufacturing process for producing vegetal-derived protein hydrolysates in Venezuela.

The search for other protein-rich plant material affordably available in large quantities worldwide led to the choice of soybean meal. Initially, the idea was to obtain the vegetal protein hydrolysate by fermentation using a koji-making process. The fungus *Aspergillus oryzae* (Su et al., 2011) was grown for several days on a culture medium made with a mash of vegetal meal. During the growing period, the fungus secreted large amounts of various degrading enzymes, which allow it to hydrolyze the proteins of various starches into sugars and amino acids. Then, the medium was diluted in water, blended and heated, at which point the enzymes began transforming complex sugars into monosaccharides and insoluble proteins into soluble peptides and amino acids. The produced soluble carbohydrates can be transformed into alcohol by the yeast *Saccharomyces cerevisiae* and distilled for other applications.

Despite the interesting biostimulant activity of the resulting vegetal protein hydrolysates, the *A. oryzae*-based production process was rejected due to the complexity of the system and the difficulties of maintaining a sterile environment in the production process and removing heat generated during fermentation. A double sugar/fiber extraction combined with commercial hydrolytic enzymes was selected as an alternative to hydrolysis of the fermented vegetal biomass using the fungal culture filtrate. This new research led to several hundred alternatives with a combination of different degrees of hydrolysis, types of proteolytic enzymes and vegetal raw materials. The product yield of different production processes at the pilot plant scale was measured together with the physical and chemical characteristics of the product prototypes (i.e., dissolved solids, solubility over time, pH, salinity, nitrogen, degree of protein hydrolysis, amino acid composition, carbohydrates and growth regulator content).

Samples of product prototypes were tested for biostimulant activity using high-throughput phenotyping platforms in a controlled environment (Paul et al., 2019a,b). Moreover, hormone-like activity of product prototypes was also determined using laboratory bioassays (Colla et al., 2014). Hundreds of field trials were conducted on crops using the best-performing product prototypes with different rates (1–5 L/ha), application methods (soil drench, seed treatment and foliar sprays) and timing (before planting, at planting, during the vegetative phase, at flowering and during fruit growth, during development and during maturity). The trials resulted in the legume-derived protein hydrolysate Trainer®, which contains 5% total nitrogen, mainly as soluble peptides (27%) with a low level of free amino acids.

Several peptides of the legume-derived protein hydrolysate have been identified as signalling molecules ('peptide hormones') that play an important role in controlling plant growth and development, flowering, fertilization, fruit growth and plant response to environmental stress. These peptide hormones have a short chain of amino acids (<50) and a specific amino acid sequence, and they exert biological activity at very low concentrations (nM). One example of a peptide hormone is a lateral root-promoting peptide having a specific 12-amino acid sequence (Colla et al., 2015a). Legume-derived protein hydrolysate application has been shown to increase photosynthesis, water-use efficiency, nutrient uptake, abiotic stress resistance and crop yield and quality in many trials (Colla et al., 2015a, 2017; Lucini et al., 2015; Rouphael et al., 2017a; Sestili et al., 2018; Di Mola et al., 2019). Moreover, recent studies (Luziatelli et al., 2019) demonstrated that foliar spraying of legume-derived protein hydrolysate Trainer® on lettuce stimulated the beneficial microflora living in the phyllosphere; the biostimulant-enhanced bacteria showed phytostimulant action and antagonistic activity against some plant pathogens. Overall, scientific articles dealing with the Trainer® product can be easily found through

Google Scholar (https://scholar.google.com/) using the key words 'Trainer + Italpollina'; this search resulted in 27 scientific papers.

The scale-up of the enzymatic production process 'LISIVEG®' for protein hydrolysates was quite complicated and time-consuming. The first production factory was established in Novara in 2012 with a production capacity of 100,000 liters of protein hydrolysate per year. Thanks to the experience gained with the first factory and the commercial success of the product, a second-generation industrial plant was constructed in 2016 in Novara with more than ten times the production capacity of the previous plant. Research activity, process optimization and new product design have been carried out continuously, allowing Italpollina to build a third-generation industrial plant in Indiana, the United States, in 2018 with increased production capacity and flexibility in the production process. Moreover, the new industrial plant was implemented with a production line of vegetal-derived protein hydrolysates in microgranule and powder forms. The solid formulations of protein hydrolysates open opportunities for new product development such as biostimulant-enriched solid water-soluble fertilizers for use in fertigation systems.

5 Future trends

Advanced technologies such as artificial intelligence and machine learning have been developed rapidly in recent years. The biostimulant sector could take advantage of these technologies in developing new products, identifying bioactive compounds, and discovering and certifying the origin of biostimulant products. Another innovative application could be the use of artificial intelligence to identify potential new raw materials or microbial strains based on those already known. Most biostimulant companies are highly specialized in their manufacturing process and the raw materials used for producing plant biostimulants. This specialization leads to an economy of scale but limits the opportunities to find new sources of biostimulants. Byproducts of the agri-industrial sector represent a potentially useful unexploited source of plant biostimulants. The use of agro-industrial byproducts for producing plant biostimulants offers new growth opportunities towards a circular economy model.

Recent studies have demonstrated a synergistic effect of combining different types of plant biostimulants (Rouphael and Colla, 2018). For instance, Rouphael et al. (2017b) demonstrated that combined application of the microbial-based tablet Asir(R) (*Rhizophagus intraradices* BEG72 and *T. atroviride* MUCL45632) and the vegetal protein hydrolysate Trainer(R) was more effective than microbial biostimulant application alone to mitigate the negative effects of salinity and alkalinity stress on lettuce growth. These findings open new opportunities for companies to develop more effective biostimulants, taking advantage of synergism between microbial and non-microbial components.

6 References

Agnolucci, M., Avio, L., Pepe, A., Turrini, A., Cristani, C., Bonini, P., Cirino, V., Colosimo, F., Ruzzi, M. and Giovannetti, M. (2018), 'Bacteria associated with a commercial mycorrhizal inoculum: community composition and multifunctional activity as assessed by Illumina sequencing and culture-dependent tools', *Frontiers in Plant Science*, 9: 1956.

Battacharyya, D., Babgohari, M. Z., Rathor, P. and Balakrishnan, P. (2015), 'Seaweed extracts as biostimulants in horticulture', *Scientia Horticulturae*, 196: 39–48.

Biological Products Industry Alliance, (2020), https://www.bpia.org.

Brown, P. and Saa, S. (2015), 'Biostimulants in agriculture', *Frontiers in Plant Science*, 6: 671.

Calvet, C., Camprubi, A., Pérez-Hernández, A. and Lovato P.E. (2013), 'Plant growth stimulation and root colonization potential of in vivo versus in vitro arbuscular mycorrhizal inocula', *HortScience* 48(7): 897-901.

Calvo, P., Nelson, L. and Kloepper, J. W. (2014), 'Agricultural uses of plant biostimulants', *Plant and Soil*, 383(1-2): 3-41.

Caradonia, F., Battaglia, V., Righi, L., Pascali, G. and La Torre, A. (2019), 'Plant biostimulant regulatory framework: prospects in Europe and current situation at international level', *Journal of Plant Growth Regulation*, 38(2): 438–48.

Castellion, G. and Markham, S. K. (2013), 'Perspective: New product failure rates: influence of argumentum ad populum and self-interest', *Journal of Product Innovation Management*, 30(5): 976–9.

Colla, G. and Rouphael, Y. (2015), 'Biostimulants in horticulture', *Scientia Horticulturae*, 196: 1–2.

Colla, G., Rouphael, Y., Canaguier, R., Svecova, E. and Cardarelli, M. (2014), 'Biostimulant action of a plant-derived protein hydrolysate produced through enzymatic hydrolysis', *Frontiers in Plant Science*, 5, article 448.

Colla, G., Nardi, S., Cardarelli, M., Ertani, A., Lucini, L., Canaguier, R. and Rouphael, R. (2015a), 'Protein hydrolysates as biostimulants in horticulture', *Scientia Horticulturae*, 196: 28–38.

Colla, G., Rouphael, Y., Di Mattia, E., El-Nakhel, C. and Cardarelli, M. (2015b), 'Co-inoculation of *Glomus intraradices* and *Trichoderma atroviride* acts as a biostimulant to promote growth, yield and nutrient uptake of vegetable crops', *Journal of the Science of Food and Agriculture*, 95(8): 1706–15.

Colla, G., Cardarelli, M., Bonini, P. and Rouphael, Y. (2017), 'Foliar applications of protein hydrolysate, plant and seaweed extracts increase yield but differentially modulate fruit quality of greenhouse tomato', *HortScience*, 52(9): 1214–20.

De Diego, N., Fürst, T., Humplík, J. F., Ugena, L., Podlešáková, K. and Spíchal, L. (2017), 'An automated method for high-throughput screening of arabidopsis rosette growth in multi-well plates and its validation in stress conditions', *Frontiers in Plant Science* 8: 1702.

De Jaeger, N., Declerck, S. and de la Providencia, I. E. (2010), 'Mycoparasitism of arbuscular mycorrhizal fungi: a pathway for the entry of saprotrophic fungi into root', *FEMS Microbiology Ecology* 73: 312–22.

Di Mola, I., Ottaiano, L., Cozzolino, E., Senatore, M., Giordano, M., El-Nakhel, C., Sacco, A., Rouphael, Y., Colla, G. and Mori, M. (2019), 'Plant-based biostimulants influence the agronomical, physiological, and qualitative responses of baby rocket leaves under diverse nitrogen conditions', *Plants*, 8: 11.

Du Jardin, P. (2015), 'Plant biostimulants: definition, concept, main categories and regulation', *Scientia Horticulturae*, 196: 3–14.

Ertani, A., Cavani, L., Pizzeghello, D., Brandellero, E., Altissimo, A., Ciavatta, C. and Nardi, S. (2009), 'Biostimulant activity of two protein hydrolyzates in the growth and nitrogen metabolism of maize seedlings', *Journal of Plant Nutrition and Soil Science*, 172: 237–44.

Estaun, V., Calvet, C. and Camprubi, A. (1994), 'Arbuscular mycorrhizae and growth enhancement of micropropagated *Prunus* rootstock in different soilless potting mixes', *Agricultural and Food Science*, 3(3): 263–7.

Fiorentino, N., Ventorino, V., Woo, S. L., Pepe, O., De Rosa, A., Gioia, L., Romano, I., Lombardi, N., Napolitano, M. Colla, G. and Rouphael, Y. (2018), 'Trichoderma-based biostimulants modulate rhizosphere microbial populations and improve N uptake efficiency, yield and nutritional quality of leafy vegetables', *Frontiers in Plant Science*, 9: 743.

Gvulru, G. and Heszxy, L. E. (1994), 'Auxin and cytokinin bioassays: a short overview', *Acta Agronomica Hungarica*, 43 (1–2): 185–97.

Lucini, L., Rouphael, Y., Cardarelli, M., Canaguier, R., Kumar, P. and Colla, G. (2015), 'The effect of a plant-derived protein hydrolysate on metabolic profiling and crop performance of lettuce grown under saline conditions', *Scientia Horticulturae* 182: 124–33.

Lucini, L., Rouphael, Y., Cardarelli, M., Bonini, P., Baffi, C. and Colla, G. (2018), 'A vegetal biopolymer-based biostimulant promoted root growth in melon while triggering brassinosteroids and stress-related compounds', *Frontiers in Plant Science*, 9: 472.

Luziatelli, F., Ficca, A. G., Colla, G., Baldassarre Švecová, E. and Ruzzi, M. (2019), 'Foliar application of vegetal-derived bioactive compounds stimulates the growth of beneficial bacteria and enhances microbiome biodiversity in lettuce', *Frontiers in Plant Science*, 10: 60.

Markets and Markets (2019), 'Biostimulants market outlook & industry demand analysis', https://www.marketsandmarkets.com.

Paul, K., Sorrentino, M., Lucini, L., Rouphael, Y., Cardarelli, M., Bonini, P., Reynaud, H., Canaguier, R., Trtílek, M., Panzarová, K. and Colla, G. (2019a) 'Understanding the biostimulant action of vegetal-derived protein hydrolysates by high-throughput plant phenotyping and metabolomics: a case study on tomato', *Frontiers in Plant Science*, 10: 47.

Paul, K., Sorrentino, M., Lucini, L., Rouphael, Y., Cardarelli, M., Bonini, P., Miras Moreno, M. B., Reynaud, H., Canaguier, R., Trtílek, M., Panzarová, K. and Colla, G. (2019b), 'A combined phenotypic and metabolomic approach for elucidating the biostimulant action of a plant-derived protein hydrolysate on tomato grown under limited water availability', *Frontiers in Plant Science*, 10: 493.

Peng, Z. Y., Zhou, X., Li, L., Yu, X., Li, H., Jiang, Z., Cao, G., Bai, M., Wang, X., Jiang, C., Lu, H., Hou, X., Qu, L., Wang, Z., Zuo, J., Fu, X., Su, Z., Li, S. and Guo, H. (2009), '*Arabidopsis* hormone database: a comprehensive genetic and phenotypic information database for plant hormone research in *Arabidopsis*', *Nucleic Acids Research*, 37(Database issue): D975–82.

Rouphael, Y. and Colla, G. (2018), 'Synergistic biostimulatory action: designing the next generation of plant biostimulants for sustainable agriculture', *Frontiers in Plant Science*, 9: 1655.

Rouphael, Y., Franken, P., Schneider, C., Schwarz, D., Giovannetti, M., Agnolucci, M., De Pascale, S., Bonini, P. and Colla, G. (2015). 'Arbuscular mycorrhizal fungi act as biostimulants in horticultural crops', *Scientia Horticulturae*, 196: 91–108.

Rouphael, Y., Colla, G., Giordano, M., El-Nakhel, C., Kyriacou, M. C. and De Pascale, S. (2017a), 'Foliar applications of a legume-derived protein hydrolysate elicit dose dependent increases of growth, leaf mineral composition, yield and fruit quality in two greenhouse tomato cultivars', *Scientia Horticulturae*, 226: 353–60.

Rouphael, Y., Cardarelli, M., Bonini, P. and Colla, G. (2017b), 'Synergistic action of a microbial-based biostimulant and a plant derived-protein hydrolysate enhances lettuce tolerance to alkalinity and salinity', *Frontiers in Plant Science*, 8: 131.

Rouphael, Y., Spíchal, L., Panzarová, K., Casa, R. and Colla, G. (2018), 'High-throughput plant phenotyping for developing novel biostimulants: from lab to field or from field to lab?', *Frontiers in Plant Science*, 9: 1197.

Sestili, F., Rouphael, Y., Cardarelli, M., Pucci, A., Bonini, P., Canaguier, R. and Colla, G. (2018), 'Protein hydrolysate stimulates growth in tomato coupled with n-dependent gene expression involved in n assimilation', *Frontiers in Plant Science*, 9: 1233.

Su, G., Ren, J., Yang, B., Cui, C. and Zhao, M. (2011), 'Comparison of hydrolysis characteristics on defatted peanut meal proteins between a protease extract from *Aspergillus oryzae* and commercial proteases', *Food Chemistry*, 126(3): 1306–11.

Xu, L. and Geelen, D. (2018), 'Developing biostimulants from agro-food and industrial by-products', *Frontiers in Plant Science*, 9: 1567.

Yakhin, O. I., Lubyanov, A. A., Yakhin, I. A. and Brown P. H. (2017), 'Biostimulants in plant science: a global perspective', *Frontiers in Plant Science*, 7: 2049.

Chapter 10

Plant biostimulants and their influence on nutrient use efficiency (NUE)

Patrick H. Brown, Douglas C. Amaral, Meerae Park, Jennifer Schmidt and Amelie Gaudin, University of California-Davis, USA

1 Introduction

In 2019 the European Union (EU) formally adopted regulations governing biostimulants and had developed a function-based description of what these products do (see Chapter 1). The EU definition, published in June of 2019,[1] states.

A plant biostimulant shall be an EU fertilising product the function of which is to stimulate plant nutrition processes independently of the product's nutrient content with the sole aim of improving one or more of the following characteristics of the plant or the plant rhizosphere:

(a) nutrient use efficiency,
(b) tolerance to abiotic stress,
(c) quality traits, or
(d) availability of confined nutrients in the soil or rhizosphere.

In this chapter we will focus on the effects of biostimulants on the 'stimulation of plant nutrition processes to ... improve nutrient use efficiency' (NUE). To meaningfully assess the existing literature on the impact of biostimulants on

1 *Official Journal of the European Union*, L 170, 25 June 2019.

http://dx.doi.org/10.19103/AS.2020.0068.13

NUE, the meaning of the term 'NUE' be clearly defined. NUE is a widely used but complicated term, the meaning of which varies depending upon the agricultural context in which it is used and also upon the understanding and intention of the user. Thus, from a grower's perspective, NUE may refer simply to the return on investments achieved from the addition of more fertilizers or biostimulant inputs, while an environmental scientist may be concerned primarily with the minimization of nutrient loss to the non-farm environment, and a plant breeder may be primarily interested in the amount of yield that is produced per unit of nutrient acquired by the plant. Each perspective will lead to a different outcome and a different rationalization for the use of a biostimulant.

Cassman et al. (2002) defined 'the NUE of a cropping system as the proportion of all N inputs that are removed in harvested crop biomass, contained in recycled crop residues, and incorporated into soil organic matter and inorganic N pools'. This definition, while being the most appropriate and holistic determination of NUE, is difficult to obtain as it requires the measurement of the current and future availability of inorganic and organic soil nutrient contents before and after experimentation, and the measurement of all nutrients in harvested and non-harvested crop parts.

A number of descriptions of NUE and the complexities of its determination have been published in recent years (Reich et al., 2015; Cassman et al., 2002). Table 1, adapted from Fixen et al. (2015), describes the two most widely utilized measures of NUE relevant to the context of biostimulant application and response. Each of the terms for NUE cited in Table 1 has specific uses and specific constraints that should be considered before conclusions of biostimulant effects on NUE can be made.

Biostimulants may influence NUE through effects on any of these descriptors of NUE. Biostimulants that enhance root growth and soil exploration increase the solubilization of soil nutrients or upregulate nutrient uptake processes that can result in an increase in agronomic nutrient use efficiency (AE) by enabling a greater amount of the total soil nutrient reserve to be acquired by the plant. This benefit can occur, however, only if both of the following conditions are met: (1) nutrient availability in the absence of the biostimulant is inadequate to effectively supply plant demands and (2) a source of the otherwise-unavailable nutrient is made available by the addition of the biostimulant; this may be a result of increased root exploration or a nutrient solubilization effect by roots, microbes or direct chemical solubilization as a result of biostimulant addition.

Agronomic efficiency (AE) can also be increased in the short term if the biostimulant acts solely to enhance shoot growth or harvested product yield and, as a consequence, enhances the total nutrient demand. A strictly growth-promoting effect of a biostimulant would not be indicative of increased internal efficiency (IE) per se, but could still be important to the overall system's NUE by

Table 1 Common NUE Terms adapted for their application to biostimulants

Term	Calculation*	Question addressed	Relevance to biostimulant Enhancement of NUE
Agronomic efficiency**	$AE = (Y - Y_0)/F$	How much productivity improvement was gained by the addition of nutrients in the presence or absence of the biostimulant?	As an integrated indicator of the impact of biostimulants on the recovery of applied and residual nutrients, typically over a single cropping season. Without measurement of soil fertility, AE cannot distinguish growth or yield promotion effects (root, shoot or harvested product) or stress mitigation effects from effects on soil nutrient availability or effects on internal efficiency (IE see below). An improvement in AE does not imply a reduced the demand for nutrients but will reduce fertilizer application.
Internal utilization efficiency (similar to physiological utilization efficiency)	$IE = Y/U$	What is the ability of the plant to transform nutrients acquired from all sources into economic yield, contrasting with and without biostimulant?	To evaluate biostimulant effects on the conversion of acquired nutrients into yield. Requires whole plant analysis of nutrients and dry matter allocation. Can distinguish growth promotion effects from improved internal utilization of acquired nutrients. An improvement in IE will reduce demand for nutrients and, as a consequence, reduce demand for fertilization.

* Y = yield of harvested portion of crop with nutrient applied; Y_0 = yield under original/control grower conditions; F = amount of nutrient applied; U = total nutrient uptake in aboveground crop biomass with nutrient applied; units are not shown in the table since the expressions are ratios on a mass basis and are therefore unitless in their standard form.
Adapted with permission from Fixen et al., 2015.

ensuring that a greater proportion of the available soil nutrients are acquired by the crop and not lost from the cropping system. This effect could also result in more efficient utilization of nutrient residues from prior crops. A biostimulant that mitigates plant stress could also enhance AE by enabling greater crop productivity under stress conditions. This potential growth stimulation-driven improvement of NUE can occur only if nutrient availability is sufficient to meet the greater plant demands induced by the growth-promoting effect.

Biostimulants may also act to directly alter the internal utilization efficiency of nutrients (IE), either through effects on within-plant allocation or remobilization of acquired nutrients or by altering the efficiency with which a given unit of the nutrient is utilized in metabolic reactions. Applied products that complex or chelate within-plant nutrients or otherwise enhance the allocation of nutrients to critical organs may, therefore, be categorized as biostimulants. Similarly, products that delay or slow the rate of leaf senescence and hence allow for greater remobilization of nutrients stored in leaf and stem organs may increase harvested product yield, nutrient storage in perennials and product nutrient content. Biostimulants may also improve 'plant nutrition processes' by altering the expression of genes involved in within-plant nutrient transport. It is also theoretically possible that a biostimulant might increase the efficiency with which a given unit of the nutrient is utilized in metabolic reactions; however, we are unaware of any explicit demonstration of this phenomenon.

In the following text, we provide an interpretation of claims of biostimulant effects on NUE in the context of these aforementioned definitions and constraints. Evidence and mechanisms of the biostimulant effect on NUE are provided for each of the major classes of biostimulants, including (1) humic substances (2) microbial biostimulants (3) seaweed extracts and (4) protein hydrolysates. While there are many additional biostimulant products including inorganic elements that are not recognized as plant essential, synthetic molecules, fermentation products, microalgae and extracts of a huge array of feedstock materials, these will not be considered here as many are not yet fully characterized and hence a mechanistic understanding cannot be determined or the body of knowledge demonstrating an effect on NUE is not sufficiently robust to draw meaningful conclusions. The lack of inclusion should not be interpreted as a lack of utility for these products.

2 Humic and fulvic substances

Humic and fulvic substances (referred to as HSs in the following text unless explicitly identified otherwise) are chemically complex, naturally occurring products of the decomposition of organisms that are present in all soils to varying degrees. HSs can be commercially extracted from a wide variety of substrates, including the organic matter of plant, animal or soil origin, peats, soft-coals and industrial or urban by-products and residues. HSs are considered to be the most abundant, naturally occurring organic molecules on the earth and have been used at all stages of agricultural production and in most crops with documented direct and indirect effects on plant growth and on NUE (specifically AE) dating back to the early 1900s (Burk et al., 1931, 1932a,b).

When applied in sufficiently great quantities, humic and fulvic rich substances (referred to generically as humic substances, HSs) can undoubtedly improve bulk soil physical and chemical characteristics and indirectly influence plant growth and NUE through the direct provision of nutrients or substrates to support beneficial soil microbial activities (Diacono and Montemurro, 2010). While broadacre applications of composts, manures, biosolids and green waste materials have long been used and clearly enhance AE, they are not relevant for this discussion as the quantity of materials that would be needed to substantively alter these gross soil characteristics in agricultural fields far exceeds common application rates of commercial biostimulants (Olk et al., 2018). Applications of HS-rich organic matter at rates sufficiently large to alter gross soil characteristics are more appropriately considered as soil amendments and not biostimulants per se. In the following text, only biostimulants applied at rates not exceeding 100 liters or kg per hectare, a quantity above which economically viable return is unlikely, will be reviewed (Rose et al., 2014).

HSs applied at low rates and targeted explicitly to the active root zone or seeding bed through drip irrigation or precision placement in the active root zone could make small but potentially meaningful contributions to soil chemical, microbial and physical conditions in the rhizosphere and hence alter nutrient availability and ultimately AE in a fashion similar to soil amendments. HSs have also been shown to directly affect plant biochemical and physiological processes. Thus HSs have dual functions, described as 'the complexing pathway (soil)' and 'the biochemical (molecular) pathway' (Olaetxea et al., 2018). The greatest amount of research on HS effects on plant nutrition has focused on Fe, N and P and mechanistic explanations for the effects of HSs on plant growth and development and has been recently reviewed (Halpern et al., 2015; Zanin et al., 2019; Olaetxea et al., 2018).

Attempts to determine the biological and agronomic function of HSs are complicated by uncertainty with nomenclature and inconsistency in commercial labeling. The source and manner in which HSs are extracted, isolated and purified are known to profoundly affect HS composition and the particular physiological effects of HSs on plants that depend on the source, concentration and molecular weight of the humic fractions applied (Calvo et al., 2014; Garcia et al., 2016). As there is no standardized methodology to quantify the 'biologically active' humic fraction present in any product, and as there are no regulatory guidelines to guarantee the composition, it is impossible to generalize results from one study to another. There is currently a critical need to develop standardized protocols for the chemical and biological functional characterization of HS (Olaetxea et al., 2018), without which the efficacy of any product claiming HS function cannot be inferred from general principles but can be derived only by rigorous and agronomically relevant experimentation with that specific product.

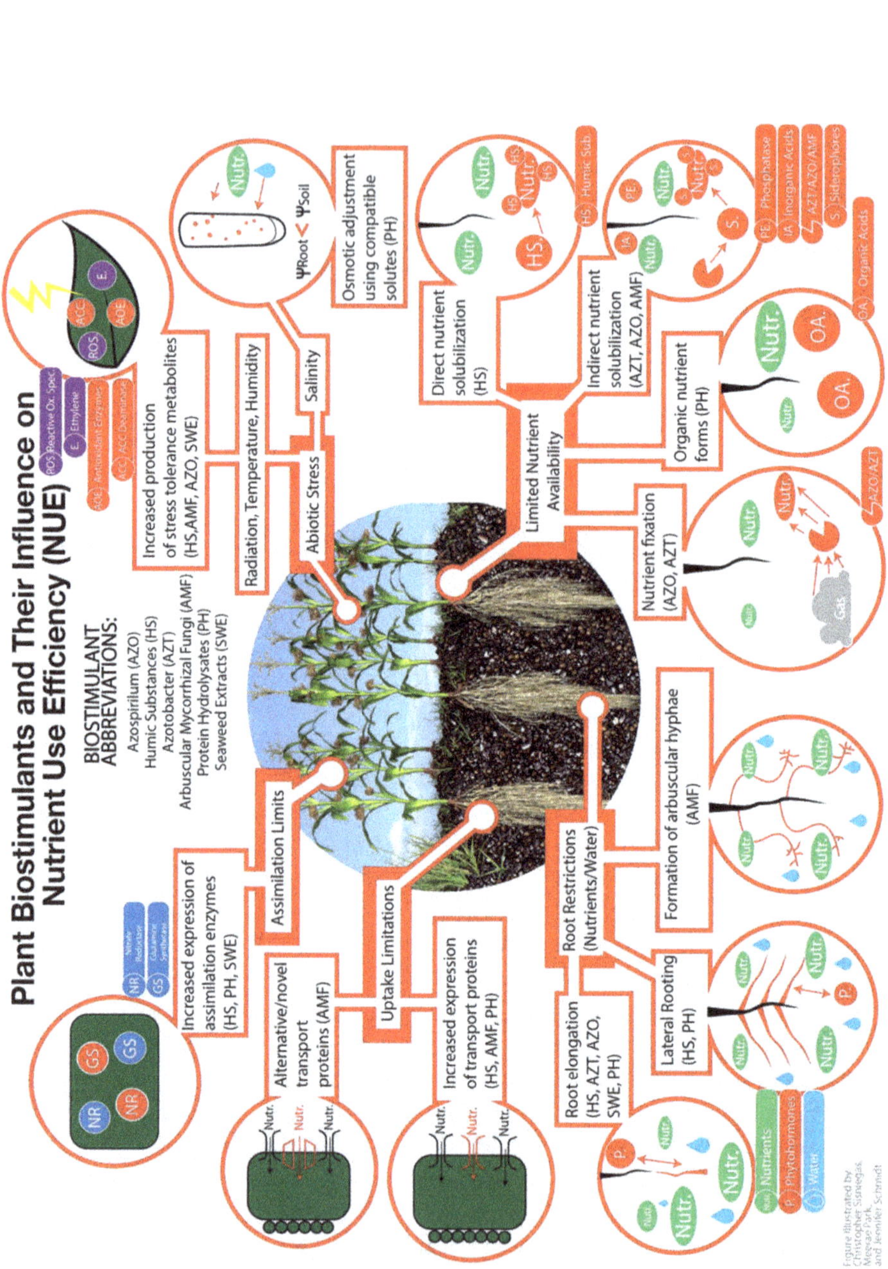

Plant Biostimulants and Their Influence on Nutrient Use Efficiency (NUE)
BIOSTIMULANT ABBREVIATIONS:
Azospirilum (AZO)
Humic Substances (HS)
Azotobacter (AZT)
Arbuscular Mycorrhizal Fungi (AMF)
Protein Hydrolysates (PH)
Seaweed Extracts (SWE)
ROS Reactive Ox. Spec.
E. Ethylene
AOE Antioxidant Enzymes
ACC ACC Deaminase
NR Nitrate Reductase
GS Glutamine Synthetase
Increased expression of assimilation enzymes (HS, PH, SWE)
Assimilation Limits
Increased production of stress tolerance metabolites (HS,AMF, AZO, SWE)
Radiation, Temperature, Humidity
Abiotic Stress
Salinity
Osmotic adjustment using compatible solutes (PH)
ΨRoot < ΨSoil
Alternative/novel transport proteins (AMF)
Uptake Limitations
Increased expression of transport proteins (HS, AMF, PH)
Limited Nutrient Availability
Direct nutrient solubilization (HS)
Indirect nutrient solubilization (AZT, AZO, AMF)
Organic nutrient forms (PH)
Nutrient fixation (AZO, AZT)
HS Humic Sub.
PE Phosphatase
IA Inorganic Acids
AZT/AZO/AMF
S Siderophores
OA Organic Acids
AZO/AZT
Gas
Root Restrictions (Nutrients/Water)
Formation of arbuscular hyphae (AMF)
Lateral Rooting (HS, PH)
Root elongation (HS, AZT, AZO, SWE, PH)
Nutr. Nutrients
P. Phytohormones
Water
Figure illustrated by Christopher Sisnegas, Meegan Park, and Jennifer Schmidt

Figure 1 ▶Representation of the various mechanisms by which the major classes of biostimulants are known to influence nutrient use efficiency in plants (NB. The figure is published under the terms of a Creative Commons Attribution 4.0 License (CC BY) (https://creativecommons.org/licenses/by/4.0/). The five main challenges or limitations that can negatively affect NUE are identified, and the consequent processes whereby biostimulants mitigate these stresses are illustrated. Abbreviations for key biostimulant categories are: PH - protein hydrolysates, HS - humic and fulvic substances, AMF - arbuscular mycorrhizal fungi, AZO - *Azospirillum*, AZT - *Azotobacter*, SWE - seaweed extracts. Key processes and their effects on agronomic efficiency (AE) and internal efficiency (IE) are described, commencing from the upper right of the diagram and proceeding in a clockwise direction. *Abiotic stress*, including high radiation or temperature and low humidity, causes oxidative damage and hence compromises carbon capture and thereby reduces shoot growth, restrict root activity and compromise nutrient demand and uptake. Alterations in carbon allocation may also alter within-plant nutrient distribution and nutrient utilization. HS, AMF, AZO and SWE mitigate abiotic stress damage through upregulation of protective mechanisms, while PHs mitigate salinity damage by stimulating compatible solute production. These processes influence both AE by altering plant demand and IE by altering nutrient allocation. *Limited nutrient availability* as a consequence of low solubility of elements including P, Mn, Fe, Zn or low absolute content (N) reduces plant nutrient acquisition, thereby lowering AE. HSs can increase availability through direct solubilization of insoluble nutrient forms, while AZT, AZO and AMF indirectly enhance availability by exudation of phosphatases, siderophores and soil acidification. PH stimulates organic acid production. AZO and AZT fix atmospheric N. *Root restrictions*, including changes in growth, architecture and activity, limit nutrient or water acquisition. HSs, AZT, AZO, SWE and PH alter root elongation, while HSs and PH alter lateral rooting and AMF enhance effective rooting volume through the formation of mycorrhizal networks. Each of these processes increases effective root volume and improved utilization of soil nutrients and enhance AE. *Uptake limitations* reduce the ability of plants to fully utilize soil nutrient resources. HSs, AMF and PH have been shown to increase the expression of transport processes, while AMF express alternative and novel transport processes. Both effects improve the effective utilization of soil nutrients and enhance AE. *Assimilation limits* reduce the efficacy with which acquired nutrients are utilized. Reduced assimilation may also cause feedback inhibition of nutrient uptake. HSs, PH and SWE have been shown to upregulate key nitrogen assimilatory enzymes, thereby improving IE.

2.1 Modes of action relevant to nutrient use efficiency

2.1.1 Influence of humic substances on plant growth

Any HS treatment that increases root or shoot growth has the potential to also increase short-term NUE as measured by AE by virtue of enhanced growth and hence enhanced use of available soil nutrient reserves. There is now a considerable body of evidence that has shown a benefit of a wide variety of HS materials on plant growth, both shoot and root (Rose et al., 2014; Yakhin et al., 2017; Abbott et al., 2018). Specifically, HSs have been reported to enhance some aspect of growth in diverse species of plants (Calvo et al., 2014), including important agricultural crops such as soybean (Chen et al., 2004), wheat (Garcia-Mina et al., 2004), maize (Canellas et al., 2002; Anjum et al., 2011) and tomatoes

(Sanchez et al., 2005). Excellent reviews of the function of HSs are provided in the reviews of Canellas et al. (2015) and Nardi et al. (2016).

The widely reported effects of HSs on plant growth have been ascribed to the presence in HSs of auxin and auxin-like functional groups that stimulate root growth and root branching. Auxin-like effects would be highly dependent on the source of HSs; HSs derived from biologically young materials (composts, earthworm castings) potentially contain plant hormones or biological analogs, while older HS sources (peats, coals, lignites, etc.) are unlikely to contain these molecules. Responses to HSs isolated from brown coal (e.g. lignite, leonardite, subbituminous coals) were observed to be less than those isolated from peats, composts or vermicomposts (Canellas et al., 2002). Root growth (increased root elongation and root biomass) and yield are the most commonly reported effects of HSs on plant growth (Garcia et al., 2016; Jannin et al., 2012; Nardi et al., 2002).

Improvements in plant growth and hence NUE may also occur as a result of HS effects on enhanced tolerance to stress. Anjum et al. (2011) conducted an experiment to establish the role of fulvic acid application in improving the performance of maize (*Zea mays* L.) under drought stress. The authors showed that the application of fulvic acid improved the performance of maize under stress, supporting the assumption that HSs may have auxin-like properties and maintain the chlorophyll contents of plants under stress conditions. However, no changes in nutrient uptake were observed. HSs are reported to have a significant influence on secondary metabolism and stress alleviation in plants (De Pascale et al., 2017), directly altering NUE. These HS effects may be the result of induced or inhibited protein synthesis, increased enzyme activities and induced morpho-functional changes in plants.

The stimulation of shoot and particularly root growth by HSs will often result in altered nutrient uptake and changes in NUE through a growth-and-demand effect or by enhancing the volume of soil that can be explored by the plant. Whether enhanced root exploration is beneficial to plant nutrient uptake will be determined by the spatial scarcity of nutrients. Thus, plants growing in nutrient solution or in highly fertilized and well-irrigated agricultural fields are unlikely to benefit from enhanced root growth, while plants growing in soil with intermittent and disperse nutrient availability may benefit from increased root exploration and root renewal. Whether these growth enhancement effects of HSs on improved AE are agronomically relevant or persistent will depend on the presence of adequate nutrients in the growth media to support the enhanced soil exploration or increased crop demand.

The vast majority of research demonstrating an effect of HSs on plant growth has been conducted in solution culture, culture media and occasionally in potted plants grown under controlled conditions and only rarely in field conditions (Rose et al., 2014), and very few studies have considered the effects of HSs on the interaction between environmental stress and NUE (Olk et al., 2018). While results from 'lab' studies may have relevance for solution- and media-grown crops, it is not evident that positive results under highly controlled conditions are predictive of those in the field, given the overwhelming abundance of native organic matter and HSs generated by crop and soil organism turnover processes as well as profound differences in environmental conditions and the availability of soil nutrients in fields in contrast to 'lab' studies. Indeed, several rigorous field studies have demonstrated that HS materials with the evidence of efficacy from controlled growth studies fail to show efficacy under field trials (Hartz and Bottoms, 2010; Mahoney et al., 2017; Lyons and Genc, 2016; Rose et al., 2014)

2.1.2 Plant nutrient uptake, nutrient solubility and utilization

In several older references (Burk et al., 1931) the 'biostimulant' effect of HSs is ascribed to its effect on enhanced root growth and the solubilization of poorly soluble Fe in the root zone. HSs applied at low rates, however, have also been shown to have direct effects on general plant biochemicals, molecular pathways or physiological processes as well as on root growth and development, plant productivity (Rose et al., 2014), nutrient uptake (Billard et al., 2014; Sanchez et al., 2005; Sánchez-Sánchez et al., 2006), chlorophyll content (Pinton et al., 1999; Anjum et al., 2011) and gene expression (Aguirre et al., 2009; Quaggiotti et al., 2004; Billard et al., 2014).

Chen et al. (2004) investigating the effects of fulvic and humic acids on nutrient solubility and uptake observed an increased chlorophyll content and enhanced nutrient uptake; however, the authors suggested that the growth enhancement of plants could be attributed to the maintenance of nutrients in solution at sufficient levels by the formation of HS-nutrient complexes. Indeed numerous studies have demonstrated that HSs, present in the nutrient solution or rooting media can not only enhance the stability and solubility of Fe in solution but stimulate Fe uptake and within-plant Fe mobility from both root and foliar-applied materials (Garcia et al., 2018; Garcia-Mina et al., 2004). While the majority of research on HSs and plant nutrition has focused on Fe, the stimulatory effects of HSs on NUE may also result from the maintenance of nutrients in solution at effective concentrations; this is relevant for any micronutrient with the ability to form complexes with HSs (e.g. Mn, Fe, Cu and Zn). The ability of HSs to bind micronutrients in a plant-available form, thus reducing precipitation reactions, will enhance crop recovery of soil reserves and hence increase AE.

In the recent mini-review, Zanin et al. (2019) summarize the multiple mechanisms of HS stimulation of Fe nutrition that have been identified as: (1) HSs naturally present in the soil or added to the root media may increase the pool of available plant Fe through the formation of HS-Fe complexes and alteration of normal soil-Fe precipitation and oxidation reactions. Stabilized Fe-HS has been shown to enhance Fe availability mediated by plant rhizosphere processes. (2) HS-Fe complexes may also act directly on plant physiological processes by enhancing within-plant transport and delivery to developing tissues and by enhancing the activity of identified root and shoot Fe-uptake mechanisms including Fe(III)-chelate reductase (LeFRO1), Fe transporter genes (*LeIRT1* and *LeIRT2*) (Tomasi et al., 2014) and leaf transporters (*CsFRO1*, *CsIRT1*, *CsNRAMP*) (Zanin et al., 2015). (3) HS materials have also been shown to induce changes in root morphology, which, depending upon existing soil conditions, may enhance not merely Fe uptake but the uptake of all nutrient elements (Zanin et al., 2019; Olaetxea et al., 2018). In a recent paper (Canellas et al., 2019) it was also suggested that (4) vermicompost-derived HA applications may have disturbed nutrient sensing, resulting in enhanced nutrient accumulation and improved growth, as indicated by the differential expression of *zmTOR*, a highly conserved gene involved in the growth and control of cell proliferation.

Studies have shown that humic acid application increases plant nutrient uptake, especially N (Jannin et al., 2012; Pinton et al., 1999), by increasing the expression of genes involved in nitrate transport (Jannin et al., 2012). Studies have also shown the effects of humic acid extract on nitrate uptake, expression of two major nitrate transporters (BnNRT1.1 and BnNRT2.1) and nitrate reductase (NR) activity in shoots and roots where an enhancement of gene expression related to N metabolism and more particularly on identified genes encoding proteins involved in uptake and assimilation was reveled. In these studies, however, no increase in nitrate accumulation in shoot or root tissues was detected, suggesting that all supplemental N taken up by the roots of treated plants was directly assimilated in relation to the growth rate without being stored. In contrast, Quaggiotti et al. (2004) showed that the nitrate content of leaves increased by 50% in seedlings treated with HSs in comparison with control plants. The authors suggest that HSs may exert direct effects on gene transcription in roots, and this may implicate a more efficient translocation of nitrate to its metabolic sinks and reflect a better internal nitrogen utilization efficiency (IE).

Palumbo et al. (2018) reported the positive effects of HSs extracted from different sources on maize plant growth and nutrient content. The authors found significant increases in leaf and root dry biomass promotion as well as enhanced nutrient content and higher concentrations of macro- and micro-elements in both leaves and roots compared to the untreated plants. Increases in several key nitrogen metabolism pathways were also observed. The paper by Palumbo et al. (2018) illustrates some of the challenges of biostimulant research. While the effects on plant growth were significant, they were limited to a very short solution culture treatment period for corn and utilized application rates that are not easily reconciled with possible field application rates and solution nutrient concentrations that differ markedly from field conditions. Remarkably, consistent increases in all measured nutrient concentrations and contents were observed for anion and cation nutrients and B. For Zn, Mn and Cu the increment in nutrient concentration resulted in substantially greater-than-required concentration values to concentrations that approach toxicity. Results such as these are hard to reconcile against established nutrient principles and may indicate that the HS effects were direct or an indirect effect of decoupling of normal constraints on crop growth and development, as suggested by Canellas et al. (2019).

2.2 Conclusion

There is now abundant evidence that HSs can alter plant nutrient uptake by processes that include the alteration of soil physical and chemical properties, stimulation of shoot and root growth that alter nutrient availability, and uptake and efficiency of internal use of nutrients. While the positive effects of HSs on NUE are well demonstrated and mechanistic explanations for these effects have been provided, there remains much that is unknown, and our ability to predict the soil and environmental conditions under which a benefit from HS application will be realized is not yet satisfactory. HSs have been shown to improve both AE and IE aspects of NUE. Given the abundance of naturally occurring HSs in soils with even modest levels of native organic matter and noting the often-superior efficacy of HSs prepared from young naturally occurring organic source materials (Garcia et al., 2019; Olaetxea et al., 2018), it remains logical to conclude that HS effects would diminish as native soil organic matter increases. While anecdotal information supports this hypothesis, there are several demonstrations of HS efficacy in OM-rich soils that would not be expected. These results may imply the presence of unknown bioactive molecules in some HS products that affect plant response even in a background of abundant, naturally occurring humic materials.

3 Microbial biostimulants

A limited number of living bacterial and fungal taxa can be classified as biostimulants under the EBIC definition, including arbuscular mycorrhizal fungi (AMF) and nitrogen-fixing bacteria such as *Azotobacter* sp. and *Azospirillum* sp. These microorganisms have a long history of use in agriculture and can increase NUE through a number of mechanisms associated with greater nutrient acquisition and plant growth. Rhizobia, which fix nitrogen in a symbiotic relationship with legume hosts, are not reviewed here since they have already been covered in extensive detail. Numerous other bacteria and fungi have been investigated as potential plant-growth-promoting microorganisms (PGPMs) that may increase plant NUE but do not fit the EBIC definition of biostimulants and are thus considered beyond the scope of this section. For instance, bacterial genera containing species known to produce phytohormones or siderophores solubilize phosphate, suppress pathogens or promote plant growth through other pathways include *Bacillus*, *Burkholderia*, *Enterobacter*, *Klebsiella*, *Microbacterium*, *Micrococcus*, *Pantoea* and *Pseudomonas* (Fuentes-Ramirez and Caballero-Mellado, 2006; Chennappa et al., 2016b). Reviews of these PGPM and their use as biofertilizers can be found elsewhere (Nakkeeran et al., 2006; Fuentes-Ramirez and Caballero-Mellado, 2006; Bhattacharyya and Jha, 2012; Hayat et al., 2010). It is worth noting that the limited list of species identified as potential biostimulants using culture-dependent methods will likely be greatly expanded with next-generation sequencing methods that facilitate the scanning of microbial genomes for relevant metabolic pathways.

3.1 Arbuscular mycorrhizal fungi (AMF)

AMF have long been recognized as plant symbionts that can contribute to productivity and NUE. The majority of land plants (70–90%) can establish symbiotic relationships with this monophyletic group (Berruti et al., 2015), which has very limited host specificity (van der Heijden et al., 2015), but the symbiosis does rely on the exchange of specific signaling molecules (e.g. strigolactones, Myc factors) and does not extend to certain agronomically important plant taxa such as the Brassicaceae (Parniske, 2008).

Interest in mycorrhizae as biostimulants is more recent than knowledge of their existence, gaining traction in the 1980s (IJdo et al., 2010). Originally based on *in vivo* cultivation with an appropriate host under nonsterile conditions, production methods for mycorrhizal biostimulants have now advanced to *in vitro* propagation under axenic conditions such as using root organ cultures (Vosátka et al., 2012). Recently, next-generation sequencing has facilitated the understanding of mycorrhizal genes involved in symbiosis and nutrient transfer (Tisserant et al., 2013). This is increasingly supporting the screening for potentially more beneficial novel strains (Igiehon and Babalola, 2017), investigation of factors influencing mycorrhizal community composition in the soil (Thonar et al., 2012) and verification of the establishment and persistence of inoculated strains over time (Pellegrino et al., 2012; Berruti et al., 2016). This knowledge presents great promise to improve inoculant formulations and effectiveness in the future. In recent decades, the use of AMF biostimulants has also expanded beyond food crops and gained traction in horticultural applications including the production of ornamental plants (Bianciotto et al., 2018).

The understanding of mycorrhizal biostimulants has also become more nuanced since the first crude inoculum-based products were developed. Rather than providing universal benefits to all host species across all environments, inoculation with AMF is now understood as the first step in establishing a resource-trading symbiosis with variable outcomes depending on the environmental context. Understanding factors regulating establishment, persistence and benefits of mycorrhizal biostimulants continues to pose a challenge and may require integrating an understanding of mycorrhizal ecology previously underemphasized in biofertilizers (Rodriguez and Sanders, 2015). Specifically, the knowledge of ecological interactions between inoculant species and native mycorrhiza (Janoušková et al., 2017), habitat niches available for colonization and priority effects controlling community establishment can be used to predict and improve the success of inoculation (Verbruggen et al., 2013).

Both single-species and multi-species mycorrhizal biostimulants are commercially available. Single-species formulations, particularly utilizing the generalist species *Rhizophagus irregularis*, *R. intraradices* and *Funneliformis mosseae*, are more common than formulations combining multiple species (Berruti et al., 2015). However, a recent meta-analysis of 902 studies revealed that the influence of mycorrhizal diversity on plant performance depends on taxonomic resolution, with family-level but not species-level diversity positively correlated with plant performance (Yang et al., 2016). Furthermore, diverse inoculants may be more effective at improving plant nutrition when host plants are simultaneously exposed to multiple abiotic stresses (Crossay et al., 2019). The effectiveness of single-strain and diverse inoculants is also related to the type of propagule (spores, extraradical hyphae or infected roots) to be used, as mycorrhizal genera vary in their ability to colonize roots from different propagules (Klironomos and Hart, 2002). *Acaulospora*, *Glomus*, *Gigaspora* and *Scutellospora* isolates are capable of colonizing roots from spores, whereas root fragments are effective propagules only for the former two genera. In addition, not all mycorrhizal genera can be produced under axenic conditions (Vosátka et al., 2012), and it is still common to use a crude inoculum formation containing trap plant roots, the surrounding soil and any included mycorrhizal propagules (Berruti et al., 2015) when developing multi-species inoculants.

3.2 Azotobacter *and* Azospirillum

Azotobacter application in agriculture is well established from a historical perspective. *A. chroococcum* has been used as a biofertilizer nearly since it was first described in 1901 (Chennappa et al., 2016b). *A. agilis*, *A. vinelandii* and *A. beijernckii* were discovered shortly thereafter in 1901–1904 and have also been investigated for potential use as biostimulants (Wani et al., 2013). Seed treatments based on *A. chroococcum* (under the name 'azotobacterin') at a density of 4–5 × 10^{10} cells ha^{-1} in a peat-nutrient formulation were widely used in the USSR in the early 20th century (Rovira, 1989; Brown, 1974).

Azospirillum spp. are gram-negative rod- or spiral-shaped bacteria (Ferreira et al., 2019). Associative N fixation can occur with the roots of a number of grasses, but *Azospirillum* spp. are also capable of forming endophytic relationships with sugarcane (Tejera et al., 2005). *Azospirillum* is increasingly used in commercial biostimulants, especially for maize (Fuentes-Ramirez and Caballero-Mellado, 2006). Peat and vermiculite formulations of *Azospirillum* sp. have a relatively long shelf life and can maintain populations of 10^7 cells g^{-1} product that successfully inoculate the target crop after 4 months or 44 weeks, respectively (Nakkeeran et al., 2006). Powder- and liquid-based formulations of *Azospirillum* sp. are most common and can be applied to seeds, transplant roots or the rhizosphere (Choudhury and Kennedy, 2004; Mehnaz, 2016).

Azospirillum spp. are a more recent addition to the group of microorganisms used as biostimulants than AMF or *Azotobacter*. *A. brasilense* was originally discovered in association with the tropical grass *Digitaria decumbens* in the second half of the 20th century (Berg et al., 1980). Extensive field inoculation experiments were conducted on a variety of crops in the 1980s and 1990s, with reported yield increases of 5–30%, leading to substantial interest in this genus as a biostimulant in Latin America in more recent years (Fuentes-Ramirez and Caballero-Mellado, 2006).

3.4 Modes of action relevant to nutrient use efficiency

3.4.1 Arbuscular mycorrhizal fungi

Mycorrhizal inoculant contributions to plant nutrition include both direct effects such as provisioning of mineral nutrients in exchange for plant-derived carbon (Casieri et al., 2013) and indirect pathways such as improved abiotic stress tolerance and protection from pathogens (Bianciotto et al., 2018; Vosátka et al., 2012). AMF enhance the plant acquisition of nutrients, including P, N, S and numerous micronutrients in most soil types (Berruti et al., 2015) and additionally of K, Ca and Mg in acidic soils (Clark and Zeto, 2000).

Direct effects of AMF on plant P status are mediated through a mycorrhizal-specific subset of phosphate (Pi) transporters in the Pht1 family (Berruti et al., 2015), which also includes nonmycorrhizal transporters present at the root–soil interface. Mycorrhizal-specific Pi transporters were first identified in potato (StPT3) (Rausch et al., 2001) and were subsequently characterized in tomato (LePT4) (Nagy et al., 2005), *Medicago truncatula* (MtPT4) (Harrison et al., 2002), rice (OsPT11) (Paszkowski et al., 2002) and other plants. Such AMF-specific transporters are localized to the periarbuscular membrane and allow the transfer of Pi from mycorrhizal hyphae to plant cortical cell cytoplasm in colonized roots. While their role is conserved across plant species, the molecular structure of these transporters varies, as, for example, in the non-orthologous transporters of rice (OsPT11) and potato (StPT3) (Paszkowski et al., 2002).

Similarly, mycorrhizal effects on N acquisition occur mainly via the transfer of NH_4^+ from the periarbuscular space to the cytoplasm of cortical cells through transporters of the AMT2 family, which are specifically induced by mycorrhizal symbiosis (Breuillin-Sessoms et al., 2015). Although transporters have been identified in tomato (*Solanum lycopersicum*) (Ruzicka et al., 2012) and *Lotus japonicus* (Guether et al., 2009) that are located exclusively in arbuscule-containing cells, other proteins involved in N transport during mycorrhizal symbiosis are inducible by AMF but also expressed in non-colonized roots (Casieri et al., 2013). Plant N nutrition is improved mainly through the transfer of NH_4^+, but induction of NO_3^- transporters and increased uptake of organic N have also been shown (Casieri et al., 2013).

Split-root experiments have shown that AMF improve plant S status in the presence of sulfate or the amino acids cysteine or methionine (Allen and Shachar-Hill, 2009). Rather than mycorrhizal-specific transporters, effects on plant S nutrition appear to proceed via the induction of existing transporters (Casieri et al., 2013). Pathways responsible for the improvement of plant micronutrient nutrition have not been as fully described as those for P, N and S, this may change with the recent publication of the *R. irregularis* genome and first characterization of metal transporters (Tamayo et al., 2014).

Mycorrhiza improves plant NUE by increasing yields and acquisition of macro- and micronutrients. Mycorrhizal fungi present in the soil colonize plant roots by entering intracellular spaces and then forming differentiated branched structures called arbuscules in root cortical cells (Harrison et al., 2002). Hyphae extend beyond plant roots and forage for soil nutrients, which can be used to meet fungal needs or transported back to host roots to be exchanged for photosynthate. The periarbuscular membrane, an extension of the plasma membrane, separates arbuscules from the cytoplasm and serves as a primary site for nutrient exchange, the primary mode of action of AMF relevant to plant nutrition and NUE (Casieri et al., 2013).

Inoculation with viable propagules of high-performing strains can reliably improve colonization in diverse field soils varying in nutrient content and composition of native microbial communities (Köhl et al., 2016). A review of 164 inoculation experiments on a variety of host plants found improved yield in 84% and plant nutrition in 92% of studies that measured these parameters (Berruti et al., 2015). That review considered plant nutrition as a single factor with a binary result (improved/not improved by colonization), making it difficult to discern nutrient-specific outcomes and effect sizes, but other studies have investigated mycorrhizal impacts on yield and plant nutrition in more detail. For example, yield increases of 9.5% have been observed for a single host plant (potato) inoculated with a single strain (*R. irregularis* DAOM 197198) in 231 field trials over 4 years (Hijri, 2016). With regard to micronutrient nutrition, meta-analyses have shown positive effects on Zn and Cu, although results are more variable for Fe and Mn. Mycorrhizal colonization improves plant Zn content of aboveground and belowground tissues as well as fruits and seeds, mediated most significantly by soil texture (Lehmann et al., 2014). Meta-analysis also revealed that plant Cu nutrition was improved by AMF by an average of 29%, but Fe nutrition was improved only in intermediate-duration studies lasting 56–112 days (Lehmann and Rillig, 2015). Mn content was increased only in herbaceous species, with mycorrhizal colonization decreasing plant Mn content by an average of 4% across all studies. Average effect sizes reported in these meta-analyses obscure substantial variability due to plant tissue of interest, plant type and soil texture, although moderating variables such as the number of strains in the inoculant, soil P status and soil pH were not found to significantly affect the outcomes of symbiosis (Lehmann and Rillig, 2015).

The establishment of mycorrhizal symbiosis and the effects of AMF on plant nutrition are context-dependent (Hoeksema et al., 2010). Host plant functional group and soil fertility influence the outcome of colonization: C_4 grasses and non-N-fixing hosts respond more positively than C_3 grasses or N-fixing plants (Hoeksema et al., 2010), and colonization is reduced under high soil P and promoted in response to N starvation (Nouri et al., 2014). Other soil properties such as K, S, Mg, Ca and Fe levels apparently do not affect colonization (Nouri et al., 2014), although pH, Mn and Zn are known drivers of mycorrhizal community composition in natural systems (Alguacil et al., 2016) and may thus affect the effectiveness and persistence of biostimulants.

3.4.2 Azotobacter *and* Azospirillum

Azotobacter increases plant growth by fixing nitrogen, solubilizing phosphate and producing beneficial compounds such as phytohormones, siderophores and pathogen-suppressing antibiotics (Chennappa et al., 2017). Strains of *Azotobacter* sp. have also been found to solubilize potassium, although this is not usually considered the primary mechanism of growth promotion as potassium is rarely the most yield-limiting factor (Verma et al., 2017). Non-symbiotic nitrogen fixation by free-living or endophytic *Azotobacter* does not provide N to the plant host in exchange for carbon, as with symbiotic rhizobia but increases rhizosphere N availability indirectly.

Phosphorus solubilization occurs via the release of inorganic acids and phosphatase enzymes (Chennappa et al., 2017). *Azotobacter* spp. also produce phytohormones including indole acetic acid (IAA) and gibberellic acid (Chennappa et al., 2016a), at rates of 2–33 μg mL^{-1} and ~0.6 μg mL^{-1}, respectively (Chennappa et al., 2014). IAA and gibberellin released due to seed treatment with *Azotobacter* sp. increase plant nutrient uptake by promoting root elongation and thus expanding the root surface area (Brown, 1974). Siderophores, low-molecular-weight organic compounds produced by microorganisms to chelate iron, increase iron availability to plants when it is limited in the soil. *A. vinelandii*, for example, is known to produce at least five different siderophores (Chennappa et al., 2017).

Azospirillum spp. contribute to plant productivity and NUE primarily through nitrogen fixation (Berg et al., 1980). This genus also solubilizes phosphate, synthesizes siderophores (Ferreira et al., 2019) and releases phytohormones, including auxins, gibberellins and cytokinins that can increase root growth, enhancing plant nutrient uptake (Fuentes-Ramirez and Caballero-Mellado, 2006). An additional mechanism contributing indirectly to plant NUE is the production of ACC deaminase, which degrades an ethylene precursor and alleviates ethylene-mediated pathways that normally reduce plant growth in response to abiotic stresses (Fuentes-Ramirez and Caballero-Mellado, 2006).

Azotobacter biostimulants have been found to improve nutrient uptake and productivity in rice, wheat, maize, barley, oat, sunflower, coffee, tea and other crops (García-Fraile et al., 2015; Mahmood et al., 2016). Inoculation with *Azotobacter* sp. can increase estimated N uptake by rice by around 11–15 kg ha^{-1}, resulting in yield increases of 7–20% in the field (Choudhury and Kennedy, 2004). However, substantial variation in N-fixing ability and growth promotion is observed even among the isolates of the same species, *A. vinelandii* (Sahoo et al., 2014). Furthermore, N fixation by *Azotobacter* requires ~1 g of a carbon source for every 10 μg N fixed (Tejera et al., 2005), so available soil carbon may ultimately impose limits on the N contributed by this genus. In addition to their potential as single-species inoculants, *Azotobacter* spp. are often applied in combination with other microorganisms that are thought to possess complementary plant-growth-promoting traits, such as AMF and P-solubilizing or N-fixing bacteria (see Chennappa et al., 2017 and references therein).

Similarly, *Azospirillum* spp. have been shown to increase nutrient uptake and yield in sorghum, wheat, maize and rice (Kannan and Ponmurugan, 2010; Choudhury and Kennedy, 2004). Inoculation of basmati rice seedlings with *A. brasilense* and *A. lipoferum* provided 47–58% of N needs and increased shoot dry weight and yield relative to an uninoculated control, although the amount of yield increase depended on host genotype (Mirza et al., 2002). In a review of 20 years of *Azospirillum* field studies, Okon and Labandera-Gonzalez (1994) found that inoculation was successful 60–70% of the time, leading to yield increases of 5–30%. They noted that inoculation success appears to be determined early on in plant development and is strongly dependent on the ability of the biostimulant formulation to maintain large, viable bacterial populations. The range of productivity gains for a variety of crops found in that review is consistent with the rice-specific value of 22% yield increase described in a more recent synthesis (Choudhury and Kennedy, 2004). In addition to its individual effects on plant growth and NUE, co-inoculation of *A. brasilense* enhances the effectiveness of the legume–rhizobia symbiosis, increasing nodulation by *Rhizobium etli* and *Rhizobium tropici* in common bean (Fuentes-Ramirez and Caballero-Mellado, 2006).

3.5 Conclusion

Among the described categories of biostimulants recognized under current EBIC regulations, AMF, rhizobia, *Azotobacter* sp. and *Azospirillum* sp. are the best researched and are the biostimulant products that most unambiguously fulfill the EBIC requirements that a biostimulant 'enhance nutrient use efficiency'. The effects of both AMF, *Azotobacter* and *Azospirillum* on NUE are restricted to improving the availability of limited soil resources and thus enhancing agronomic efficiency (AE), and very little evidence exists to suggest that these organisms alter within-plant nutrient transport, the efficiency with which acquired nutrients are converted to biomass or the physiological or internal use efficiency (IE). AMF, *Rhizobia*, *Azotobacter* and *Azospirillum* may also indirectly enhance the NUE of all nutrient elements by reducing constraints to production (nutrient, water or other stress) and thus enhancing plant growth and resource use efficiency. With the exception of the effects caused by nitrogen fixation, any NUE benefit from the use of these microbial biostimulants can occur only if there is a source of the nutrient in the soil or water and no benefit can be expected in soils that are wholly depleted of an essential plant nutrient.

While AMF, *Rhizobia* sp., *Azotobacter* sp. and *Azospirillum* sp. have a long history of use in agriculture and their use and benefits for productivity have been clearly demonstrated, questions remain regarding the interactions of these microorganisms with other soil biota. It is also unknown if applications of other non-microbial biostimulants influence the behavior of AMF, *Rhizobia* sp., *Azotobacter* sp. and *Azospirillum* sp. or whether the efficacy of some biostimulant products is indeed mediated by effects on AMF, *Rhizobia* sp., *Azotobacter* sp. or *Azospirillum* sp.

4 Seaweeds and algae

Seaweeds have been used in agriculture for centuries and were among the first source materials used to produce agricultural biostimulants (Craigie, 2011; Khan et al., 2009). Since the introduction to commercial agriculture, seaweed extracts (SWEs) have been subject to the extensive characterization of critical components and potential mode(s) of action (Abbott et al., 2018; Shukla et al., 2019; Halpern et al., 2015; Ertani et al., 2018). The determination of SWE composition and the identification of biological effects are, however, greatly complicated by the substantial diversity in sources of available SWE and the manner in which SWE extracts are prepared (Khan et al., 2009). Even within seaweed species, studies have found a widely varying composition of amino acids, secondary/exogenous metabolites, hormones and minerals and consequently have observed varied effects on plant growth, stress tolerance and mineral acquisition. Recognizing this diversity of composition and effect, several recent reviews have appropriately focused on the analysis of the composition and function of individual species of seaweeds rather than attempting to provide a unified explanation for the function of SWEs as a whole (Shukla et al., 2019; Ertani et al., 2018). The complexity and varied nature of plant responses to SWEs was nicely illustrated in the recent work of Ertani et al. (2018) who examined six commercial SWEs from *Ascophyllum nodosum* and *Laminaria* sp. While all SWEs tested resulted in varied but significant increases in root growth and esterase activity, the effects on mineral concentrations and sugar contents were inconsistent (Ertani et al., 2018).

Despite the wide diversity in species, origin and preparation for various SWE products, it is widely observed that SWEs enhance root growth and less frequently shoot growth, with consequent effects on nutrient content and concentration and hence AE-NUE. Given the effects on root and shoot growth, it is frequently unclear if the effects of SWEs on NUE are a consequence of increased nutrient demand, increased root exploration, increased activity of root transport processes and/or more efficient internal utilization of nutrients.

4.1 Modes of action relevant to nutrient use efficiency

4.1.1 Root growth

SWEs have long been observed to enhance root growth with explanations, including the beneficial effects of SWEs on soil structure by virtue of improved aggregate formation, hydrophilic characteristics of alginates and other gel-like components and positive effects on the soil microbial composition and function (Shukla et al., 2019). These beneficial effects are frequently greater when applied to young plants or during periods of drought stress and when applied as root drenches or through targeted deposition in the root rhizosphere through drip irrigation (Khan et al., 2009). Given the small volumes of SWEs typically applied under commercial field conditions and recognizing that SWEs are highly biologically available and hence would rapidly be degraded (Craigie, 2011), it is perhaps not surprising that applications need to be carefully timed and precisely placed to achieve positive effects. While the SWE would not be expected to persist long in field soils, the benefits of the enhanced root growth and improved establishment of tolerance of stress may have long-term benefits.

Investigations of the mechanism by which SWEs improve root growth have largely been conducted under highly controlled environment assays, a condition that effectively eliminates the possibility of benefits to soil health and soil structure of soil microbial dynamics. Nevertheless, under these conditions, benefits to root production are frequently observed, suggesting that SWEs positively influence endogenous phytohormone production (Craigie, 2011; Ertani et al., 2018; Shukla et al., 2019). Positive effects from SWE application are not always seen, which may be a consequence of the concentrations used; dose dependency is also a characteristic of many phytohormone responses (Wally et al., 2013). Despite this apparent contradiction, a majority of literature supports that SWEs increase root growth across various phenological stages (Jannin et al., 2013; Sharma et al., 2014).

Because of the similarity in plant response to SWEs, with responses seen as a result of phytohormone applications, it has long been suggested that SWEs function by virtue of the phytohormone content within SWE extracts (Craigie, 2011; Khan et al., 2009; Kurepin et al., 2014). More recent examinations have suggested, however, that at the rate of SWEs used and given the frequently modest levels of phytohormone concentrations it is unlikely that phytohormone content in SWEs is responsible for any observed phenotypic changes; rather, it appears likely that SWEs stimulate phytohormone pathways through a process that is not well understood (Shukla et al., 2019; Ertani et al., 2018), perhaps by virtue of other molecules within SWEs that interact with endogenous phytohormone production.

4.1.2 Nutrient transport

SWE-induced root growth responses are likely a *primary* driver of enhanced AE as a consequence of increased root absorptive surface area and hence better utilization of soil reserves; however, it has also been shown that some formulations of *Ascophyllum nodosum* can enhance nutrient uptake through mechanisms other than root stimulation (Ertani et al., 2018). Indeed, there are many reports of altered nutrient content and nutrient concentration in experiments conducted in well-stirred solution culture with the abundant provision of soluble nutrients, conditions under which root growth would not be expected to have a substantial effect on nutrient acquisition.

Modulation of transporter gene expression has been observed in a number of experiments (Jannin et al., 2013; Billard et al., 2014). On occasions, however, changes in gene expression have not resulted in predictable changes in tissue nutrient concentrations.

Despite no significant differences in shoot N concentration of *Brassica napus*, microarray analysis revealed that both BnNRT1.1 and BnNRT 2.2 were significantly upregulated by 68 fold and 16 fold, respectively, relative to control (Jannin et al., 2013).

While there are a large number of reports of SWE effects on nutrient concentration and content, many such reports have not adequately controlled the interactive effects of enhanced growth on nutrient dilution, both within plant and growth media on the final measured nutrient concentration. The frequent reports of simultaneous increase in the concentration (often incorrectly reported as content) of a diverse range of nutrients, both monovalent and divalent and cationic and anionic for which there would have to be simultaneous upregulation of a wide diversity of transporter and assimilation processes, suggest that many observed effects of SWEs on NUE are likely secondary in nature occurring as a consequence of growth promotion.

4.1.3 Nutrient assimilation and storage

Nitrate reductase (NR), putatively a key rate-limiting step in nitrogen assimilation, is increased by SWE application (Billard et al., 2014; Chouliaras et al., 2009; Jannin et al., 2013; Rouphael et al., 2017c, 2018; Sabir et al., 2014; Zodape et al., 2011; Suzuki et al., 2010). Increases in NR activity, however, do not always result in predictable changes in plant nitrogen status, perhaps due to growth-induced nutrient dilution. SWEs have also been shown to increase the expression of glutamine synthetase (GS1) in spinach (Fan et al., 2013), which plays an important role in the regulation of plant nitrogen cycling (Masclaux-Daubresse et al., 2010).

A large body of research has established that SWEs can effectively mitigate plant stress caused by salinity, heat or water deficit (Shukla et al., 2019). Enhanced plant senescence caused by stress will reduce the duration during which nutrient remobilization can occur and hence will reduce physiological NUE (Uauy et al., 2006; Erenoglu et al., 2011; Masclaux-Daubresse et al., 2010). Positive effects of SWEs on stress-induced senescence, which have been widely observed, will therefore have beneficial effects on internal NUE and realize a greater yield (Brown and Saa, 2015). This is a promising but underexplored use of SWEs.

4.2 Conclusion

SWEs have been clearly demonstrated to improve NUE under both field and controlled growth conditions. The most compelling mechanism by which this could occur is through stimulatory effects on root growth and thus enhanced exploration of the soil volume which would result in improved agronomic NUE. While SWEs also clearly influence the expression of nutrient transport and assimilation processes, there is an inconsistency between observed upregulation of these processes and tissue nutrient content, and it remains uncertain if these effects are direct or a consequence of gross stimulatory effects of SWEs on plant metabolism.

The greatest promise for SWE use to positively influence NUE in crop plants is likely to occur through the enhanced establishment of roots in seedlings and the mitigation of plant stress events that compromise root uptake and result in enhanced tissue senescence and hence compromised remobilization of nutrients. Since SWEs are likely to be rapidly broken down in the environment, there is an imperative to time the application of SWEs to mitigate these stressors and to place SWEs precisely in the root zone or as a targeted foliar application.

As is the case with many classes of biostimulants, there is dramatic diversity in source and preparation techniques utilized to prepare commercial SWE biostimulants; given this diversity, it is unlikely that a single unified explanation for the beneficial effects of SWEs will be identified, and hence there is an imperative that each commercial SWE undergo careful well-replicated field testing.

5 Protein hydrolysates

Protein hydrolysates (PHs) are recycled plant/animal waste products that may enhance plant growth and improve plant mineral nutrition. Given the diversity and inconsistency of source materials used for these products, validation of these products as a class faces considerable challenges. Wide variations in processing, application method, cropping systems use, management practices and analytical methodology further complicate this analysis. PHs can be divided into two general categories: animal (typically derived from connective tissues) and plant (typically derived from unharvested plant parts of by-products of agricultural processing). The use of agricultural and industrial by-products as a source of plant biostimulants has been proposed as a potentially important aspect of the circular-economy concept (Xu and Geelen, 2018). A distinguishing characteristic of PHs is their relatively high concentration of soluble proteins, peptides and both essential/non-essential and free or peptide-bound amino acids. Additionally, PHs frequently contain significant concentrations of carbohydrates but negligible quantities of mineral nutrients, phenols, phytohormones and other organic compounds (Colla et al., 2017a; Tejada et al., 2018; Vesela and Friedrich, 2009).

Despite the wide array of source materials and preparation methods, PH studies have generally shown that PHs can affect mineral nutrition in plants. Results are, however, frequently inconsistent, perhaps due to differences in experimental designs, crop, methodology, environmental conditions or PH composition. The variation in responses is observed in both types of PHs (i.e. animal and plant-derived) and across a wide range of application rate/method.

Regardless of the underlying cause(s), these 'outliers' should be seriously considered (Lucini et al., 2015; Cerdán et al., 2009; Tsouvaltzis et al., 2014).

Examples of enhancements of physiological processes from PH application are sparse and primarily limited to measurements of photosynthetic parameters such as chlorophyll concentration and quantum efficiency. PH interaction with photosynthetic parameters is not fully understood, given the variability in findings across studies, with approximately 40% of the surveyed studies finding no significant difference in chlorophyll concentration (Colla et al., 2014; Ertani et al., 2013; Grabowska et al., 2012; Lucini et al., 2015; Paradikovic et al., 2011; Rouphael et al., 2017a, 2018; Sestili et al., 2018).

5.1 Modes of action relevant to nutrient use efficiency

5.1.1 Direct effects of nutrient form in soils

Depending upon environmental conditions, plants can extract nutrients from variable forms in the soil. Several studies have postulated that organic nutrient forms present in PHs or resulting from PH application may explain the observed increases in NUE. To test this hypothesis, Teixeira et al. (2018) investigated the effect of isolates and combinations of cysteine, phenylalanine, glycine, glutamate on nitrate and total N uptake. In both PH and amino acid combinations, they found results similar to those reported earlier, notably no consistent effect on leaf NR activity but an increase in leaf nitrate concentration, suggesting that amino acids may be involved in the modulation of assimilation/storage networks observed with PH application. This relationship is still inconclusive, given that PH and amino acid isolates are yet to be tested on full cropping cycles (Teixeira et al., 2018).

At the rates of PHs utilized in most commercial applications, the total N applied and the content of putative nutrient complexing amino acids are far too low to influence gross nutrient dynamics, particularly for full-growth cycle experiments. PH applications directly to seeds or transplants or injected precisely into root rhizosphere may indirectly increase full-season nutrient use by stimulating early phenological states where minute amounts of nutrients can significant effects.

5.1.2 Root growth/morphology

Root growth increase is a commonly reported effect of PHs. Increases in root growth translate to increased surface area and root zone exploration for nutrient absorption – effectively increasing the total uptake of nutrients and improving AE. Field-based and season-long studies are, however, limited, with most studies being conducted for short time frames under solution culture and in greenhouse studies. PH applications increased root growth in maize and lettuce under both optimal and limiting conditions (Rouphael et al., 2017b; Lucini et al., 2015; Santi et al., 2017; Ertani et al., 2009, 2013). Studies in tomato suggest that the root growth response interacts with the mode of application (i.e. root vs foliar application) but was unaffected by nutrient conditions (Sestili et al., 2018).

Microarray analyses suggest that the effects of PHs on root growth are related to an array of factors, including the expression of transcription factors controlling lateral root formation, cell wall components and hormone metabolism/synthesis/signaling (Colla et al., 2017b; Paul et al., 2019). Regardless of the specific pathway(s), microarray analyses have demonstrated that PHs are mechanistically distinct from isolated amino acid components in affecting root growth and ultimately distinct in how they affect nutrient acquisition (Santi et al., 2017).

In addition to root growth stimulation, it has been shown that soybean-derived PHs can modulate root morphology through root hair-promoting peptides (RHPPs) that activate root hair formation by increasing the density of trichoblasts (root hair cells) and atrichoblasts (root hairless cells). RHPPs increased the root surface area 16.6 times relative to control and altered root morphology through mechanisms physiologically distinct to other plant hormones (Matsumiya and Kubo, 2011; Matsubayashi and Sakagami, 2006). A subtilisin-like alkaline protease in *B. circulans* HA12 produces RHPP from soybean meal, which is also a component in the commercial PH 'Trainer' with several sources corroborating its beneficial root growth effects (Matsumiya and Kubo, 2011; Rouphael et al., 2018; Lucini et al., 2015; Colla et al., 2017a, 2014; Sestili et al., 2018). In these studies, PH 'Trainer' significantly increased both shoot and root growth – ultimately increasing nutrient uptake, which would result in increased AE. Inconsistency in the effects of PHs on nutrient concentration in plants suggests that either limiting factors exist upstream from uptake or nutrient uptake is diluted by growth – the latter likely the predominant factor (Cerdán et al., 2009; Colla et al., 2014, 2017a; Ertani et al., 2009, 2013; Lucini et al., 2015; Rouphael et al., 2017a,b, 2018; Santi et al., 2017; Sestili et al., 2018; Tejada et al., 2018).

Several research groups have demonstrated that PHs can modulate salt stress response through enhanced production of compatible solutes of low molecular weight that have been documented as tools for osmotic adjustment and/or involved in stress signaling during salt stress (Colla et al., 2017b; Van Oosten et al., 2017; Ertani et al., 2013; Lucini et al., 2015; Petrozza et al., 2014; Rouphael et al., 2017a). Microarray analyses have found that stress-related transcripts are shared and differentially expressed between PH and equivalent amino acid treatments, suggesting that PH stress-related response is partially linked to free amino acid components yet also due to different components (Santi et al., 2017).

Improved growth under salt or drought stress conditions have been observed and will generally increase use/uptake efficiency as continued plant growth results in greater nutrient uptake and hence an improvement in agronomic nitrogen use efficiency (AE).

5.1.3 Nutrient transporter and assimilation

Microarray analyses have shown that PHs affect the gene expression of transporters and related transcription factors (TFs). In maize roots, it was found that the TCP family, putatively related to the regulation of dual-affinity NRT1.1 expression, was differentially expressed between PHs, equivalent amino acid treatment and equivalent inorganic N treatment (Sestili et al., 2018; Santi et al., 2017). These findings suggest that different forms of N differentially affect TF expression, thereby modulating transporter expression and nutrient uptake. However, the analysis of transporter/TF expression must consider that intrafamily regulation of transporter proteins and posttranslational modifications prevent causal extrapolation of expression variation to observed differences in nutrient uptake. Putative effects on transporter networks are yet to be proven across multiple species, suggesting that a general mechanism for enhancing nutrient transport is yet to be proven for higher plants. Furthermore, modulation of transporter expression is network specific. This distinction is evident in the observed down-regulation of high-affinity nitrate transporters, NRT 2.1 and NRT 2.3, in tomato roots across the levels of N deficiency/sufficiency. Additionally, under N sufficiency, the root application of PHs was associated with increased expression of ammonium transporters while under N deficiency, the same treatment was associated with increased expression of amino acid transporters. These observations suggest that the PH not only modulates transporter expression but is so selectively dependent on conditions of nutrient availability (Santi et al., 2017; Sestili et al., 2018)

Concomitant with observed increases in nitrate and ammonium concentration of tomato leaves, NR in leaves was decreased after legume-derived PH application (Sestili et al., 2018; Colla et al., 2018). Contrary to these results, increases in nitrate assimilation were observed in roots under N sufficiency regardless of the application method (Sestili et al., 2018), and PHs were observed to decrease leaf nitrate levels in a variety of vegetables (Colla et al., 2018). Interestingly, Ertani et al. (2009) found that under N-sufficiency nitrate levels significantly decreased in a dose-dependent fashion in maize roots, 14 days after seeding and 2 days after treatment with both alfalfa and meat-flour PHs. Observed decreases in root nitrate corresponded to increased NR and glutamine synthetase (GS) activity, also in a dose-dependent fashion. Furthermore, leaf nitrate concentration increased despite counteracting increases in biomass. These trends confirmed those found in tomato by Sestili et al. (2018) and suggest there may be a common mechanism for modulating assimilation/storage networks between legume-derived and animal-derived PHs (Ertani et al., 2009).

Given that both ammonium and nitrate concentration of leaves increased across multiple studies, N deficiency/sufficiency and the mode of application, it is likely that observed increases in total N are generally attributable to root growth effects and to a lesser extent depend upon assimilation/storage networks as explained earlier.

While the specific effects on N uptake and assimilation described earlier and non-specific increases on general nutrient uptake as a result of root and shoot growth have been widely reported, specific effects on the uptake or transport of other nutrients have not been widely observed. In iron-limited conditions induced by liming, Cerdán et al. (2009) found that PHs from plant origin applied as a root drench can increase the concentration of iron without increasing shoot or root growth of tomato. This increase in uptake efficiency was associated with increased Fe(III) chelate reducatase (FCR) activity. In contrast to observed positive effects from soil drench, Cerdan et al. (2013) found that foliarly applied PHs from plant origin and all modes of application of PHs from animal origin significantly decreased iron concentration.

5.2 Conclusion

Given the wealth of studied components and positive effects associated with PH application, it is likely that the PH enhances NUE via multiple mechanisms. With the exception of consistent effects of PH on root growth response, there is a wide variability of findings with regards to the effects on specific uptake processes (transporters and transport regulation), the activity of assimilatory pathways and effects on tissue nutrient content and concentration. This may be a result of the great diversity in PH products present in the marketplace and may suggest that some PH components have specific effects on critical metabolic pathways. As with each of the biostimulants reviewed in this chapter, the effects of PH on root growth are clearly a primary driver of enhanced NUE.

6 Conclusion and future trends

In the context of plant nutrients, the current definition of biostimulants enacted by the EU mandates that biostimulants act through their ability to 'stimulate plant nutrition processes independently of the products nutrient content', specifically by improving NUE. A review of the available literature suggests clearly that many biostimulants indeed improve the AE by enhancing root growth and soil exploration, increasing the solubilization of soil nutrients or upregulating nutrient uptake processes, thereby enabling a greater amount of the total soil nutrient reserve to be acquired by the plant. Agronomic NUE will also be increased if a biostimulant increases stress tolerance, thereby maintaining plant growth and nutrient uptake from soils under conditions that otherwise compromise plant growth. In many agronomic contexts, an improvement in AE would be of great practical value as it allows for the greater exploitation of added fertilizers and improved recovery of residual nutrients. Given the relatively low efficiency with which nutrients are used in agriculture (Cassman et al., 2002), and concerns with N and P pollution and depletion of limited global P reserves, this potential benefit of biostimulants is significant.

Though the potential for beneficial effects of biostimulants on AE represents an exciting opportunity to enhance agricultural efficiency, these positive effects on NUE can occur only if a source of the otherwise-unavailable nutrient is made available by the addition of the biostimulant or if there was a pre-existing excess of that nutrient in the root zone that is now utilized in response to enhanced crop growth. The difficulty in truly assessing the benefits of biostimulants on NUE is complicated by the preponderance of studies that have been conducted under non-field or non-commercial conditions, often in very short-term experiments. While valuable information on mechanistic explanations for biostimulant effects can be gained from controlled experimentation, a true measure of AE-NUE can be gained only by examining the full cropping cycle in the context of the environment in which the crop would be commercially grown.

While evidence for improved AE with the use of biostimulants is clear, there is much less evidence to demonstrate that biostimulants alter the internal or physiological nutrient use efficiency (IE) by increasing the productivity of a crop for a given quantity of acquired nutrient. An increase in IE is most frequently associated with (a) improved internal allocation of limiting nutrients acquired by the plant to satisfy actively growing tissues; (b) increased within-plant remobilization to satisfy nutrient demands of tuber, fruit or seed development; or (c) enhanced nutrient remobilization to storage tissues during senescence in perennial species. The ability of many biostimulants to enhance plant stress tolerance would be expected to improve IE if it favorably alters plant senescence and nutrient remobilization. The identification of biostimulants that explicitly improve IE would be of substantial value to agriculture.

The recently published EU definition of biostimulants as products that improve plant nutrition processes has provided a useful regulatory framework for the differentiation of these products and has helped focus research and development in the biostimulant industry. Under this definition, any compound that enhances plant growth but is not a recognized plant nutrient, plant hormone or pest control product could reasonably be classified as a biostimulant. There are, however, a number of challenges with this definition that will need to be addressed in future refinements of the regulatory framework.

Under the current definition materials such as soil structural amendments (gypsum, organic residues, composts and manures, etc.) that enhance water penetration and root health, or surface amendments that provide favorable growth conditions could all be classified as biostimulants since they will enhance crop growth and root exploration and will, therefore, stimulate plant nutrition processes. The differentiation between an amendment and a biostimulant is however arbitrary; for example, while the broadacre application of tons of compost would not be considered a biostimulant application, a water-soluble 'tea' extract applied at a hundred liter per acre, while a concentrated extract applied at one liter per acre, would clearly might be considered a biostimulant if it positively impacted plant growth or NUE. In this context, the term 'biostimulant' has come to be defined as much by the quantity of materials required as the observed effect.

A second major constraint of the current regulatory framework is the explicit limitation of microbial biostimulants to *Rhizobia*, *Mycorrhiza*, *Azotobacter* and *Azospirillum*, a narrow classification that ignores many microbial species that are well known to explicitly influence plant growth, nutrient availability and internal NUE. Increasing evidence also suggests that many biostimulants may function by virtue of their ability to stimulate specific microbial consortia (Tejada et al., 2011; Qiao et al., 2019; Hellequin et al., 2018; Woo and Pepe, 2018), and hence a clear separation of biostimulants by the class of source materials is artificial and does not recognize the full potential of an integrated approach to biostimulant formulation and function (Rouphael et al., 2017a; Xu and Geelen, 2018).

7 References

Abbott, L. K., Macdonald, L. M., Wong, M. T. F., Webb, M. J., Jenkins, S. N. and Farrell, M. 2018. Potential roles of biological amendments for profitable grain production - a review. *Agriculture, Ecosystems and Environment* 256, 34–50. doi:10.1016/j.agee.2017.12.021.

Aguirre, E., Lemenager, D., Bacaicoa, E., Fuentes, M., Baigorri, R., Zamarreno, A. M. and Garcia-Mina, J. M. 2009. The root application of a purified leonardite humic acid modifies the transcriptional regulation of the main physiological root responses to Fe deficiency in Fe-sufficient cucumber plants. *Plant Physiology and Biochemistry: PPB* 47(3), 215–23. doi:10.1016/j.plaphy.2008.11.013.

Alguacil, M. D. M., Torres, M. P., Montesinos-Navarro, A. and Roldán, A. 2016. Soil characteristics driving arbuscular mycorrhizal fungal communities in semiarid Mediterranean soils. *Applied and Environmental Microbiology* 82(11), 3348–56. doi:10.1128/AEM.03982-15.

Allen, J. W. and Shachar-Hill, Y. 2009. Sulfur transfer through an arbuscular mycorrhiza. *Plant Physiology* 149(1), 549–60. doi:10.1104/pp.108.129866.

Anjum, S. A., Wang, L., Farooq, M., Xue, L. and Ali, S. 2011. Fulvic acid application improves the maize performance under well-watered and drought conditions. *Journal of Agronomy and Crop Science* 197(6), 409–17. doi:10.1111/j.1439-037X.2011.00483.x.

Berg, R. H., Tyler, M. E., Novick, N. J., Vasil, V. and Vasil, I. K. 1980. Biology of Azospirillum-Sugarcane Association: enhancement of nitrogenase activity. *Applied and Environmental Microbiology* 39(3), 642–9. doi:10.1128/AEM.39.3.642-649.1980.

Berruti, A., Lumini, E., Balestrini, R. and Bianciotto, V. 2015. Arbuscular mycorrhizal fungi as natural biofertilizers: let's benefit from past successes. *Frontiers in Microbiology* 6, 1559. doi:10.3389/fmicb.2015.01559.

Berruti, A., Lumini, E. and Bianciotto, V. 2016. AMF components from a microbial inoculum fail to colonize roots and lack soil persistence in an arable maize field. *Symbiosis* 72(1), 73–80. doi:10.1007/s13199-016-0442-7.

Bhattacharyya, P. N. and Jha, D. K. 2012. Plant growth-promoting rhizobacteria (PGPR): emergence in agriculture. *World Journal of Microbiology and Biotechnology* 28(4), 1327–50. doi:10.1007/s11274-011-0979-9.

Bianciotto, V., Victorino, I., Scariot, V. and Berruti, A. 2018. Arbuscular mycorrhizal fungi as natural biofertilizers: current role and potential for the horticulture industry. *Acta Horticulturae* 1191(1191), 207–16. doi:10.17660/ActaHortic.2018.1191.29.

Billard, V., Etienne, P., Jannin, L., Garnica, M., Cruz, F., Garcia-Mina, J. M., Yvin, J. C. and Ourry, A. 2014. Two biostimulants derived from algae or humic acid induce similar responses in the mineral content and gene expression of winter oilseed rape (Brassica napus L.). *Journal of Plant Growth Regulation* 33(2), 305–16. doi:10.1007/s00344-013-9372-2.

Breuillin-Sessoms, F., Floss, D. S., Gomez, S. K., Pumplin, N., Ding, Y., Levesque-Tremblay, V., Noar, R. D., Daniels, D. A., Bravo, A., Eaglesham, J. B., Benedito, V. A., Udvardi, M. K. and Harrison, M. J. 2015. Suppression of arbuscule degeneration in *Medicago truncatula phosphate transporter4* mutants is dependent on the ammonium transporter 2 family protein AMT2;3. *The Plant Cell* 27(4), 1352–66. doi:10.1105/tpc.114.131144.

Brown, M. E. 1974. Seed and root bacterization. *Annual Review of Phytopathology* 12(1), 181–97. doi:10.1146/annurev.py.12.090174.001145.

Brown, P. and Saa, S. 2015. Biostimulants in agriculture. *Frontiers in Plant Science* 6, 671. doi:10.3389/fpls.2015.00671.

Burk, D., Lineweaver, H., Horner, C. K. and Allison, F. E. 1931. The relation between iron, humic acid and organic matter in the nutrition and stimulation of plant growth. *Science* 74(1925), 522–4. doi:10.1126/science.74.1925.522.

Burk, D., Lineweaver, H. and Horner, C. K. 1932a. Iron in relation to the stimulation of growth by humic acid. *Soil Science* 33(6), 413–54. doi:10.1097/00010694-193206000-00002.

Burk, D., Lineweaver, H. and Horner, C. K. 1932b. The physiological nature of humic acid stimulation of Azotobacter growth. *Soil Science* 33(6), 455–88. doi:10.1097/00010694-193206000-00003.

Calvo, P., Nelson, L. and Kloepper, J. W. 2014. Agricultural uses of plant biostimulants. *Plant and Soil* 383(1-2), 3-41. doi:10.1007/s11104-014-2131-8.

Canellas, L. P., Olivares, F. L., Okorokova-Facanha, A. L. and Facanha, A. R. 2002. Humic acids isolated from earthworm compost enhance root elongation, lateral root emergence, and plasma membrane H+-ATPase activity in maize roots. *Plant Physiology* 130(4), 1951–7. doi:10.1104/pp.007088.

Canellas, L. P., Olivares, F. L., Aguiar, N. O., Jones, D. L., Nebbioso, A., Mazzei, P. and Piccolo, A. 2015. Humic and fulvic acids as biostimulants in horticulture. *Scientia Horticulturae* 196, 15–27. doi:10.1016/j.scienta.2015.09.013.

Canellas, L. P., Canellas, N. O. A., Soares, T. S. and Olivares, F. L. 2019. Humic acids interfere with nutrient sensing in plants owing to the differential expression of TOR. *Journal of Plant Growth Regulation* 38(1), 216–24. doi:10.1007/s00344-018-9835-6.

Casieri, L., Lahmidi, N. A., Doidy, J., Veneault-Fourrey, C., Migeon, A., Bonneau, L., Courty, P. E., Garcia, K., Charbonnier, M., Delteil, A., Brun, A., Zimmermann, S., Plassard, C. and Wipf, D. 2013. Biotrophic transportome in mutualistic plant-fungal interactions. *Mycorrhiza* 23(8), 597–625. doi:10.1007/s00572-013-0496-9.

Cassman, K. G., Dobermann, A. and Walters, D. T. 2002. Agroecosystems, nitrogen-use efficiency, and nitrogen management. *Ambio* 31(2), 132–40. doi:10.1579/0044-7447-31.2.132.

Cerdán, M., Sánchez-Sánchez, A., Oliver, M., Juárez, M. and Sánchez-Andreu, J. J. 2009. *Effect of foliar and root applications of amino acids on iron uptake by tomato plants.* International Society for Horticultural Science (ISHS), Leuven, Belgium, 481–8. doi:10.17660/ActaHortic.2009.830.68.

Cerdan, M., Sanchez-Sanchez, A., Jorda, J. D., Juarez, M. and Sanchez-Andreu, J. 2013. Effect of commercial amino acids on iron nutrition of tomato plants grown under lime-induced iron deficiency. *Journal of Plant Nutrition and Soil Science* 176(6), 859–66. doi:10.1002/jpln.201200525.

Chen, Y., Clapp, C. E. and Magen, H. 2004. Mechanisms of plant growth stimulation by humic substances: the role of organo-iron complexes. *Soil Science and Plant Nutrition* 50(7), 1089–95. doi:10.1080/00380768.2004.10408579.

Chennappa, G., Adkar-Purushothama, C. R., Suraj, U., Tamilvendan, K. and Sreenivasa, M. Y. 2014. Pesticide tolerant *Azotobacter* isolates from paddy growing areas of northern Karnataka, India. *World Journal of Microbiology and Biotechnology* 30(1), 1–7. doi:10.1007/s11274-013-1412-3.

Chennappa, G., Naik, M. K., Adkar-Purushothama, C. R., Amaresh, Y. S. and Sreenivasa, M. Y. 2016a. PGP potential, abiotic stress tolerance and antifungal activity of Azotobacter strains isolated from paddy soils. *Indian Journal of Experimental Biology* 54(5), 322–31.

Chennappa, G., Naik, M. K. and Sreenivasa, M. Y. 2016b. Azotobacter: PGPR activities with special reference to effect of pesticides and biodegradation. *Microbial Inoculants in Sustainable Agricultural Productivity*. SpringerLink.

Chennappa, G., Naik, M. K., Amaresh, Y. S., Nagaraja, H. and Sreenivasa, M. Y. 2017. Azotobacter: a potential biofertilizer and bioinoculants for sustainable agriculture. In: Panpatte, D. G., Jhala, Y. K., Vyas, R. V. and Shelat, H. N. (Eds), *Microorganisms for Green Revolution*. SpringerLink.

Choudhury, A. T. M. A. and Kennedy, I. R. 2004. Prospects and potentials for systems of biological nitrogen fixation in sustainable rice production. *Biology and Fertility of Soils* 39(4), 219–27. doi:10.1007/s00374-003-0706-2.

Chouliaras, V., Tasioula, M., Chatzissavvidis, C., Therios, I. and Tsabolatidou, E. 2009. The effects of a seaweed extract in addition to nitrogen and boron fertilization on productivity, fruit maturation, leaf nutritional status and oil quality of the olive (*Olea europaea* L.) cultivar Koroneiki. *Journal of the Science of Food and Agriculture* 89(6), 984–8. doi:10.1002/jsfa.3543.

Clark, R. B. and Zeto, S. K. 2000. Mineral acquisition by arbuscular mycorrhizal plants. *Journal of Plant Nutrition* 23(7), 867–902. doi:10.1080/01904160009382068.

Colla, G., Rouphael, Y., Canaguier, R., Svecova, E. and Cardarelli, M. 2014. Biostimulant action of a plant-derived protein hydrolysate produced through enzymatic hydrolysis. *Frontiers in Plant Science* 5, 448. doi:10.3389/fpls.2014.00448.

Colla, G., Cardarelli, M., Bonini, P. and Rouphael, Y. 2017a. Foliar applications of protein hydrolysate, plant and seaweed extracts increase yield but differentially modulate fruit quality of greenhouse tomato. *Hortscience* 52(9), 1214–20. doi:10.21273/HORTSCI12200-17.

Colla, G., Hoagland, L., Ruzzi, M., Cardarelli, M., Bonini, P., Canaguier, R. and Rouphael, Y. 2017b. Biostimulant action of protein hydrolysates: unraveling their effects on plant physiology and microbiome. *Frontiers in Plant Science* 8, 2202. doi:10.3389/fpls.2017.02202.

Colla, G., Kim, H. J., Kyriacou, M. C. and Rouphael, Y. 2018. Nitrate in fruits and vegetables. *Scientia Horticulturae* 237, 221–38. doi:10.1016/j.scienta.2018.04.016.

Craigie, J. S. 2011. Seaweed extract stimuli in plant science and agriculture. *Journal of Applied Phycology* 23(3), 371–93. doi:10.1007/s10811-010-9560-4.

Crossay, T., Majorel, C., Redecker, D., Gensous, S., Medevielle, V., Durrieu, G., Cavaloc, Y. and Amir, H. 2019. Is a mixture of arbuscular mycorrhizal fungi better for plant growth than single-species inoculants? *Mycorrhiza* 29(4), 325–39. doi:10.1007/s00572-019-00898-y.

De Pascale, S., Rouphael, Y. and Colla, G. 2017. Plant biostimulants: innovative tool for enhancing plant nutrition in organic farming. *European Journal of Horticultural Science* 82(6), 277–85. doi:10.17660/eJHS.2017/82.6.2.

Diacono, M. and Montemurro, F. 2010. Long-term effects of organic amendments on soil fertility. A review. *Agronomy for Sustainable Development* 30(2), 401–22. doi:10.1051/agro/2009040.

Erenoglu, E. B., Kutman, U. B., Ceylan, Y., Yildiz, B. and Cakmak, I. 2011. Improved nitrogen nutrition enhances root uptake, root-to-shoot translocation and remobilization of zinc (65Zn) in wheat. *The New Phytologist* 189(2), 438–48. doi:10.1111/j.1469-8137.2010.03488.x.

Ertani, A., Cavani, L., Pizzeghello, D., Brandellero, E., Altissimo, A., Ciavatta, C. and Nardi, S. 2009. Biostimulant activity of two protein hydrolyzates in the growth and nitrogen metabolism of maize seedlings. *Journal of Plant Nutrition and Soil Science-Zeitschrift Fur Pflanzenernahrung Und Bodenkunde* 172, 237–44.

Ertani, A., Schiavon, M., Muscolo, A. and Nardi, S. J. P. 2013. Alfalfa plant-derived biostimulant stimulate short-term growth of salt stressed Zea mays L. plants. *Plant and Soil* 364(1-2), 145–58. doi:10.1007/s11104-012-1335-z.

Ertani, A., Francioso, O., Tinti, A., Schiavon, M., Pizzeghello, D. and Nardi, S. 2018. Evaluation of seaweed extracts from laminaria and *Ascophyllum nodosum* spp. as biostimulants in *Zea mays* L. using a combination of chemical, biochemical and morphological approaches. *Frontiers in Plant Science* 9, 428. doi:10.3389/fpls.2018.00428.

Fan, D., Hodges, D. M., Critchley, A. T. and Prithiviraj, B. 2013. A commercial extract of brown macroalga (*Ascophyllum nodosum*) affects yield and the nutritional quality of spinach in vitro. *Communications in Soil Science and Plant Analysis* 44(12), 1873–84. doi:10.1080/00103624.2013.790404.

Ferreira, C. M. H., Soares, H. M. V. M. and Soares, E. V. 2019. Promising bacterial genera for agricultural practices: an insight on plant growth-promoting properties and microbial safety aspects. *The Science of the Total Environment* 682, 779–99. doi:10.1016/j.scitotenv.2019.04.225.

Fixen, P. E., Brentrup, F., Bruulsema, T., Garcia, F., Norton, R. and Zingore, S. 2015. Nutrient/fertilizer use efficiency: measurement, current situation and trends. In: IFA, I., IPNI and IPI (Eds), *Managing Water and Fertilizer for Sustainable Agricultural Intensification*. IFA, Horgan, Switzerland.

Fuentes-Ramirez, L. E. and Caballero-Mellado, J. 2006. Bacterial biofertilizers. In: Siddiqui, Z. A. (Ed.), *PGPR: Biocontrol and Biofertilization*. Springer.

Garcia, A. C., De Souza, L. G. A., Pereira, M. G., Castro, R. N., Garcia-Mina, J. M., Zonta, E., Lisboa, F. J. G. and Berbara, R. L. L. 2016. Structure-property-function relationship in humic substances to explain the biological activity in plants. *Scientific Reports* 6, 20798. doi:10.1038/srep20798.

Garcia, A. C., Tavares, O. C. H., Balmori, D. M., Almeida, V. D., Canellas, L. P., Garcia-Mina, J. M. and Berbara, R. L. L.. 2018. Structure-function relationship of vermicompost humic fractions for use in agriculture. *Journal of Soils and Sediments* 18(4), 1365–75. doi:10.1007/s11368-016-1521-3.

Garcia, A. C., De Castro, T. A. V., Santos, L. A., Tavares, O. C. H., Castro, R. N., Berbara, R. L. L. and Garcia-Mina, J. M. 2019. Structure-property-function relationship of humic substances in modulating the root growth of plants: a review. *Journal of Environmental Quality* 48(6), 1622–32. doi:10.2134/jeq2019.01.0027.

García-Fraile, P., Menéndez, E. and Rivas, R. 2015. Role of bacterial biofertilizers in agriculture and forestry. *AIMS Bioengineering* 2(3), 183–205. doi:10.3934/bioeng.2015.3.183.

Garcia-Mina, J. M., Antolin, M. C. and Sanchez-Diaz, M. 2004. Metal-humic complexes and plant micronutrient uptake: a study based on different plant species cultivated in diverse soil types. *Plant and Soil* 258(1), 57–68. doi:10.1023/B:PLSO.0000016509.56780.40.

Grabowska, A., Kunicki, E., Sękara, A., Kalisz, A. and Wojciechowska, R. 2012. The effect of cultivar and biostimulant treatment on the carrot yield and its quality. *Vegetable Crops Research Bulletin* 77(1), 37–48. doi:10.2478/v10032-012-0014-1.

Guether, M., Neuhauser, B., Balestrini, R., Dynowski, M., Ludewig, U. and Bonfante, P. 2009. A mycorrhizal-specific ammonium transporter from Lotus japonicus acquires nitrogen released by arbuscular mycorrhizal fungi. *Plant Physiology* 150(1), 73–83. doi:10.1104/pp.109.136390.

Halpern, M., Bar-Tal, A., Ofek, M., Minz, D., Muller, T. and Yermiyahu, U. 2015. The use of biostimulants for enhancing nutrient uptake. *Advances in Agronomy* 130(130), 141–74. doi:10.1016/bs.agron.2014.10.001.

Harrison, M. J., Dewbre, G. R. and Liu, J. 2002. A phosphate transporter from Medicago truncatula involved in the acquisition of phosphate released by arbuscular mycorrhizal fungi. *The Plant Cell* 14(10), 2413–29. doi:10.1105/tpc.004861.

Hartz, T. K. and Bottoms, T. G. 2010. Humic substances generally ineffective in improving vegetable crop nutrient uptake or productivity. *HortScience* 45(6), 906–10. doi:10.21273/HORTSCI.45.6.906.

Hayat, R., Ali, S., Amara, U., Khalid, R. and Ahmed, I. 2010. Soil beneficial bacteria and their role in plant growth promotion: a review. *Annals of Microbiology* 60(4), 579–98. doi:10.1007/s13213-010-0117-1.

Hellequin, E., Monard, C., Quaiser, A., Henriot, M., Klarzynski, O. and Binet, F. 2018. Specific recruitment of soil bacteria and fungi decomposers following a biostimulant application increased crop residues mineralization. *PLoS ONE* 13(12), e0209089. doi:10.1371/journal.pone.0209089.

Hijri, M. 2016. Analysis of a large dataset of mycorrhiza inoculation field trials on potato shows highly significant increases in yield. *Mycorrhiza* 26(3), 209–14. doi:10.1007/s00572-015-0661-4.

Hoeksema, J. D., Chaudhary, V. B., Gehring, C. A., Johnson, N. C., Karst, J., Koide, R. T., Pringle, A., Zabinski, C., Bever, J. D., Moore, J. C., Wilson, G. W. T., Klironomos, J. N. and Umbanhowar, J. 2010. A meta-analysis of context-dependency in plant response to inoculation with mycorrhizal fungi. *Ecology Letters* 13(3), 394–407. doi:10.1111/j.1461-0248.2009.01430.x.

Igiehon, N. O. and Babalola, O. O. 2017. Biofertilizers and sustainable agriculture: exploring arbuscular mycorrhizal fungi. *Applied Microbiology and Biotechnology* 101(12), 4871–81. doi:10.1007/s00253-017-8344-z.

Ijdo, M., Cranenbrouck, S. and Declerck, S. 2010. Methods for large-scale production of AM fungi: past, present, and future. *Mycorrhiza* 21(1), 1–16. doi:10.1007/s00572-010-0337-z.

Jannin, L., Arkoun, M., Ourry, A., Philippe, L., Goux, D., Garnica, M., Fuentes, M., Francisco, S. S., Baigorri, R., Cruz, F., Houdusse, F., Garcia-Mina, J., Yvin, J.-C. and Etienne, P. 2012. Microarray analysis of humic acid effects on *Brassica napus* growth: involvement of N, C and S metabolisms. *Plant and Soil* 359(1–2), 297–319. doi:10.1007/s11104-012-1191-x.

Jannin, L., Arkoun, M., Etienne, P., Laine, P., Goux, D., Garnica, M. and Fuentes, M. San Francisco, S., Baigorri, R., Cruz, F., Houdusse, F., Garcia-Mina, J. M., Yvin, J. C. and Ourry, A. 2013. Brassica napus growth is promoted by *Ascophyllum nodosum* (L.) Le Jol. seaweed extract: microarray analysis and physiological characterization of N, C, and S metabolisms. *Journal of Plant Growth Regulation* 32, 31–52.

Janoušková, M., Krak, K., Vosátka, M., Püschel, D. and Štorchová, H. 2017. Inoculation effects on root-colonizing arbuscular mycorrhizal fungal communities spread beyond directly inoculated plants. *PLOS ONE* 12(7), e0181525. doi:10.1371/journal.pone.0181525.

Kannan, T. and Ponmurugan, P. 2010. Response of paddy (Oryza sativa L.) varieties to Azospirillum brasilense inoculation. *Journal of Phytology* 2, 8–13.

Khan, W., Rayirath, U. P., Subramanian, S., Jithesh, M. N., Rayorath, P., Hodges, D. M., Critchley, A. T., Craigie, J. S., Norrie, J. and Prithiviraj, B. 2009. Seaweed extracts as biostimulants of plant growth and development. *Journal of Plant Growth Regulation* 28(4), 386–99. doi:10.1007/s00344-009-9103-x.

Klironomos, J. N. and Hart, M. M. 2002. Colonization of roots by arbuscular mycorrhizal fungi using different sources of inoculum. *Mycorrhiza* 12(4), 181–4. doi:10.1007/s00572-002-0169-6.

Köhl, L., Lukasiewicz, C. E. and Van Der Heijden, M. G. 2016. Establishment and effectiveness of inoculated arbuscular mycorrhizal fungi in agricultural soils. *Plant, Cell and Environment* 39(1), 136–46. doi:10.1111/pce.12600.

Kurepin, L. V., Zaman, M. and Pharis, R. P. 2014. Phytohormonal basis for the plant growth promoting action of naturally occurring biostimulators. *Journal of the Science of Food and Agriculture* 94(9), 1715–22. doi:10.1002/jsfa.6545.

Lehmann, A. and Rillig, M. C. 2015. Arbuscular mycorrhizal contribution to copper, manganese and iron nutrient concentrations in crops - a meta-analysis. *Soil Biology and Biochemistry* 81, 147–58. doi:10.1016/j.soilbio.2014.11.013.

Lehmann, A., Veresoglou, S. D., Leifheit, E. F. and Rillig, M. C. 2014. Arbuscular mycorrhizal influence on zinc nutrition in crop plants - a meta-analysis. *Soil Biology and Biochemistry* 69, 123–31. doi:10.1016/j.soilbio.2013.11.001.

Lucini, L., Rouphael, Y., Cardarelli, M., Canaguier, R., Kumar, P. and Colla, G. 2015. The effect of a plant-derived biostimulant on metabolic profiling and crop performance of lettuce grown under saline conditions. *Scientia Horticulturae* 182, 124–33. doi:10.1016/j.scienta.2014.11.022.

Lyons, G. and Genc, Y. 2016. Commercial humates in agriculture: real substance or smoke and mirrors? *Agronomy* 6(4). doi:10.3390/agronomy6040050.

Mahmood, A., Turgay, O. C., Farooq, M. and Hayat, R. 2016. Seed biopriming with plant growth promoting rhizobacteria: a review. *FEMS Microbiology Ecology* 92(8), 1–14. doi:10.1093/femsec/fiw112.

Mahoney, K. J., Mccreary, C. M., Depuydt, D. and Gillard, C. L. 2017. Fulvic and humic acid fertilizers are ineffective in dry bean. *Canadian Journal of Plant Science* 97, 202–5. doi:10.1139/CJPS-2016-0143.

Masclaux-Daubresse, C., Daniel-Vedele, F., Dechorgnat, J., Chardon, F., Gaufichon, L. and Suzuki, A. 2010. Nitrogen uptake, assimilation and remobilization in plants: challenges for sustainable and productive agriculture. *Annals of Botany* 105(7), 1141–57. doi:10.1093/aob/mcq028.

Matsubayashi, Y. and Sakagami, Y. 2006. Peptide hormones in plants. *Annual Review of Plant Biology* 57, 649–74. doi:10.1146/annurev.arplant.56.032604.144204.

Matsumiya, Y. and Kubo, M. 2011. *Soybean Peptide: Novel Plant Growth Promoting Peptide from Soybean*.

Mehnaz, S. 2016. An overview of globally available bioformulations. In: Arora, N. K., Mehnaz, S. and Balestrini, R. (Eds), *Bioformulations: for Sustainable Agriculture*. SpringerLink.

Mirza, M. S., Rasul, G., Mehnaz, S., Ladha, J. K., So, R. B., Ali, S. and Malik, K. A. 2002. Beneficial effects of inoculated nitrogen-fixing bacteria on rice. In: Kadha, J. K. and Reddy, P. M. (Eds), *The Quest for Nitrogen Fixation in Rice*.

Nagy, R., Karandashov, V., Chague, V., Kalinkevich, K., Tamasloukht, M. B., Xu, G., Jakobsen, I., Levy, A. A., Amrhein, N. and Bucher, M. 2005. The characterization of novel mycorrhiza-specific phosphate transporters from *Lycopersicon esculentum* and *Solanum tuberosum* uncovers functional redundancy in symbiotic phosphate transport in solanaceous species. *The Plant Journal: for Cell and Molecular Biology* 42(2), 236–50. doi:10.1111/j.1365-313X.2005.02364.x.

Nakkeeran, S., Dilantha Fernando, W. G. and Siddiqui, Z. A. 2006. Plant growth promoting rhizobacteria formulations and its scope in commercialization for the management of pests and diseases. In: Siddiqui, Z. A. (Ed.), *PGPR: Biocontrol and Biofertilization*. SpringerLink.

Nardi, S., Pizzeghello, D., Muscolo, A. and Vianello, A. 2002. Physiological effects of humic substances on higher plants. *Soil Biology and Biochemistry* 34(11), 1527–36. doi:10.1016/S0038-0717(02)00174-8.

Nardi, S., Pizzeghello, D., Schiavon, M. and Ertani, A. 2016. Plant biostimulants: physiological responses induced by protein hydrolyzed-based products and humic substances in plant metabolism. *Scientia Agricola* 73(1), 18–23. doi:10.1590/0103-9016-2015-0006.

Nouri, E., Breuillin-Sessoms, F., Feller, U. and Reinhardt, D. 2014. Phosphorus and nitrogen regulate arbuscular mycorrhizal symbiosis in *Petunia hybrida*. *PLOS ONE* 9(6), e90841. doi:10.1371/journal.pone.0090841.

Okon, Y. and Labandera-Gonzalez, C. A. 1994. Agronomic applications of Azospirillum: an evaluation of 20 years worldwide field inoculation. *Soil Biology and Biochemistry* 26(12), 1591–601. doi:10.1016/0038-0717(94)90311-5.

Olaetxea, M., De Hita, D., Garcia, C. A., Fuentes, M., Baigorri, R., Mora, V., Garnica, M., Urrutia, O., Erro, J., Zamarreno, A. M., Berbara, R. L. and Garcia-Mina, J. M. 2018. Hypothetical framework integrating the main mechanisms involved in the promoting action of rhizospheric humic substances on plant root- and shoot-growth. *Applied Soil Ecology* 123, 521–37. doi:10.1016/j.apsoil.2017.06.007.

Olk, D. C., Dinnes, D. L., Scoresby, J. R., Callaway, C. R. and Darlington, J. W. 2018. Humic products in agriculture: potential benefits and research challenges-a review. *Journal of Soils and Sediments* 18(8), 2881-91. doi:10.1007/s11368-018-1916-4.

Palumbo, G., Schiavon, M., Nardi, S., Ertani, A., Celano, G. and Colombo, C. M. 2018. Biostimulant potential of humic acids extracted From an amendment obtained via combination of olive Mill wastewaters (OMW) and a pre-treated organic material derived from municipal solid waste (MSW). *Frontiers in Plant Science* 9, 1028. doi:10.3389/fpls.2018.01028.

Paradikovic, N., Vinkovic, T., Vrcek, I. V., Zuntar, I., Bojic, M. and Medic-Saric, M. 2011. Effect of natural biostimulants on yield and nutritional quality: an example of sweet yellow pepper (*Capsicum annuum* L.) plants. *Journal of the Science of Food and Agriculture* 91(12), 2146–52. doi:10.1002/jsfa.4431.

Parniske, M. 2008. Arbuscular mycorrhiza: the mother of plant root endosymbioses. *Nature Reviews: Microbiology* 6(10), 763–75. doi:10.1038/nrmicro1987.

Paszkowski, U., Kroken, S., Roux, C. and Briggs, S. P. 2002. Rice phosphate transporters include an evolutionarily divergent gene specifically activated in arbuscular mycorrhizal symbiosis. *Proceedings of the National Academy of Sciences of the United States of America* 99(20), 13324–9. doi:10.1073/pnas.202474599.

Paul, K., Sorrentino, M., Lucini, L., Rouphael, Y., Cardarelli, M., Bonini, P., Reynaud, H., Canaguier, R., Trtilek, M., Panzarova, K. and Colla, G. 2019. Understanding the biostimulant action of vegetal-derived protein hydrolysates by high-throughput plant phenotyping and metabolomics: a case study on tomato. *Frontiers in Plant Science* 10, 47. doi:10.3389/fpls.2019.00047.

Pellegrino, E., Turrini, A., Gamper, H. A., Cafa, G., Bonari, E., Young, J. P. and Giovannetti, M. 2012. Establishment, persistence and effectiveness of arbuscular mycorrhizal fungal inoculants in the field revealed using molecular genetic tracing and measurement of yield components. *The New Phytologist* 194(3), 810-22. doi:10.1111/j.1469-8137.2012.04090.x.

Petrozza, A., Santaniello, A., Summerer, S., Di Tommaso, G., Di Tommaso, D., Paparelli, E., Piaggesi, A., Perata, P. and Cellini, F. 2014. Physiological responses to Megafol (R) treatments in tomato plants under drought stress: a phenomic and molecular approach. *Scientia Horticulturae* 174, 185–92. doi:10.1016/j.scienta.2014.05.023.

Pinton, R., Cesco, S., Iacolettig, G., Astolfi, S. and Varanini, Z. 1999. Modulation of NO3- uptake by water-extractable humic substances: involvement of root plasma membrane H(+)ATPase. *Plant and Soil* 215(2), 155–61. doi:10.102 3/A:1004752531903.

Qiao, C. C., Penton, C. R., Xiong, W., Liu, C., Wang, R. F., Liu, Z. Y., Xu, X., Li, R. and Shen, Q. R. 2019. Reshaping the rhizosphere microbiome by bio-organic amendment to enhance crop yield in a maize-cabbage rotation system. *Applied Soil Ecology* 142, 136–46. doi:10.1016/j.apsoil.2019.04.014.

Quaggiotti, S., Ruperti, B., Pizzeghello, D., Francioso, O., Tugnoli, V. and Nardi, S. 2004. Effect of low molecular size humic substances on nitrate uptake and expression of genes involved in nitrate transport in maize (Zea mays L.). *Journal of Experimental Botany* 55(398), 803–13. doi:10.1093/jxb/erh085.

Rausch, C., Daram, P., Brunner, S., Jansa, J., Laloi, M., Leggewie, G., Amrhein, N. and Bucher, M. 2001. A phosphate transporter expressed in arbuscule-containing cells in potato. *Nature* 414(6862), 462–70. doi:10.1038/35106601.

Reich, M., Aghajanzadeh, T. and De Kok, L. J. 2015. Physiological basis of plant nutrient use efficiency - concepts, opportunities and challenges for its improvement. In: Hawkesford, M. J., Kopriva, S. and De Kok, L. J. (Eds), *Nutrient Use Efficiency in Plants*. Springer-Verlag, Groningen.

Rodriguez, A. and Sanders, I. R. 2015. The role of community and population ecology in applying mycorrhizal fungi for improved food security. *The ISME Journal* 9(5), 1053–61. doi:10.1038/ismej.2014.207.

Rose, M. T., Patti, A. F., Little, K. R., Brown, A. L., Jackson, W. R. and Cavagnaro, T. R. 2014. A meta-analysis and review of plant-growth response to humic substances: practical implications for agriculture. *Advances in Agronomy* 124(124), 37–89.

Rouphael, Y., Cardarelli, M., Bonini, P. and Colla, G. 2017a. Synergistic action of a microbial-based biostimulant and a plant derived-protein hydrolysate enhances lettuce tolerance to alkalinity and salinity. *Frontiers in Plant Science* 8, 131. doi:10.3389/fpls.2017.00131.

Rouphael, Y., Colla, G., Giordano, M., El-Nalchel, C., Kyriacou, M. C. and De Pascale, S. 2017b. Foliar applications of a legume-derived protein hydrolysate elicit dose-dependent increases of growth, leaf mineral composition, yield and fruit quality in two greenhouse tomato cultivars. *Scientia Horticulturae* 226, 353–60. doi:10.1016/j.scienta.2017.09.007.

Rouphael, Y., De Micco, V., Arena, C., Raimondi, G., Colla, G. and De Pascale, S. 2017c. Effect of Ecklonia maxima seaweed extract on yield, mineral composition, gas exchange, and leaf anatomy of zucchini squash grown under saline conditions. *Journal of Applied Phycology* 29(1), 459–70. doi:10.1007/s10811-016-0937-x.

Rouphael, Y., Giordano, M., Cardarelli, M., Cozzolino, E., Mori, M., Kyriacou, M. C., Bonini, P. and Colla, G. 2018. Plant- and seaweed-based extracts increase yield but differentially modulate nutritional quality of greenhouse spinach through biostimulant action. *Agronomy* 8(7). doi:10.3390/agronomy8070126.

Rovira, A. D. 1989. Rhizosphere research - 85 years of progress and frustration. In: Keister, D. L. and Cregan, P. B. (Eds), *The Rhizosphere and Plant Growth*. SpringerLink, Beltsville, MD.

Ruzicka, D. R., Hausmann, N. T., Barrios-Masias, F. H., Jackson, L. E. and Schachtman, D. P. 2012. Transcriptomic and metabolic responses of mycorrhizal roots to nitrogen patches under field conditions. *Plant and Soil* 350(1–2), 145–62. doi:10.1007/s11104-011-0890-z.

Sabir, A., Yazar, K., Sabir, F., Kara, Z., Yazici, M. A. and Goksu, N. 2014. Vine growth, yield, berry quality attributes and leaf nutrient content of grapevines as influenced by seaweed extract (*Ascophyllum nodosum*) and nanosize fertilizer pulverizations. *Scientia Horticulturae* 175, 1–8. doi:10.1016/j.scienta.2014.05.021.

Sahoo, R. K., Ansari, M. W., Dangar, T. K., Mohanty, S. and Tuteja, N. 2014. Phenotypic and molecular characterisation of efficient nitrogen-fixing Azotobacter strains from rice fields for crop improvement. *Protoplasma* 251(3), 511–23. doi:10.1007/s00709-013-0547-2.

Sanchez, A. S., Juarez, M., Sanchez-Andreu, J., Jorda, J. and Bermudez, D. 2005. Use of humic substances and amino acids to enhance iron availability for tomato plants from applications of the chelate FeEDDHA. *Journal of Plant Nutrition* 28(11), 1877–86. doi:10.1080/01904160500306359.

Sánchez-Sánchez, A., Sánchez-Andreu, J., Juárez, M., Jordá, J. and Bermúdez, D. 2006. Improvement of iron uptake in table grape by addition of humic substances. *Journal of Plant Nutrition* 29(2), 259–72. doi:10.1080/01904160500476087.

Santi, C., Zamboni, A., Varanini, Z. and Pandolfini, T. 2017. Growth stimulatory effects and genome-wide transcriptional changes produced by protein hydrolysates in maize seedlings. *Frontiers in Plant Science* 8, 433. doi:10.3389/fpls.2017.00433.

Sestili, F., Rouphael, Y., Cardarelli, M., Pucci, A., Bonini, P., Canaguier, R. and Colla, G. 2018. Protein hydrolysate stimulates growth in tomato coupled With N-dependent gene expression involved in N assimilation. *Frontiers in Plant Science* 9, 1233. doi:10.3389/fpls.2018.01233.

Sharma, H. S. S., Fleming, C., Selby, C., Rao, J. R. and Martin, T. 2014. Plant biostimulants: a review on the processing of macroalgae and use of extracts for crop management to reduce abiotic and biotic stresses. *Journal of Applied Phycology* 26(1), 465–90. doi:10.1007/s10811-013-0101-9.

Shukla, P. S., Mantin, E. G., Adil, M., Bajpai, S., Critchley, A. T. and Prithiviraj, B. 2019. Ascophyllum nodosum-based biostimulants: sustainable applications in agriculture for the stimulation of plant growth, stress tolerance, and disease management. *Frontiers in Plant Science* 10, 655. doi:10.3389/fpls.2019.00655.

Suzuki, A., Chardon, F., Daniel-Vedele, F., Dechorgnat, J., Gaufichon, L. and Masclaux-Daubresse, C. 2010. Nitrogen uptake, assimilation and remobilization in plants: challenges for sustainable and productive agriculture. *Annals of Botany* 105(7), 1141–57. doi:10.1093/aob/mcq028.

Tamayo, E., Gómez-Gallego, T., Azcón-Aguilar, C. and Ferrol, N. 2014. Genome-wide analysis of copper, iron and zinc transporters in the arbuscular mycorrhizal fungus *Rhizophagus irregularis*. *Frontiers in Plant Science* 5, 547. doi:10.3389/fpls.2014.00547.

Teixeira, W. F., Fagan, E. B., Soares, L. H., Soares, J. N., Reichardt, K. and Neto, D. D. 2018. Seed and foliar application of amino acids improve variables of nitrogen metabolism and productivity in soybean crop. *Frontiers in Plant Science* 9, 396. doi:10.3389/fpls.2018.00396.

Tejada, M., Benitez, C., Gomez, I. and Parrado, J. 2011. Use of biostimulants on soil restoration: effects on soil biochemical properties and microbial community. *Applied Soil Ecology* 49, 11–7. doi:10.1016/j.apsoil.2011.07.009.

Tejada, M., Rodriguez-Morgado, B., Paneque, P. and Parrado, J. 2018. Effects of foliar fertilization of a biostimulant obtained from chicken feathers on maize yield. *European Journal of Agronomy* 96, 54–9. doi:10.1016/j.eja.2018.03.003.

Tejera, N., Lluch, C., Martìnez-Toledo, M. V. and Gonzàlez-López, J. 2005. Isolation and characterization of *Azotobacter* and *Azospirillum* strains from the sugarcane rhizosphere. *Plant and Soil* 270(1), 223–32. doi:10.1007/s11104-004-1522-7.

Thonar, C., Erb, A. and Jansa, J. 2012. Real-time PCR to quantify composition of arbuscular mycorrhizal fungal communities – marker design, verification, calibration and field validation. *Molecular Ecology Resources* 12(2), 219–32. doi:10.1111/j.1755-0998.2011.03086.x.

Tisserant, E., Malbreil, M., Kuo, A., Kohler, A., Symeonidi, A., Balestrini, R., Charron, P., Duensing, N., Frei Dit Frey, N., Gianinazzi-Pearson, V., Gilbert, L. B., Handa, Y., Herr, J. R., Hijri, M., Koul, R., Kawaguchi, M., Krajinski, F., Lammers, P. J., Masclaux, F. G., Murat, C., Morin, E., Ndikumana, S., Pagni, M., Petitpierre, D., Requena, N., Rosikiewicz, P., Riley, R., Saito, K., San Clemente, H., Shapiro, H., Van Tuinen, D., Bécard, G., Bonfante, P., Paszkowski, U., Shachar-Hill, Y. Y., Tuskan, G. A., Young, J. P. W., Sanders, I. R., Henrissat, B., Rensing, S. A., Grigoriev, I. V., Corradi, N., Roux, C. and Martin, F. 2013. Genome of an arbuscular mycorrhizal fungus provides insight into the oldest plant symbiosis. *Proceedings of the National Academy of Sciences of the United States of America* 110(50), 20117–22. doi:10.1073/pnas.1313452110.

Tomasi, N., Mimmo, T., Terzano, R., Alfeld, M., Janssens, K., Zanin, L., Pinton, R., Varanini, Z. and Cesco, S. 2014. Nutrient accumulation in leaves of Fe-deficient cucumber plants treated with natural Fe complexes. *Biology and Fertility of Soils* 50(6), 973–82. doi:10.1007/s00374-014-0919-6.

Tsouvaltzis, P., Koukounaras, A. and Siomos, A. 2014. Application of amino acids improves lettuce crop uniformity and inhibits nitrate accumulation induced by the supplemental inorganic nitrogen fertilization. *International Journal of Agriculture and Biology* 16, 951–5.

Uauy, C., Distelfeld, A., Fahima, T., Blechl, A. and Dubcovsky, J. 2006. A NAC gene regulating senescence improves grain protein, zinc, and iron content in wheat. *Science* 314(5803), 1298–301. doi:10.1126/science.1133649.

Van Der Heijden, M. G. A., Martin, F. M., Selosse, M. A. and Sanders, I. R. 2015. Mycorrhizal ecology and evolution: the past, the present, and the future. *The New Phytologist* 205(4), 1406–23. doi:10.1111/nph.13288.

Van Oosten, M. J., Pepe, O., De Pascale, S., Silletti, S. and Maggio, A. 2017. The role of biostimulants and bioeffectors as alleviators of abiotic stress in crop plants. *Chemical and Biological Technologies in Agriculture* 4(1). doi:10.1186/s40538-017-0089-5.

Verbruggen, E., Van Der Heijden, M. G., Rillig, M. C. and Kiers, E. T. 2013. Mycorrhizal fungal establishment in agricultural soils: factors determining inoculation success. *The New Phytologist* 197(4), 1104–9. doi:10.1111/j.1469-8137.2012.04348.x.

Verma, P., Yadav, A. N., Khannam, K. S., Saxena, A. K. and Suman, A. 2017. Potassium-solubilizing microbes: diversity, distribution, and role in plant growth promotion. In: Panpatte, D. G., Jhala, Y. K., Vyas, R. V. and Shelat, H. N. (Eds), Microorganisms for Green Revolution. SpringerLink.

Vesela, M. and Friedrich, J. 2009. Amino acid and soluble protein cocktail from waste keratin hydrolysed by a fungal keratinase of *Paecilomyces marquandii*. *Biotechnology and Bioprocess Engineering* 14(1), 84–90. doi:10.1007/s12257-008-0083-7.

Vosátka, M., Látr, A., Gianinazzi, S. and Albrechtová, J. 2012. Development of arbuscular mycorrhizal biotechnology and industry: current achievements and bottlenecks. *Symbiosis* 58(1–3), 29–37. doi:10.1007/s13199-012-0208-9.

Wally, O. S. D., Critchley, A. T., Hiltz, D., Craigie, J. S., Han, X. M., Zaharia, L. I., Abrams, S. R. and Prithiviraj, B. 2013. Regulation of phytohormone biosynthesis and accumulation in Arabidopsis following treatment with commercial extract from the marine macroalga *Ascophyllum nodosum*. *Journal of Plant Growth Regulation* 32(2), 324–39. doi:10.1007/s00344-012-9301-9.

Wani, S. A., Chand, S. and Ali, T. 2013. Potential use of Azotobacter chroococcum in crop production: an overview. *Current Agriculture Research Journal* 1(1), 35-8. doi:10.12944/CARJ.1.1.04.

Woo, S. L. and Pepe, O. 2018. Microbial consortia: promising probiotics as plant biostimulants for sustainable agriculture. *Frontiers in Plant Science* 9, 1801. doi:10.3389/fpls.2018.01801.

Xu, L. and Geelen, D. 2018. Developing biostimulants From agro-food and industrial By-Products. *Frontiers in Plant Science* 9, 1567. doi:10.3389/fpls.2018.01567.

Yakhin, O. I., Lubyanov, A. A., Yakhin, I. A. and Brown, P. H. 2017. Biostimulants in plant science: a global perspective. *Frontiers in Plant Science* 7, 2049. doi:10.3389/fpls.2016.02049.

Yang, H., Zhang, Q., Koide, R. T., Hoeksema, J. D., Tang, J., Bian, X., Hu, S. and Chen, X. 2016. Taxonomic resolution is a determinant of biodiversity effects in arbuscular mycorrhizal fungal communities. *Journal of Ecology* 105(1), 219-28. doi:10.1111/1365-2745.12655.

Zanin, L., Tomasi, N., Rizzardo, C., Gottardi, S., Terzano, R., Alfeld, M., Janssens, K., De Nobili, M., Mimmo, T. and Cesco, S. 2015. Iron allocation in leaves of Fe-deficient cucumber plants fed with natural Fe complexes. *Physiologia Plantarum* 154(1), 82-94. doi:10.1111/ppl.12296.

Zanin, L., Tomasi, N., Cesco, S., Varanini, Z. and Pinton, R. 2019. Humic substances contribute to plant iron nutrition acting as chelators and biostimulants. *Frontiers in Plant Science* 10, 675. doi:10.3389/fpls.2019.00675.

Zodape, S. T., Gupta, A., Bhandari, S. C., Rawat, U. S., Chaudhary, D. R., Eswaran, K. and Chikara, J. 2011. Foliar application of seaweed sap as biostimulant for enhancement of yield and quality of tomato (*Lycopersicon esculentum* Mill.). *Journal of Scientific and Industrial Research* 70, 215-9.

Chapter 11

Combining plant biostimulants and precision agriculture

Raffaele Casa, University of Tuscia, Italy; Davide Cammarano, Purdue University, USA; Domenico Ronga, CREA, Italy; and Giuseppe Cillo, University of Teramo, Italy

1 Introduction

Precision agriculture (PA) can encompass precision management of both plant and animal production, that is, precision crop management and precision livestock farming. PA can be considered as a technology-driven agricultural revolution with the potential of markedly changing the way farming is practised. Several definitions have been proposed in recent decades, but one of the best known definitions describes it as an approach to managing the agricultural production process by 'doing the right thing at the right time at the right place' (Gebbers and Adamchuk, 2010). A more complete and recent definition is: 'a management strategy that gathers, processes and analyzes temporal, spatial and individual data and combines it with other information to support management decisions according to estimated variability for improved resource use efficiency, productivity, quality, profitability and sustainability of agricultural production' (ISPA, 2019). This definition effectively summarizes the philosophy and the main objectives of PA:

- to take into account the variability in time and space of factors that affect the agricultural production process, and

http://dx.doi.org/10.19103/AS.2020.0068.14

- to improve efficiency in use of management inputs in agricultural production.

Improving efficiency means using fewer resources to achieve the same result or getting a better result with the same use of inputs (e.g. water, fertilizers, plant protection products etc.). There is a close link between the philosophy on which PA is based and the need to increase sustainability and reduce the use of inputs that is potentially harmful for the environment (Stafford, 2018), including the need to reduce the impact of agriculture on global warming (Soto et al., 2019). To achieve this aim, PA makes use of the best that technology can offer in terms of capacity to monitor spatial and temporal variability of soils and crops, and to implement management strategies that can take into account this variability, that is, site-specific or variable-rate management. In addition to environmental benefits, another objective of site-specific management is increasing cost-effectiveness in the application of inputs, such as seeds, fertilizers, plant regulators and plant-protection products. It should be noted that the economics of PA should take account of externalities, related, for example, to the reduction of the pollution of aquifers which reduces potential harm and cost for other users of water resources.

Biostimulants include a wide range of products, from non-microbial natural substances such as humic acids (HA), protein hydrolysates (PH) and seaweed extracts (SWE), to microbial-based products, for example, plant growth-promoting rhizobacteria (PGPR) (Rouphael and Colla, 2018). These are examined in detail in other chapters of this book. Their use, as soil or seed inoculants or in foliar spraying of vegetable or field crops, is increasingly considered as a sustainable way to promote plant growth, reduce abiotic stresses and increase nutrient and water-use efficiency (Calvo et al., 2014).

Biostimulants typically act on specific plant physiological processes. Seaweeds, for example, contain cytokinins and auxins or other hormone-like substances which stimulate plant hormones (Bulgari et al., 2015), whereas microbial inoculants stimulate plant growth and nutrient uptake by interacting with processes in the rhizosphere (Calvo et al., 2014). These mechanisms are subjected to a high degree of variability in time and space across agricultural fields, because of the differences in soil properties and plant phenotype interactions with the surrounding environment. The efficacy of biostimulants can therefore be highly dependent on application timing and placement (Preininger et al., 2018). Additionally the cost of biostimulants is generally high when compared, for example, to standard mineral fertilizers (Ronga et al., 2019a), so it would make sense for farmers to employ efficient strategies for their optimal application. PA is therefore an opportunity to enhance the cost-effectiveness of applying biostimulants by using variable-rate technology in different management zones. Some of the technologies developed in precision crop farming for the site-specific application of other inputs such as fertilizers, herbicides and pesticides, could therefore be used in the application of biostimulants.

However, research and operational examples of the use of biostimulants in precision agriculture are critically lacking. The wide variety of biostimulants and their differing modes of action make it more difficult, compared to other standard farm inputs, to identify the correct application doses, timing and placing of individual products. This chapter reviews the research on monitoring of soil and crop variables which allows the use of PA in the application of biostimulants.

2 Monitoring spatial variability in soil and plants

In order to develop precision crop-management strategies, it is necessary to be able to monitor soil and crop properties at the field scale. There have been impressive advances in the tools available to monitor and map the spatial and temporal variability of crops and their status or soils, such as those based on remote and proximal sensing (Jones and Vaughan, 2010; Viscarra Rossel et al., 2011). The former use airborne or space-borne sensors, whereas the latter are defined as techniques in which the sensors are placed within 2 m or so of the target, typically on a ground-based platform such as a tractor. The advantage of remote and proximal sensing, as compared to traditional destructive methods in which crops are sampled from the field and taken to a laboratory for further analysis, is that they provide diagnostic information in a more rapid and cost-effective way and in a spatial context. It has been shown in many studies that remote and proximal sensing tools can detect plant abiotic and biotic stresses (Mahlein et al., 2019; Zarco-Tejada et al., 2012).

Optical sensors measure reflectance in the visible (0.4–0.7 μm), near-infrared (NIR, 0.7–1.3 μm) and shortwave infrared (SWIR, 1.3–2.5 μm) part of the spectrum, whereas thermal sensors detect heat emission in the thermal infrared spectrum (TIR, 7–20 μm). For the monitoring of plant stress or of traits linked with photosynthesis, there is increasing interest in the use of hyperspectral sensing (Transon et al., 2018) and of fluorescence (Meroni et al., 2009). Hyperspectral sensing, in which hundreds of narrow spectral bands are used, as compared to typically 4–13 wide bands used in multispectral sensors, was until recently, only possible using non-imaging ground-based spectroradiometers, but it is now becoming available on unmanned aerial vehicles (UAVs) (Aasen et al., 2018) and on satellite platforms (Pignatti et al., 2015) with imaging spectroscopy systems. Sensing the fluorescence of chlorophyll, or of other pigments, has been shown to be extremely useful for monitoring the photosynthetic process and the occurrence of abiotic or biotic stresses (Mahlein et al., 2019; Zarco-Tejada et al., 2012).

Remote sensing for water-stress detection has been generally done using the Crop Water Stress Index (CWSI) which uses TIR data (Jackson et al., 1988).

The CWSI is calculated using the sensed difference in canopy and ambient air temperatures, and vapour pressure deficit derived from meteorological information. There are crop stresses that can cause elevated plant temperature and are correlated with both crop water status and yield reduction (Pinter et al., 2003).

Proximal soil sensing aims at mapping soil variability and is generally based on several approaches such as the use of electrical and electromagnetic sensors, optical and radiometric sensors, sensors based on mechanical interactions between sensors and soil, and electrochemical sensors that can quantify the activity of specific ions or molecules (Viscarra Rossel et al., 2011).

Sensors can acquire measurements at specific times, for example, in order to explore spatial variability in the field (e.g. measuring soil electrical conductivity before sowing) or be stationary in the field, such as, for example, with wireless sensors networks (WSN) (Di Gennaro et al., 2017) and therefore quantify the temporal variability of given crop/soil properties. An important feature of most sensing systems is that the relationship between the actual measurement (e.g., voltage) and the soil/plant property is not directly measured by the sensor (Schirrmann et al., 2013). For example, a farmer might be interested in sensing the soil to quantify the amount of organic matter, but the sensor used to scan the soil will only detect the ability of the soil to conduct an electrical charge. Therefore, to quantify the relationship between the soil property and the sensor's output, some kind of site-specific inference is needed to relate the measurement to a given property.

In the context of precision application of biostimulants, for example, for those products especially targeted at abiotic stress mitigation or nutrient uptake improvement, the capability to monitor spatially and accurately the occurrence of these stresses, or of nutritional deficiencies, is crucial. Given that diagnosis and detection might be non-specific, it will be often desirable to combine multiple sensors, using data fusion approaches (Nawar et al., 2017; Viscarra Rossel et al., 2011). For example, Tsoulias et al. (2019) used apparent soil electrical conductivity measurements (ECa) as an indicator for soil water availability, and hence water stress, in an apple orchard and a tractor-based LiDAR scanning system to estimate the leaf area. A significant correlation was found between ECa and leaf area per tree. Trees in lower ECa areas had reduced leaf area and enhanced water stress. This kind of information enables precision application of biostimulants targeting water stress mitigation.

3 Site-specific management based on uniform management zones

Traditionally, agronomic management uses a whole-field approach where each field is considered as a homogenous area and spatial variability in the field is not considered. While this method is common among farmers because it is easy to implement, it is not efficient in optimizing input application. If a fertilizer is applied uniformly, some may be wasted (e.g. because crops already have adequate nutrition) and some be unused. The unused nutrients in some areas of the field are lost through leaching, runoff and gaseous emissions, resulting in an economic loss for the farmer and an environmental cost for the society (Basso et al., 2016). Yield is the result of interactions between factors such as soil, weather, cultivar, agronomic management, pest and diseases. Soil, for example, is highly variable across a field. Factors such as topography, land use and soil-plant-atmosphere interactions will also impact crop growth and development (Nawar et al., 2017).

Site-specific management (SSM) aims to manage soil and crops while taking into account spatial variability within a cultivated field. There are two main ways to carry out site-specific management:

1. constantly varying inputs (e.g. fertilizers, pesticides etc.) for each point of the field, in response to specific needs detected and quantified by sensors, or translated into previously produced maps using different data sources, and
2. dividing the plot into homogeneous sub-areas within which the inputs are applied uniformly within a site-specific management zone (MZ).

The latter approach, which requires the preparation of maps of uniform zones into which the field is divided, is a kind of compromise between conventional management and the first site-specific approach cited above, but is often more feasible than the former.

MZs are defined as sub-regions that present similar combinations of yield-limiting factors (Nawar et al., 2017; Miao et al., 2018). In such zones, a given agronomic input is applied at a given rate to maximize its efficiency. MZs can be generated by overlaying data of different sources including yield maps, topography, soil sampling data, aerial photographs, apparent electrical conductivity (ECa) and canopy images assessing crop growth variability (Maestrini and Basso, 2018a,b; Miao et al., 2018). Basso et al. (2011) have demonstrated how the subdivision of the field into MZs, combined with long-term weather information, has helped in quantifying optimal trade-offs between economic and environmental fertilization objectives. Delgado et al. (2005) have shown how the subdivision of fields into MZs helped to increase yield and nitrogen-use efficiency as well as reduce leaching.

Nowadays, the availability of data from different sources, such as yield maps, soil and crop sensing, topography, weather, and new geostatistical methods allows a better delineation of MZs (Nawar et al., 2017). These developments have improved the performance of MZ, as compared to traditional MZ delineation methods that were based only on soil or crop properties. In addition, the adoption of new sensing technologies for characterizing spatial variability in the soil, for example, gamma ray (Priori et al., 2016), electromagnetic induction (EMI) and visible and near-infrared (vis–NIR) spectroscopy (Viscarra Rossel et al., 2011) has helped to improve measurement accuracy. Measurement of variability in crop growth with proximal or satellite sensors to delineate MZs based on spectral indices (Pascucci et al., 2018) has meant that a wide range of yield-limiting factors in soil and crops can be measured rapidly at finer resolutions.

4 Site-specific application of farming inputs

There is an increased demand, especially in the most affluent societies, for food produced in a more sustainable way (Dias et al., 2015). At the same time, the increasing demand for food from a growing population is putting enormous pressure on agricultural productivity (Foley et al., 2011). The world is also facing many environmental challenges, increasing the need for more environmentally sustainable production practices in agriculture (Foley et al., 2011). Farmers are thus called on to increase agricultural production while reducing their impact on the environment (Cammarano et al., 2019). Researchers need to develop innovative products and technologies that are able to increase crop yields and quality while decreasing their agricultural carbon footprint (Dias et al., 2015; Ronga et al., 2019b).

In this context, the combined use of innovative eco-friendly biostimulants and precision agriculture approaches could promote soil health and optimize crop nutrition management. Biostimulants are able to synchronize nutrient availability and crop uptake, allowing more targeted and efficient fertilization. Biostimulants based on organic matrices have the additional benefit of closing the loop 'from the land to the land', because these organic compounds come from other agricultural sources. Biostimulants can influence various physicochemical aspects of soil such as decreased erosion, increased water availability and infiltration, improved soil structure, decreased leaching and stimulation of rhizosphere biological activity (Calvo et al., 2014; Rouphael and Colla, 2018).

In addition, microbial consortia, in particular AMF and PGPR, are ubiquitous organisms naturally present in the cultivated environment that interact with most plant roots and provide several benefits to the host plant. These include improved water and nutrition uptake (of P, K, Mg, N and micronutrients), enhanced tolerance to environmental stresses, reduced sensitivity to heavy metals and better soil structure (Bernardo et al., 2017; Calvo et al., 2014; du Jardin, 2015). The European Union is encouraging the use of innovative biostimulants containing organic recycled by-products in line with a circular economy model (Rouphael and Colla, 2018).

The benefits of the biostimulants can be further enhanced by adapting their application to precision agriculture site-specific techniques, that is, by varying their amount spatially within the field, for example, in different management zones. This will depend on both:

1 the ability to obtain maps that are able to show how to vary application rates of biostimulant products; and
2 the availability of suitable machinery to accomplish variable rate applications (VRA) of biostimulants.

The first requirement relies heavily on sensing, for example, of nutrient deficiencies and abiotic stresses and on the exploitation of multi-source information layers for the definition of MZ, as discussed earlier. The second aspect is examined in the following section.

5 Precision application techniques for biostimulants

The ability to implement variable rate application (VRA) of biostimulants depends on the way the product needs to be applied, which generally varies according to the biostimulant category (Table 1). Biostimulant application methods include: soil application, soil drench (i.e. the process of adding diluted products directly to the base of a plant), foliar spray and seed inoculation. Seed or tuber treatments do not lend themselves to VRA. Soil application also depends on whether the product is a solid (e.g. granular form) or a liquid, which is a more common scenario. If the biostimulant is available in liquid form or can be diluted in a solution before application, conventional sprayer machines can be used.

Sprayers are also used for foliar applications of many categories of biostimulants (Preininger et al., 2018). When applied to soil (rather than the plant itself), macronutrients and micronutrients may not satisfy plant requirements due to physical, biochemical and chemical processes regulating their transport and uptake. Because they reach the plant directly, foliar spray application is a better way of supplying plant nutrient requirements that are essential to achieve high yields (Shaaban, 2001). Using foliar sprays, plants can be treated throughout the whole season, by timely administration of efficient formulations with appropriate spraying equipment, as reviewed by Preininger et al. (2018) for microbial biostimulants. VRA technologies for sprayers are now available and allow precise regulation of application according to a prescription map, allowing a considerable saving in the amount of products used as compared to uniform blanket applications (Castaldi et al., 2017; Tona et al., 2018).

The possibility of developing prescription maps for site-specific biostimulant applications, for example, using VRA sprayers, will also depend on

Table 1 Main agronomic effects and application methods of the main categories of biostimulants

Biostimulant category	Main agronomic effects reported	Application method	References
Humic and fulvic acids	• increased uptake of macronutrients and micronutrients (N, P, Mn, Cu, Zn, Fe); • water stress protection	soil application; in irrigation water	du Jardin (2015), Halpern et al. (2015)
Seaweed extracts	• increased uptake of macronutrients and micronutrients (N, P, K, Ca, Mg, Fe, Zn, S)	foliar spray; soil application	du Jardin (2015), Halpern et al. (2015)
Protein hydrolysates and other N-containing compounds	• tolerance to salinity; • tolerance to drought; • increased N uptake	foliar spray; soil drench; seed treatment	Colla et al. (2017)
Plant-growth-promoting bacteria (PGPB)	• increased uptake of macronutrients and micronutrients (P, N, Fe);	Seed inoculation; soil application	Halpern et al. (2015)
Beneficial fungi (e.g. AMF)	• drought resistance • increased macronutrients and micronutrients uptake (N, S, B, Cu, K, Zn, Ca, Mg, Na, Mn, Fe, Al and Si,) • protection against oxidative stress • mineralization of organic nutrients	Seed and potato tuber inoculation; soil drench	Drobek et al. (2019), Hart and Trevors (2005), Hijri (2016)

which nutrient and other requirements are being addressed (Table 1). They are likely to be easier for biostimulants targeting macronutrient uptake, especially nitrogen (N) or phosphorous (P), as well as for those helping to reduce abiotic stresses such as water stress. Many tools have been developed to diagnose N deficiency, based on proximal or remote sensing (Xia et al., 2016), mostly exploiting chlorophyll content as a proxy of N status (Osborne et al., 2002). Narrow-band indices obtained from hyperspectral sensors have been shown to have the potential for detecting macro- and micronutrient concentrations in vineyards, leading to different fertilization treatments based on *Ascophyllum nodosum* seaweed and chelates (Gil-Pérez et al., 2010)

Remote assessment of plant canopy P deficiency has been more challenging. Spectro-radiometric sensors for the rapid assessment of soil P have also been proposed. Although the content of P in the soil has no direct effect on its spectral reflectance, Mouazen and Kuang (2016) showed that a spectroradiometric system, using chemometric models and embedded in the back of a subsoiler, can provide mapping of spatial variability of phosphorus in the soil which is comparable with laboratory analysis.

Soil geophysical mapping systems, such as apparent electrical conductivity (ECa) measurements, have also been shown to provide a potential means for assessing the spatial variability of vesicular-arbuscular mycorrhizal (VAM) fungi across fields (Johnson et al., 2004). Biological indicators of VAM presence and activity could be related to specific soil physical and chemical characteristics and crop yields, due to significant partitioning of each of these parameters among ECa zones (Johnson et al., 2004).

Welbaum et al. (2004) have discussed new approaches that involve the precision management of microorganisms in the agroecosystem. According to the authors, the current difficulty in resolving soil microbial-community structure dynamics is our limited ability to examine soil ecosystems in real time. However with the continuing evolution of molecular genetic techniques, computer technology and nanotechnology (Mura et al., 2013), it will be possible to precisely manage the rhizosphere while monitoring soil and plant health in real time.

Adaptation of environmental probes based on nanosensors for agroecosystem monitoring (Giraldo et al., 2019), would allow the development of 'smart fields'. In smart fields, simple though essential environmental factors such as temperature, pH, water, mineral nutrients and gas composition (e.g. oxygen, carbon dioxide and ethylene) could be monitored in real time and at multiple locations in the field. Such probes could be integrated into arrays and linked to a computer-based 'monitoring/command' system that turns on specifically prescribed response inputs to correct for imbalances that might arise. Changes to the rhizosphere could be rapidly made through a subsurface irrigation system. Diverse inputs, such as water, mineral nutrients, microbial inoculants, crop protection chemicals, gases (e.g. O_2 or H_2) and organic nutrients (e.g. sugars and/or signal molecules) that preferentially feed soil and plant beneficial microbes could readily be delivered, creating both a beneficial rhizosphere and healthy plant stands (Welbaum et al., 2004).

Ma (2019) has reviewed microbial seed-coating technology and its contribution to sustainable precision agriculture. Seed coatings can deliver seeds in a larger, rounder, heavier and more uniform form, which can be placed individually with enhanced spacing and depth control. Biological seed coatings with PGPM (plant growth-promoting microorganisms) have been suggested as an alternative to conventional seed coatings (e.g. with fertilizer and protection products). The survival and colonization of these microbes in the rhizosphere can, however, be an issue. Survival is determined by environmental factors such as soil pH, nutrient and moisture content; plant genotype and physiological status; and the presence of other microbial species (Ma, 2019). There is a pressing need to understand the behaviour of introduced microbes via seed coating at the root level and their function in the field under different environmental conditions.

Crop production is often affected by abiotic stresses such as water deficiency, salinization of soils, low and high temperatures. Biostimulants can play a role in reducing abiotic stresses and restoring damage already caused by unfavourable environmental conditions (Van Oosten et al., 2017). For example, microalgal extracts (ME) contain macronutrients and micronutrients essential for crop growth and development (Calvo et al., 2014; Ronga et al., 2019a). The application of ME provides potential protection against abiotic stresses in plants. As an example, extracts from *Dunaliella* spp. and *Phaeodactylum* spp. alleviated salt stress during the seed germination stage of bell pepper (*Capsicum annuum* L.) (Guzmán-Murillo et al., 2013). Abd El-Baky et al. (2010) showed that the application of ME to wheat (*Triticum aestivum* L.) could improve wheat salinity tolerance. The spatial and temporal variability monitoring of water stress, required in case of VRA application of these biostimulants, can be performed by TIR sensing, for example, using the CWSI (Jackson et al., 1988) from UAV-based thermal cameras. Soil salinity variability can be assessed using geoelectric proximal sensing (Corwin and Lesch, 2005).

6 Case study: biostimulants in precision viticulture

Studies in which biostimulant application has been coupled with precision agriculture management approaches are just starting to be undertaken, and published data are still lacking. Preliminary tests were carried out in Central Italy (Cantalamessa, 2018; Cillo, 2017), in the context of precision viticulture, the sector in which precision crop management shows the highest adoption potential in Italy (Casa and Pisante, 2016).

The VitisL@b project took place in 2017 at the Argiano winery (Lat. 42° 58′ N, Long. 11° 25′ E), in Tuscany (Montalcino, Siena, Italy), which produces premium quality wine Brunello di Montalcino DOCG and is managed according to an organic-farming regime (Cillo, 2017). Following the acquisition of information related to variables such as soil pedology and vigour of the crop, space-time variability was identified in order to delineate MZs in a 4.5 ha San Giovese vineyard.

Data on soil properties, combined with data obtained by remotely monitoring the spatial variability of crop vigour of the vineyard, were used to identify homogeneous areas. Information layers used included:

1 an apparent electrical conductivity map obtained from the use of an EMI geoelectric sensor; and
2 the NDVI and Green vegetation NDVI obtained from RapidEye and Quickbird satellites, respectively.

The different information layers were then used to obtain a classification of the vineyard in zones showing homogeneous crop vigour, by applying a fuzzy-c-means classification, using the Management Zone Analyst software (Fridgen et al., 2004). The vineyard was divided into three MZs with high, low and average vigour, respectively.

In each of the 3 areas a plot consisting of 12 plants (4 plants on each of 3 consecutive rows) was selected. Foliar spraying of a biostimulant obtained from hydrolysis of alfalfa, *Ascophyllum nodosum* and molasses (Basfoliar Planate BIO SL, Compo Expert, Italy) was carried out at different rates. The doses of the product distributed, differentiated between the different rows, were:

1 standard dosage suggested by the company Compo Expert: 5 mL product per litre of water;
2 double dose: 10 mL product per litre of water;
3 half dose: 2.5 mL of product per litre of water; and
4 Control: water only.

The product application dates were July 25, August 4 and August 14 of 2017, and the application was carried out using a backpack sprayer (Fig. 1). After about 72 hours after each treatment, grapes from each identified plant were sampled, in order to be able to determine the relative °Brix by refractometry. The variation of this variable (expressing sugar content) was then correlated with the dose of product administered in the three different areas with different degrees of vigour and assessed by statistical analysis (ANOVA).

The results showed a significant difference in the sugar content of grapes sampled from the different zones (Fig. 1). However, the differences among

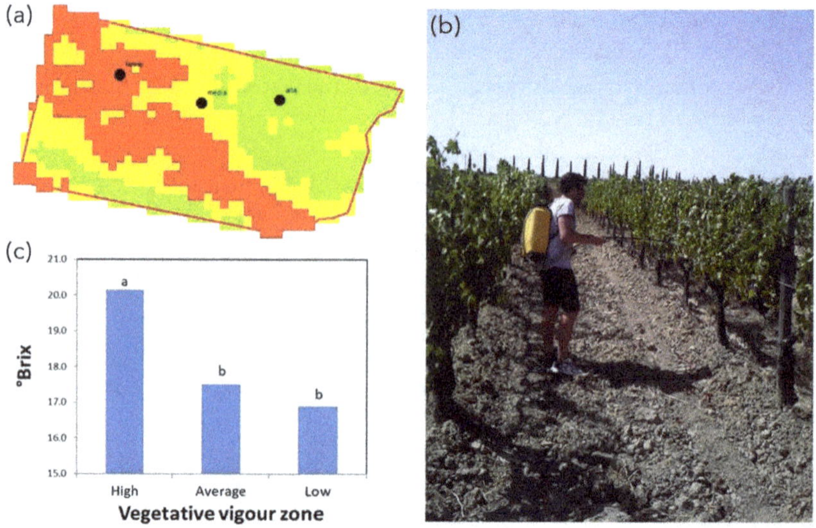

Figure 1 Experiment on the application of biostimulants in a precision viticulture context carried out in 2017 at the Argiano winery (Cillo, 2017): (a) uniform vigour zones; (b) foliar application of biostimulant in test area (points in (a)); (c) grape-sugar content (Brix degrees) results for the three zones. Different letters indicate statistically significant differences (P<0.05).

application rates of biostimulants, and their interaction with the vigour zones, were not statistically significant. It should be noted that these are preliminary results and might have been influenced by a particularly hot weather recorded in August 2017 in the area. In fact, some studies have reported statistically significant effects of the application of *Ascophyllum nodosum* extract on grape yield, canopy chlorophyll, berry N content and anthocyanin extractability (Taskos et al., 2019).

Another experiment was carried out at Luigi Valori's farm (3 ha) in Controguerra (Teramo), Central Italy, producing Montepulciano d'Abruzzo DOC wine, under an organic farming regime, in 2018 (Cantalamessa, 2018). Different information layers, including apparent electrical conductivity (ECa) using a Geonics EM38-MK2 sensor, UAV multispectral and ground-based spectroradiometric measurements, were used to delineate homogeneous areas in terms of crop vigour. The biostimulant administered was a formulation based on auxins and cytokines of vegetable origin, extracted from *Ecklonia maxima* algae (Basfoliar Kelp BIO SL). The treatment was carried out every 15 days in the months of July–August, at the uniform dose of 300 g L^{-1}, in order to be able to verify its effect and therefore the increase of anthocyanin in grapes, relative to the spatial variability of vineyard. The results revealed significant differences in anthocyanin pigment concentration, with higher values in high

ECa (typically associated with higher clay content and water retention) which were also high-vigour areas as assessed from spectral indices.

7 Conclusion and future trends

Currently, plant nutrition is mainly based on the use of synthetic fertilizers with typically low efficiency and considerable environmental impact (Basso et al., 2016). The application of biostimulants to soil or plants using variable-rate technologies might help to solve these problems. The combined use of biostimulants and precision agricultural techniques allow such benefits as:

1. the reduction in the use of external inputs (synthetic fertilizers and plant protection products),
2. an increase in soil soluble phosphorus solubilization,
3. improved soil biodiversity,
4. improved soil protection,
5. an increase in crop yield and quality,
6. a reduction of the risk of leaching nutrients, and
7. an improvement in crop tolerance to abiotic stresses.

In addition, integrated use of precision agriculture and crop modelling could allow the identification of the main cause of yield reduction through different areas and periods (Basso et al., 2011).

In order to allow the development of methodologies for the application of biostimulants in a precision agriculture context, much research has still to be carried out, since this area is still almost completely unexplored. The physiological mechanism of action of different biostimulants is currently the subject of much research and is becoming clearer (Bernardo et al., 2017; Colla et al., 2017; Guzmán-Murillo et al., 2013). However, agronomic field trials, which could provide some guidance on the best timing and rates of application of the different categories of biostimulants, are still needed for the different cropping systems. It will be essential to consider the interaction with environmental factors and other agronomic practices such as fertilization (Taskos et al., 2019). This kind of information is essential in order to develop prescription maps for biostimulant application.

Research on plant stress-monitoring technologies, in particular for drought and salinity, which are among the most important limiting factors for crop production, is still needed, especially in order to distinguish different abiotic and biotic stresses, since their symptoms are often very similar. Hyperspectral and fluorescence sensing (Aasen et al., 2018; Transon et al., 2018) as well as UAV-based thermal imaging (Zarco-Tejada et al., 2012) have great potential in this context. Soil imaging spectroscopy (Ben-Dor et al., 2009) and geophysical sensing, for example, gamma ray techniques (Priori et al., 2016), are also very active areas of research in order to provide methods and technologies for the mapping of soil properties linked to nutrient availability (N, P) and soil salinity. This will allow the development of methods for precise application of biostimulants that specifically target these stresses or which improve the uptake of the specific nutrients. A current trend in research in the context of remote and proximal sensing is on the development of data-fusion approaches, in which data from different sensors are combined (Casa et al., 2013). The increasing progress in the field of artificial intelligence (i.e. machine learning algorithms) will also have an increasing importance in this context (Upreti et al., 2019).

8 Where to look for further information

Currently, there is almost no information available in the literature on the combination of precision agriculture and biostimulants. There are some recent books which cover in detail all the principles and technologies employed in precision agriculture (Castrignano et al., 2020; Oliver et al., 2018; Stafford, 2018).

The scientific journals which are the best sources for up-to-date information include:

- Computer and Electronics in Agriculture (Elsevier)
- Precision Agriculture (Springer)
- Remote Sensing of the Environment (Elsevier)
- Remote Sensing (MDPI)
- Sensors (MDPI)

The main international conferences on precision agriculture are the series of European Conferences on Precision Agriculture (ECPA), held every 2 (odd) years in Europe and the International Conference on Precision Agriculture (ICPA) held every 2 (even) years in the United States.

A good source of information on news and events, including conferences, is the website of the International Society for Precision Agriculture (ISPA), https://www.ispag.org which is the main scientific society on this topic.

9 References

Aasen, H., Honkavaara, E., Lucieer, A. and Zarco-Tejada, P. 2018. Quantitative remote sensing at ultra-high resolution with uav spectroscopy: a review of sensor technology, measurement procedures, and data correction workflows. *Remote Sens.* 10, 1091. https://doi.org/10.3390/rs10071091.

Abd El-Baky, H. H., El-Baz, F. K. and El Baroty, G. S. 2010. Enhancing antioxidant availability in wheat grains from plants grown under seawater stress in response to microalgae extract treatments. *J. Sci. Food Agric.* 90, 299–303. https://doi.org/10.1002/jsfa.3815.

Basso, B., Ritchie, J. T., Cammarano, D. and Sartori, L. 2011. A strategic and tactical management approach to select optimal N fertilizer rates for wheat in a spatially variable field. *Eur. J. Agron.* 35, 215–22. https://doi.org/10.1016/J.EJA.2011.06.004.

Basso, B., Dumont, B., Cammarano, D., Pezzuolo, A., Marinello, F. and Sartori, L. 2016. Environmental and economic benefits of variable rate nitrogen fertilization in a nitrate vulnerable zone. *Sci. Total Environ.* 545–546, 227–35. https://doi.org/10.1016/J.SCITOTENV.2015.12.104.

Ben-Dor, E., Chabrillat, S., Demattê, J. A. M., Taylor, G. R., Hill, J., Whiting, M. L. and Sommer, S. 2009. Using Imaging Spectroscopy to study soil properties. *Remote Sens. Environ.* 113, S38–55. https://doi.org/10.1016/J.RSE.2008.09.019.

Bernardo, L., Morcia, C., Carletti, P., Ghizzoni, R., Badeck, F. W., Rizza, F., Lucini, L. and Terzi, V. 2017. Proteomic insight into the mitigation of wheat root drought stress by arbuscular mycorrhizae. *J. Proteomics* 169, 21–32. https://doi.org/10.1016/j.jprot.2017.03.024.

Bulgari, R., Cocetta, G., Trivellini, A., Vernieri, P. and Ferrante, A. 2015. Biostimulants and crop responses: a review. *Biol. Agric. Hortic.* 31, 1–17. https://doi.org/10.1080/01448765.2014.964649.

Calvo, P., Nelson, L. and Kloepper, J. W. 2014. Agricultural uses of plant biostimulants. *Plant Soil* 383, 3–41. https://doi.org/10.1007/s11104-014-2131-8.

Cammarano, D., Ceccarelli, S., Grando, S., Romagosa, I., Benbelkacem, A., Akar, T., Al-Yassin, A., Pecchioni, N., Francia, E. and Ronga, D. 2019. The impact of climate change on barley yield in the Mediterranean basin. *Eur. J. Agron.* 106, 1–11. https://doi.org/10.1016/j.eja.2019.03.002.

Cantalamessa, S. 2018. Indici vegetazionali e contenuto di polifenoli durante la fase di maturazione delle uve. BSc Thesis. University of Teramo.

Casa, R. and Pisante, M. 2016. Introduzione. In: Casa, R. (Ed.), *Agricoltura Di Precisione*. New Business Media, Milano, Italy, pp. 1–8.

Casa, R., Castaldi, F., Pascucci, S., Basso, B. and Pignatti, S. 2013. Geophysical and hyperspectral data fusion techniques for in-field estimation of soil properties. *Vadose Zoo. J.* 12. https://doi.org/10.2136/vzj2012.0201.

Castaldi, F., Pelosi, F., Pascucci, S. and Casa, R. 2017. Assessing the potential of images from unmanned aerial vehicles (UAV) to support herbicide patch spraying in maize. *Precis. Agric.* 18, 76–94. https://doi.org/10.1007/s11119-016-9468-3.

Castrignano, A., Buttafuoco, G., Khosla, R., Mouazen, A. M., Moshou, D. and Naud, O. 2020. *Agricultural Internet of Things and Decision Support for Precision Smart Farming*. Elsevier Academic Press, London, UK.

Cillo, G. 2017. Progetto VitisL@b: La fertilizzazione di precisione in viticoltura a sostegno della qualità, in regime biologico. Thesis, Master in Precision Agriculture. University of Teramo.

Colla, G., Hoagland, L., Ruzzi, M., Cardarelli, M., Bonini, P., Canaguier, R. and Rouphael, Y. 2017. Biostimulant action of protein hydrolysates: unraveling their effects on plant physiology and microbiome. *Front. Plant Sci.* https://doi.org/10.3389/fpls.2017.02202.

Corwin, D. L. and Lesch, S. M. 2005. Apparent soil electrical conductivity measurements in agriculture. *Comput. Electron. Agric.* 46, 11–43. https://doi.org/10.1016/j.compag.2004.10.005.

Delgado, J. A., Khosla, R., Bausch, W. C., Westfall, D. G. and Inman, D. J. 2005. Nitrogen fertilizer management based on site-specific management zones reduces potential for nitrate leaching. *J. Soil Water Conserv.* 60, 402–10.
Dias, V. da V., Schultz, G., Schuster, M. da S., Talamini, E., Révillion, J. P., Dias, V. da V., Schultz, G., Schuster, M. da S., Talamini, E. and Révillion, J. P. 2015. The organic food market: a quantitative and qualitative overview of international publications. *Ambient. Soc.* 18, 155–74. https://doi.org/10.1590/1809-4422ASOC841V1812015en.
Di Gennaro, S. F., Matese, A., Gioli, B., Toscano, P., Zaldei, A., Palliotti, A. and Genesio, L. 2017. Multisensor approach to assess vineyard thermal dynamics combining high-resolution unmanned aerial vehicle (UAV) remote sensing and wireless sensor network (WSN) proximal sensing. *Sci. Hortic. (Amsterdam).* 221, 83–7. https://doi.org/10.1016/J.SCIENTA.2017.04.024.
Drobek, M., Frąc, M. and Cybulska, J. 2019. Plant biostimulants: importance of the quality and yield of horticultural crops and the improvement of plant tolerance to abiotic stress-a review. *Agronomy* 9. https://doi.org/10.3390/agronomy9060335.
du Jardin, P. 2015. Plant biostimulants: Definition, concept, main categories and regulation. *Sci. Hortic. (Amsterdam)* 196, 3–14. https://doi.org/10.1016/j.scienta.2015.09.021.
Foley, J. A., Ramankutty, N., Brauman, K. A., Cassidy, E. S., Gerber, J. S., Johnston, M., Mueller, N. D., O'Connell, C., Ray, D. K., West, P. C., Balzer, C., Bennett, E. M., Carpenter, S. R., Hill, J., Monfreda, C., Polasky, S., Rockström, J., Sheehan, J., Siebert, S., Tilman, D. and Zaks, D. P. M. 2011. Solutions for a cultivated planet. *Nature* 478, 337–42. https://doi.org/10.1038/nature10452.
Fridgen, J. J., Kitchen, N. R., Sudduth, K. A., Drummond, S. T., Wiebold, W. J. and Fraisse, C. W. 2004. Management Zone Analyst (MZA). *Agron. J.* 96, 100–8. https://doi.org/10.2134/AGRONJ2004.1000.
Gebbers, R. and Adamchuk, V. I. 2010. Precision agriculture and food security. *Science* 327, 828–31. https://doi.org/10.1126/science.1183899.
Gil-Pérez, B., Zarco-Tejada, P. J., Corrêa-Guimarães, A., Relea-Gangas, E., Navas-Gracia, L. M., Hernandéz-Navarro, S., Sanz-Requena, J. F., Berjon, A. and Martín-Gil, J. 2010. Remote sensing detection of nutrient uptake in vineyards using narrow-band hyperspectral imagery. *Vitis* 49, 167–73.
Giraldo, J. P., Wu, H., Newkirk, G. M. and Kruss, S. 2019. Nanobiotechnology approaches for engineering smart plant sensors. *Nat. Nanotechnol.* 14, 541–53. https://doi.org/10.1038/s41565-019-0470-6.
Guzmán-Murillo, M. A., Ascencio, F. and Larrinaga-Mayoral, J. A. 2013. Germination and ROS detoxification in bell pepper (*Capsicum annuum* L.) under NaCl stress and treatment with microalgae extracts. *Protoplasma* 250, 33–42. https://doi.org/10.1007/s00709-011-0369-z.
Halpern, M., Bar-Tal, A., Ofek, M., Minz, D., Muller, T. and Yermiyahu, U. 2015. The use of biostimulants for enhancing nutrient uptake. *Advances in Agronomy* 130, 141–74. https://doi.org/10.1016/bs.agron.2014.10.001.
Hart, M. M., Trevors, J. T. 2005. Microbe management: application of mycorrhyzal fungi in sustainable agriculture. Front. Ecol. Environ. 3, 533–9. https://doi.org/10.1890/1540-9295(2005)003[0533:MMAOMF]2.0.CO;2.
Hijri, M. 2016. Analysis of a large dataset of mycorrhiza inoculation field trials on potato shows highly significant increases in yield. *Mycorrhiza* 26, 209–14. https://doi.org/10.1007/s00572-015-0661-4.

ISPA. 2019. Precision Ag Definition [WWW Document]. Int. Soc. Precis. Agric. website. URL https://www.ispag.org/about/definition (accessed on 20 September 2019).

Jackson, R. D., Kustas, W. P. and Choudhury, B. J. 1988. A reexamination of the crop water stress index. *Irrig. Sci.* 9, 309–17. https://doi.org/10.1007/BF00296705.

Johnson, C. K., Wienhold, B. J., Doran, J. W., Drijber, R. A. and Wright, S. F. 2004. Linking microbial-scale findings to farm-scale outcomes in a dryland cropping system. *Precis. Agric.* 5, 311–28. https://doi.org/10.1023/B:PRAG.0000040803.35346.b2.

Jones, H. G., Vaughan, R. A. 2010. *Remote Sensing of Vegetation : Principles, Techniques, and Applications*. Oxford University Press, Oxford, UK.

Ma, Y. 2019. Seed coating with beneficial microorganisms for precision agriculture. *Biotechnol. Adv.* 37(7), 107423. https://doi.org/10.1016/j.biotechadv.2019.107423.

Maestrini, B. and Basso, B. 2018a. Drivers of within-field spatial and temporal variability of crop yield across the US Midwest. *Sci. Rep.* 8, 14833. https://doi.org/10.1038/s41598-018-32779-3.

Maestrini, B. and Basso, B. 2018b. Predicting spatial patterns of within-field crop yield variability. *Field Crops Res.* 219, 106–12. https://doi.org/10.1016/J.FCR.2018.01.028.

Mahlein, A.-K., Alisaac, E., Al Masri, A., Behmann, J., Dehne, H.-W. and Oerke, E.-C. 2019. Comparison and combination of thermal, fluorescence, and hyperspectral imaging for monitoring fusarium head blight of wheat on spikelet scale. *Sensors* 19, 2281. https://doi.org/10.3390/s19102281.

Meroni, M., Rossini, M., Guanter, L., Alonso, L., Rascher, U., Colombo, R. and Moreno, J. 2009. Remote sensing of solar-induced chlorophyll fluorescence: review of methods and applications. *Remote Sens. Environ.* 113, 2037–51. https://doi.org/10.1016/J.RSE.2009.05.003.

Miao, Y., Mulla, D. J. and Robert, P. C. 2018. An integrated approach to site-specific management zone delineation. *Front. Agric. Sci. Eng.* https://doi.org/10.15302/J-FASE-2018230.

Mouazen, A. M. and Kuang, B. 2016. On-line visible and near infrared spectroscopy for in-field phosphorous management. *Soil Tillage Res.* 155, 471–7. https://doi.org/10.1016/J.STILL.2015.04.003.

Mura, S., Seddaiu, G., Bacchini, F., Roggero, P. P. and Greppi, G. F. 2013. Advances of nanotechnology in agro-environmental studies. *Ital. J. Agron.* 8, 18. https://doi.org/10.4081/ija.2013.e18.

Nawar, S., Corstanje, R., Halcro, G., Mulla, D. and Mouazen, A. M. 2017. Delineation of soil management zones for variable-rate fertilization: a review. *Advances in Agronomy* 143, 175–245. https://doi.org/10.1016/bs.agron.2017.01.003.

Oliver, M. A., Bishop, T. F. A. and Marchant, B. P. 2018. *Precision Agriculture for Sustainability and Environmental Protection*. Taylor and Francis, Oxford, UK. https://doi.org/10.4324/9780203128329

Osborne, S. L., Schepers, J. S., Francis, D. D. and Schlemmer, M. R. 2002. Detection of phosphorus and nitrogen deficiencies in corn using spectral radiance measurements. *Agron. J.* 94, 1215. https://doi.org/10.2134/agronj2002.1215.

Pascucci, S., Carfora, M., Palombo, A., Pignatti, S., Casa, R., Pepe, M. and Castaldi, F. 2018. A comparison between standard and functional clustering methodologies: application to agricultural fields for yield pattern assessment. *Remote Sens.* 10, 585. https://doi.org/10.3390/rs10040585.

Pignatti, S., Acito, N., Amato, U., Casa, R., Castaldi, F., Coluzzi, R., De Bonis, R., Diani, M., Imbrenda, V., Laneve, G., Matteoli, S., Palombo, A., Pascucci, S., Santini, F., Simoniello,

T., Ananasso, C., Corsini, G. and Cuomo, V. 2015. Environmental products overview of the Italian hyperspectral PRISMA mission: the SAP4PRISMA project. In: *2015 IEEE International Geoscience and Remote Sensing Symposium (IGARSS)*. IEEE, pp. 3997-4000. https://doi.org/10.1109/IGARSS.2015.7326701.

Pinter Jr., P. J., Hatfield, J. L., Schepers, J. S., Barnes, E. M., Moran, M. S., Daughtry, C. S. T. and Upchurch, D. R. 2003. Remote sensing for crop management. *Photogramm. Eng. Remote Sens.* 69, 647-64. https://doi.org/10.14358/PERS.69.6.647.

Preininger, C., Sauer, U., Bejarano, A. and Berninger, T. 2018. Concepts and applications of foliar spray for microbial inoculants. *Appl. Microbiol. Biotechnol.* 102, 7265-82. https://doi.org/10.1007/s00253-018-9173-4.

Priori, S., Fantappiè, M., Bianconi, N., Ferrigno, G., Pellegrini, S. and Costantini, E. A. C. 2016. Field-scale mapping of soil carbon stock with limited sampling by coupling gamma-ray and vis-NIR spectroscopy. *Soil Sci. Soc. Am. J.* 80, 954. https://doi.org/10.2136/sssaj2016.01.0018.

Ronga, D., Biazzi, E., Parati, K., Carminati, D., Carminati, E. and Tava, A. 2019a. Microalgal biostimulants and biofertilisers in crop productions. *Agronomy* 9, 1-22. https://doi.org/10.3390/agronomy9040192.

Ronga, D., Gallingani, T., Zaccardelli, M., Perrone, D., Francia, E., Milc, J. and Pecchioni, N. 2019b. Carbon footprint and energetic analysis of tomato production in the organic vs the conventional cropping systems in Southern Italy. *J. Clean. Prod.* 220, 836-45. https://doi.org/10.1016/J.JCLEPRO.2019.02.111.

Rouphael, Y. and Colla, G. 2018. Synergistic biostimulatory action: designing the next generation of plant biostimulants for sustainable agriculture. *Front. Plant Sci.* 871, 1-7. https://doi.org/10.3389/fpls.2018.01655.

Schirrmann, M., Gebbers, R. and Kramer, E. 2013. Performance of automated near-infrared reflectance spectrometry for continuous in situ mapping of soil fertility at field scale. *Vadose Zoo. J.* 12. https://doi.org/10.2136/vzj2012.0199.

Shaaban, M. M. 2001. Green microalgae water extract as foliar feeding to wheat plants. *Pakistan J. Biol. Sci.* 4, 628-32. https://doi.org/10.3923/pjbs.2001.628.632.

Soto, I., Barnes, A., Balafoutis, A., Beck, B., Sánchez, B., Vangeyte, J., Fountas, S., Van der Wal, T., Eory, V. and Gómez-Barbero, M. 2019. The contribution of precision agriculture technologies to farm productivity and the mitigation of greenhouse gas emissions in the EU. https://doi.org/10.2760/016263.

Stafford, J. 2018. *Precision Agriculture for Sustainability* (1st edn.). Burleigh Dodds Science Publishing. https://doi.org/10.1201/9781351114592.

Taskos, D., Stamatiadis, S., Yvin, J. C. and Jamois, F. 2019. Effects of an *Ascophyllum nodosum* (L.) Le Jol. extract on grapevine yield and berry composition of a Merlot vineyard. *Sci. Hortic. (Amsterdam)* 250, 27-32. https://doi.org/10.1016/j.scienta.2019.02.030.

Tona, E., Calcante, A. and Oberti, R. 2018. The profitability of precision spraying on specialty crops: a technical-economic analysis of protection equipment at increasing technological levels. *Precis. Agric.* 19, 606-29. https://doi.org/10.1007/s11119-017-9543-4.

Transon, J., d'Andrimont, R., Maugnard, A. and Defourny, P. 2018. Survey of hyperspectral earth observation applications from space in the Sentinel-2 context. *Remote Sens.* 10, 157. https://doi.org/10.3390/rs10020157.

Tsoulias, N., Paraforos, D. S., Fountas, S. and Zude-Sasse, M. 2019. Calculating the water deficit spatially using LiDAR laser scanner in an apple orchard. In: Stafford, J. V. (Ed.),

Precision Agriculture '19. Wageningen Academic Publishers, the Netherlands, pp. 115–21. https://doi.org/10.3920/978-90-8686-888-9_13.

Upreti, D., Huang, W., Kong, W., Pascucci, S., Pignatti, S., Zhou, X., Ye, H. and Casa, R. 2019. A comparison of hybrid machine learning algorithms for the retrieval of wheat biophysical variables from Sentinel-2. *Remote Sens.* 11, 481. https://doi.org/10.3390/rs11050481.

Van Oosten, M. J., Pepe, O., De Pascale, S., Silletti, S. and Maggio, A. 2017. The role of biostimulants and bioeffectors as alleviators of abiotic stress in crop plants. *Chem. Biol. Technol. Agric.* 4, 5. https://doi.org/10.1186/s40538-017-0089-5.

Viscarra Rossel, R. A., Adamchuk, V. I., Sudduth, K. A., McKenzie, N. J. and Lobsey, C. 2011. Proximal soil sensing: an effective approach for soil measurements in space and time. *Adv. Agron.* 113, 237–82. https://doi.org/10.1016/B978-0-12-386473-4.00010-5.

Welbaum, G. E., Sturz, A. V., Dong, Z. and Nowak, J. 2004. Managing soil microorganisms to improve productivity of agro-ecosystems. *Crit. Rev. Plant Sci.* 23, 175–93. https://doi.org/10.1080/07352680490433295.

Xia, T., Miao, Y., Wu, D., Shao, H., Khosla, R. and Mi, G. 2016. Active optical sensing of spring maize for in-season diagnosis of nitrogen status based on nitrogen nutrition index. *Remote Sens.* 8, 605. https://doi.org/10.3390/rs8070605.

Zarco-Tejada, P. J., González-Dugo, V. and Berni, J. A. J. 2012. Fluorescence, temperature and narrow-band indices acquired from a UAV platform for water stress detection using a micro-hyperspectral imager and a thermal camera. *Remote Sens. Environ.* 117, 322–37. https://doi.org/10.1016/J.RSE.2011.10.007.

Index

CPSIA information can be obtained
at www.ICGtesting.com
Printed in the USA
BVHW052249250522
638037BV00003B/75